DICTIONARY OF ELECTRICAL ENGINEERING

DICTIONARY OF ELECTRICAL ENGINEERING

K. G. JACKSON

LONDON

NEWNES-BUTTERWORTHS

THE BUTTERWORTH GROUP

ENGLAND
Butterworth & Co (Publishers) Ltd
London: 88 Kingsway, WC2B 6AB

AUSTRALIA
Butterworths Pty Ltd
Sydney: 586 Pacific Highway, NSW 2067
Melbourne: 343 Little Collins Street, 3000
Brisbane: 240 Queen Street, 4000

CANADA
Butterworth & Co (Canada) Ltd
Scarborough: 2265 Midland Avenue,
Ontario M1P 4S1

NEW ZEALAND
Butterworths of New Zealand Ltd
Wellington: 26-28 Waring Taylor Street, 1

SOUTH AFRICA
Butterworth & Co (South Africa) (Pty) Ltd
Durban: 152-154 Gale Street

First published in 1965 by George Newnes Ltd
Revised and enlarged impression published in
1973 by Newnes-Butterworths, an imprint of
the Butterworth Group
Second impression 1975

© Butterworth & Co. (Publishers) Ltd. 1973

ISBN O 6004 1060 9

Printed in England by Lewis Reprints Ltd
Member of Brown Knight & Truscott Group
London and Tonbridge

PREFACE

An electrical engineer is a man of various talents. He may install lighting and power supplies in people's homes; he may erect and service electric motors in a factory. He may be concerned with the laying and jointing of cables; he may be more at home with a soldering iron in a laboratory. He may erect overhead power supplies, or design high-voltage insulators. He may provide supplies for electric traction, or develop new forms of motor for traction. He may build generators, or design nuclear power stations. He may be seeking new methods of generating electricity, or using old methods of testing equipment. He may write and read papers for the Institution of Electrical Engineers, or his reading may be limited to the Wiring Regulations. Lifts and cranes, insulators and conductors, power transformers and instrument transformers, switching stations and miniature circuit-breakers, 400 kV supplies and 6 V batteries—any one of these may come within his purview. Heating and lighting, welding and melting, mining and flying, clocks and computers—an electrical engineer is involved in each of these fields.

It is not surprising, therefore, that there is a need for a comprehensive dictionary of electrical engineering. Though the present work does not claim to be complete, it does contain something for each of the activities listed in the first paragraph. What has been attempted is to include terms that are of general use among engineers, as well as a selection of the more common or more important terms from each aspect of the subject. It is hoped that, in this way, the book will be of value to all, from the apprentice or student to the experienced engineer trying to keep abreast of his field.

It has been difficult to decide what terms to omit. Some of the entries might be classified as electronics. The justification in these cases is the frequency with which they are encountered by the electrical engineer, or their similarity to other, electrical, terms. Mathematical principles also have been described briefly, where they are of importance to electrical calculations.

Since the dictionary first appeared in 1965, much of the world has changed or is changing to S.I. units, and this is necessarily reflected in this new impression although the units in common use to denote electrical quantities are based on the M.K.S.µ system and are unaltered in the International System. There have also been considerable further advances in solid-state electronics, and a short glossary of semiconductor terms has been added to the Appendix.

An effort has been made to keep the definitions concise, without omitting any necessary information. This has been done with the help of extensive cross-referencing. Key words printed in small capitals refer the reader to other entries that will provide additional necessary or supplementary information. Additional tabulated data will be found in the Appendix.

I acknowledge gratefully the help received from sources too numerous to detail. They include contributors to *Newnes Concise Encyclopaedia of Electrical Engineering* and the press officers of many leading manufacturers, who have, almost without exception, responded readily to my requests for information. I would also like to thank Mr. P. Chapman for critically reviewing the work and suggesting a number of improvements.

K.G.J.

1973

A

A.C. Balancer. See STATIC BALANCER.

A.C. Bridge. A BRIDGE circuit employing alternating current for the measurement of circuit parameters. Balance is more complex in a.c. than in d.c. bridges, because both the magnitude and the phase angle of impedances enter into the balance condition. However, this fact makes possible the precise measurement of reactive components (capacitors and inductors) as well as of resistors.

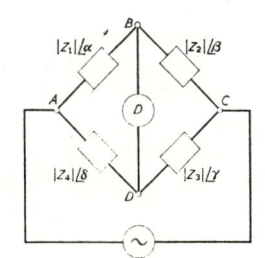

FIG. A–1. FOUR-ARM A.C. BRIDGE.

In the arrangement of Fig. A–1, by analogy with the WHEATSTONE BRIDGE, balance is obtained when

$$Z_1 Z_3 = Z_2 Z_4$$

that is, when simultaneously

$$|Z_1 Z_3| = |Z_2 Z_4| \quad \text{and} \quad \alpha + \gamma = \beta + \delta$$

Let $Z_1 = R_1 + jX_1$ be the unknown impedance. Then

$$R_1 + jX_1 = (Z_2/Z_3) \cdot Z_4 = (Z_2 Z_4) \cdot Y_3.$$

In the former relation, Z_1 is balanced by a variable impedance Z_4 multiplied by a *ratio* (Z_2/Z_3); in the latter, it is balanced by an admittance Y_3 multiplied by a *product* $(Z_2 Z_4)$. Examples of four-arm product bridges are MAXWELL BRIDGE, HAY BRIDGE and SCHERING BRIDGE. Available for either type of measurement is the UNIVERSAL BRIDGE. Six-arm bridges are the ANDERSON BRIDGE and the PARALLEL-T BRIDGE (see also HEAVYSIDE–CAMPBELL BRIDGE).

A.C. Calculating Table. A form of NETWORK ANALYSER.

A.C. Potentiometer

A.C. Potentiometer. Instrument for comparing alternating voltages. A.C. potentiometers differ from d.c. types as no alternating voltage standard comparable to a standard cell exists. In practice the a.c. potentiometer is usually balanced on direct current against a standard cell, using a transfer instrument which is commonly a precision electrodynamic galvanometer or milliammeter equally accurate on direct or alternating current. This is then used to adjust the a.c. side for balance before use on a.c. measurements. Two general forms exist: the polar, and the quadrature or rectangular co-ordinate. An example of the polar type is the DRYSDALE POTENTIOMETER, and of the quadrature type is the PEDERSEN POTENTIOMETER.

A.C. Resistance. Alternative name for EFFECTIVE RESISTANCE (q.v.).

A-Class Insulation. One of seven classes of insulating materials for electrical machinery and apparatus defined in B.S. 2757 on the basis of thermal stability in service. Class A insulation is assigned a temperature of 105°C. It consists of materials such as cotton, silk or paper, suitably impregnated or coated.

A-Connection. A three-phase transformer connection between phases which provides a neutral point.

Abbreviations. See Appendix.

Absolute Ampere. See AMPERE.

Absolute Permeability. The ratio of the magnetic flux density in a medium or material to the magnetizing force. Absolute permeability $\mu = \mu_r \mu_0$, where μ_r is the relative permeability of the material and μ_0 is the absolute permeability of free space (magnetic space constant). μ_0 is $4\pi/10^7$ in the rationalized M.K.S. system and unity in the C.G.S. system.

Absolute Permittivity. The capacitance of a unit cube of a dielectric or of space. Absolute permittivity $\epsilon = \epsilon_r \epsilon_0$, where ϵ_r is the relative permittivity of the dielectric, and ϵ_0 is the absolute permittivity of free space (electric space constant). $\epsilon_0 = (1/36\pi) 10^{-9}$ F/m in the rationalized M.K.S. system, $(1/36\pi) 10^{-11}$ F/cm in the "practical" system, and unity in the electrostatic C.G.S. system.

Absolute Unit. The unit of any system that includes length, mass and time in its fundamental units. Systems of absolute electrical units require an additional unit besides the basic mechanical units. This unit represents either the magnetic or the electric property of space. The three systems of absolute electrical units are: M.K.S., electromagnetic C.G.S. and electrostatic C.G.S. (see C.G.S. SYSTEM, M.K.S. SYSTEM also SYSTÈME INTERNATIONAL).

Absorptiometer. Instrument for determining the concentration of a solution. A beam of light of selected wavelength is transmitted through the solution. The light absorbed is proportional to concentration and is measured by means of one or two photocells.

Absorption. The reduction in the intensity of a beam of radiation (light, X-rays, etc.) during its passage through matter (see also DIELECTRIC ABSORPTION).

Absorption Dynamometer. A type of testing brake for measuring the output of electric motors and internal-combustion engines. It absorbs the output as well as measuring it.

The simplest form is the friction brake; the force required to restrain the braking device and the distance of its line of action from the axis of the shaft are measured, and the resulting restraining torque, equal to the output torque, is calculated. Alternatively, a magnetic brake can be used in which a metal disc rotates between the poles of electromagnets. For larger outputs the hydraulic dynamometer is suitable and comprises a rotor, fitted with vanes, moving in an enclosed stator also fitted with vanes; admission of water to the device results in turbulence which gives a high braking torque.

Absorption Factor. (1) The ratio of radiation absorbed in a material to the radiation incident on it. This is also known as the *absorption coefficient*.

(2) An allowance made in interior lighting calculations for light absorbed before reaching the working plane. In clear atmospheres the factor is unity, but it is less when smoke or steam is present.

Absorption Inductor. A centre-tapped inductor connected between the two star points of a double three-phase transformer supplying a mercury-arc rectifier. The centre-tap forms the d.c. negative terminal. It is also called an *interphase reactor* or *phase equalizer*.

Accelerating Machinery. See PARTICLE ACCELERATOR.

Accelerating Relay. A device which controls the time interval between the closing of successive resistor-short-circuiting contactors in starting a motor, so providing automatic acceleration.

Acceleration

Such relays may be adjustable for time delay (*timing relay*), or of the current-operated type which closes when the current passing through it drops to a predetermined value.

Acceleration. See COUNTER-E.M.F. ACCELERATION, CURRENT-CONTROL ACCELERATION, TIME-CONTROL ACCELERATION.

Acceptor Circuit. A tuned circuit which will accept a signal of one frequency more readily than those of other frequencies.

Acceptor Impurity. An element introduced into the material of a SEMICONDUCTOR, having a lower valency than the semiconductor. It captures electrons to complete its valency bonding, producing a positive charge carrier or *hole*. This gives rise to p-type conductivity. Cf. DONOR IMPURITY.

Access Time. The time taken to extract an item of information (a number or an instruction) from the store of a DIGITAL COMPUTER.

Accumulator. A device for receiving and storing energy, and discharging it, by chemical action. Also known as a *storage cell* or *secondary cell*, the accumulator is reversible, i.e. it can, after discharging, be brought back to a full state of charge by passing a reverse current through it.

A storage cell consists essentially of two electrodes (plates) immersed in an electrolyte in a suitable container. In practice, multiple plates may be used in the cell. There are two basic types: LEAD–ACID CELL and STEEL–ALKALINE CELL.

Acheson Furnace. An electric furnace for producing silicon carbide. A heavy current is passed through a central core of coke, around which the coke and sand which make the silicon carbide are packed.

Acid Dip. An acid into which chemically cleaned metals are dipped before electroplating in order to remove films of oxide, etc., which would interfere with the plating process.

Acoustic Controller. A form of automatic control by noise level, comprising a microphone, an amplifier, a high-pass filter to segregate the significant noise from low-frequency external noise, and a relay adjusted to operate in accordance with fluctuations in noise above and below a set level. An application is control of the feed motors in a pulverized fuel mill, by noise level inside the pulverizing drums.

Acoustic Sounding. A system of determining the depth of the sea-bed or an immersed object, by time-interval measurements of the echo of acoustic waves, using piezoelectric oscillators and recording instruments.

Acrylic Resin. Thermoplastic synthetic resin made from acrylic and methacrylic acids, used as insulating material. Polymethylmethacrylate (Perspex, Diakon, etc.) is the most important; it is a rigid, clear, transparent resin which can be obtained in flat or curved sheets, rods, tubes, and mouldings made from powder. It is of value for high-frequency use, but has a low softening point which limits it to low-temperature applications.

Actino-therapy. Ultra-violet therapy, i.e. treatment of disease by radiation with ultra-violet waves (4,000–2,500 Å).

Active Component. The component of an alternating current or voltage (considered as vector quantities) which is in phase with the current or voltage (see also ACTIVE VOLT–AMPERES).

Active Electrode. An electrode of an ELECTROSTATIC PRECIPITATOR. The active (discharge) electrode is insulated, and the other (receiving) electrode, usually positive and earthed, collects the precipitated particles.

Active Volt–Amperes. Product of the current and ACTIVE COMPONENT of the voltage or of the voltage and active component of the current.

Actuator. An electromechanical device for performing a mechanical action, usually with a linear rather than a rotary motion. The motion is speed- and position-controlled in the same way as in a SERVOMECHANISM.

Adaptor. A device which may be inserted into a SOCKET–OUTLET to receive one or more plugs which may or may not be of the same type as that for which the socket–outlet is intended. The name is also sometimes applied to accessories for insertion into a lampholder.

Admittance. The ratio current/voltage in a circuit. It is the reciprocal of impedance and is measured in reciprocal-ohms or mhos ($\mho$).

Advancer. See PHASE ADVANCER.

After-Glow. The emission of light from a phosphor after the cessation of bombardment by an electron beam.

Ageing. The gradual change that takes place with time in the useful properties of a material.

Air. A mixture of gases, water and various pollutions comprising the atmosphere. It is an important insulating medium. The main constituents of clean dry air, by volume, are nitrogen, 78·08%; oxygen, 20·95%; argon, 0·93%; carbon dioxide, 0·03%. Its electric strength is 30 kV/cm (76 V/mil) at 0°C and 760 mm

Air-Blast Circuit-Breaker

pressure, increasing with pressure. The sparkover voltage is 330 V d.c. minimum, at atmospheric pressure.

Air-Blast Circuit-Breaker. A CIRCUIT-BREAKER in which the arc is extinguished by a blast of air. It may use compressed air to force the contacts apart as well as to blow out the resulting arc. Very high breaking capacities are possible at voltages from 6,600 V a.c. to 380 kV. In all forms of air-blast circuit-breaker, high-pressure air causes the arcing contacts to open, by pressure either on the contacts themselves, or on an actuating piston. A jet of air sweeps the arc into a low-pressure zone where it is de-ionized and extinguished at a current zero. Arcing times of less than one cycle are possible. Axial-blast, radial-blast and cross-blast jets are employed, the last being used with arc chutes, for the lower voltages only. Arc shutes are not required for axial and radial-blast circuit-breakers, but cooling vanes may be built into the exhaust chambers. See Fig. A–2.

Air-Break Circuit-Breaker. A CIRCUIT-BREAKER in which the circuit is opened in air. It utilizes "free" air at atmospheric pressure to extinguish the arc and to provide insulation, thereby avoiding the need for processing of the dielectric. Free-air circuit-breakers are manufactured for voltages up to 3,000 V d.c. and 3,300 V a.c. in three distinct forms: low-breaking-capacity a.c. and d.c. circuit-breakers, high-speed d.c. circuit-breakers and high-breaking-capacity a.c. circuit-breakers.

Air Capacitor. A CAPACITOR in which the dielectric is air.

Air Conditioning. The production of an artificial atmosphere suited to particular requirements. It may include cleaning, humidifying, dehumidifying, cooling or heating, air movement and possibly ozonizing.

Air-Cooling. The carrying away, by means of air, of the heat caused by inevitable losses in the operation of a motor, transformer or other machine. An open-type machine may be cooled by natural convection, a protected type by fan or fans mounted on the shaft, and an enclosed type either by a fan mounted internally or externally, or by air blown through ingoing and outgoing ventilation pipes.

Air Core. Inductor core that is non-ferromagnetic and (usually) non-metallic.

Air-Gap. (1) Part of the magnetic circuit of rotating electrical machines and electromagnets, requiring the largest proportion of the magnetomotive force; for example, the space between

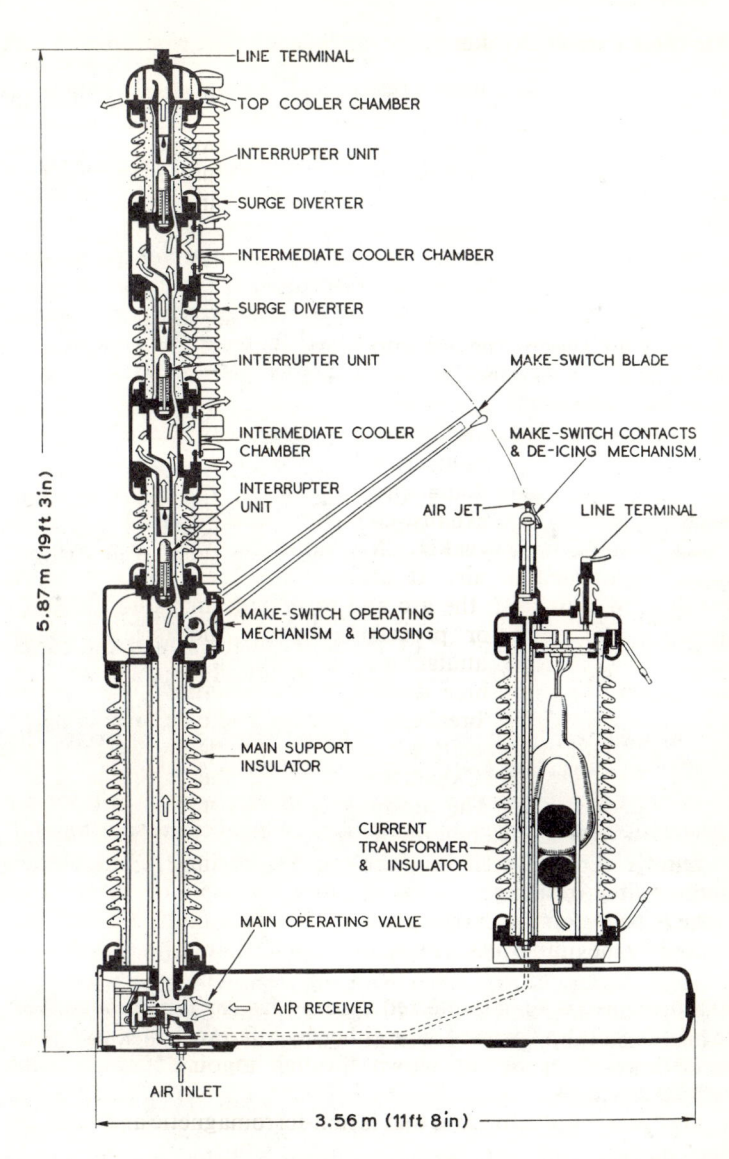

FIG. A–2. SECTIONAL DRAWING OF SINGLE-POLE 165 kV AIR-BLAST CIRCUIT-BREAKER.
(*The English Electric Co. Ltd.*)

Air Termination

the rotor and stator of a motor, or between the armature and core of an electromagnet.

(2) A point of discontinuity in an overhead conductor for electric traction (see SECTION GAP).

Air Termination. Part of a lightning-conductor system designed to collect discharges from the atmosphere or to distribute charge into the atmosphere.

Alcomax. Trade name for nickel–aluminium permanent-magnet materials exhibiting anisotropic characteristics. The properties in the preferred direction of magnetization are obtained by heat treatment in a magnetic field, and are very superior to the properties in other directions. See NICKEL ALLOYS.

Alive. See LIVE.

Alkaline Cell. See STEEL–ALKALINE CELL.

Alkyd Resin. Synthetic resin, the condensation products of polybasic acids with polyhydric alcohols. They are used (non-hardening resins) as adhesives, or (thermosetting types) as finishes or for bonding insulating materials such as asbestos or mica.

All-In Tariff. An electricity supply TARIFF which does not involve any consideration of the purpose for which the supply is to be used. A domestic supply is usually on a TWO-PART TARIFF.

All-Insulated. Having all external surfaces protected by coverings consisting entirely of insulating material.

Allan Cell. American name for an electrolytic cell for the production of hydrogen by electrolysis of an alkaline solution.

Allen's Loop Test. A fault-localization test, suitable for finding high-resistance faults in short lengths of cable. It is similar to VARLEY'S LOOP TEST.

Alni, Alnico. Trade names for nickel–aluminium permanent-magnet materials exhibiting isotropic characteristics. Alni was the first commercially used alloy, and is now partially displaced. Alnico has the highest magnetic energy of the isotropic alloys. See NICKEL ALLOYS.

Alpha Particle. Nucleus of a helium atom. It consists of two protons and two neutrons. It has a positive electric charge and may be expelled at high velocity from a disintegrating nucleus of greater complexity.

Alternating Current, Voltage. A current or voltage varying with time in a cyclic manner, the current direction and the voltage polarity reversing periodically.

Alternator. An a.c. GENERATOR having its field windings excited by means of direct current. Alternators may be divided into two types depending on the arrangement of the field system: (*a*) the *turbo type* where the surface of the rotor is a smooth cylinder and the field winding is placed in slots machined longitudinally in the steel (see TURBO-ALTERNATOR), and (*b*) the *salient-pole type* having pole pieces, surrounded by the field coils, projecting outwards from a hub.

Alumel. A nickel-base alloy containing manganese, aluminium and silicon. It is stable at temperatures up to 1,200°C and is used in thermocouples and as an electrical resistance alloy. See NICKEL ALLOYS.

Aluminium. A bluish-white metal, atomic number 13 and atomic weight 27. It possesses properties of value in electrical applications. Pure aluminium is about one-third of the weight of iron, copper, zinc or their alloys. Its electrical conductivity is high, and its alloys have high tensile strength. It is corrosion-resisting.

Steel-cored aluminium conductors are widely used for medium- and long-span lines and for very long individual spans. Aluminium wire, usually square, is sometimes used for winding the coils of large lifting magnets to save weight, and it can be insulated to withstand high temperatures. In the form of flat bar, rod or tube, aluminium is frequently used for busbars. Other applications of pure aluminium are for electricity meter discs, and, as pressed or drawn sheet, for electrostatic screening.

Aluminium Alloy. Aluminium combined with another metal or metals, principally copper, magnesium, zinc, silicon, manganese or nickel to improve its characteristics. Alloy with 5–12% silicon is used for busbar casings, since it is non-magnetic, light, and has good casting and reasonable machining properties. High-tensile alloys such as Duralumin are used for the moving parts of switchgear, and result in a reduction of inertia effects. Silicon alloys are used for castings for cage rotors, the percentage of silicon varying between 5% and 12%, depending on the resistivity required. Aluminium alloys containing 2·5–5% copper are used for the collector bows used with overhead wires by electric trolleys and locomotives.

Aluminium Rectifier

Aluminium Rectifier. A type of ELECTROLYTIC RECTIFIER comprising cells each with an electrolyte of ammonium phosphate, with an aluminium anode and an inert lead-plate cathode. Such a cell is an undirectional conductor. The action is chemical as it depends on the forming and reforming of a film on the anode surface. There is appreciable reverse current and the efficiency is low. Only low voltages and currents can be handled by a single cell, but a number may be connected in series or parallel.

American Wire Gauge. See BROWN AND SHARP GAUGE.

Ammeter. An ampere-meter, an instrument for measuring current, with a scale and moving element which carries a pointer. It is connected in series with the circuit carrying the current to be measured and is of low electrical resistance or impedance, in order that only a small voltage drop shall be caused in the circuit and a minimum of power absorbed from it. Ammeters may be MOVING-IRON, MOVING-COIL, ELECTRODYNAMIC, HOT-WIRE, INDUCTION or RECTIFIER INSTRUMENTS.

Ammeter Shunt. One or more flat strips of metal alloy, of very low temperature coefficient (e.g. manganin), and of definite resistance, connected in the main circuit and across the terminals of a moving-coil millivoltmeter which is calibrated as an ammeter.

Ammonium Dihydrogen Phosphate. A material, generally known as ADP, showing pronounced piezoelectric effects. It has been used as a transducer element for ultrasonic depth measurement and in submarine detection systems.

Amortisseur. Damping winding of a machine (see DAMPER).

Ampere. The practical unit of electric current; also the absolute unit in the M.K.S. system (abbreviation: A, symbol: I). An *absolute ampere* is that steady current which, when maintained in two parallel rectilinear conductors of infinite length and negligible cross section and separated by a distance of 1m in free space, produces between them a force of 2×10^{-7} newton per metre length. The ampere is one of the seven basic S.I. units. See also INTERNATIONAL AMPERE, SYSTÈME INTERNATIONAL.

Ampere Balance. A laboratory current-measuring device. A sliding weight balances the force between a pair of coils carried on the weight beam and a system of fixed coils through which the same current is passing.

Ampere-Hour. A unit of electric quantity. It is the amount of electricity which has passed when one ampere has been flowing for one hour. 1 Ah = 3,600 coulombs.

Ampere-Hour Capacity. The capacity of a battery expressed in ampere-hours, usually based on continuous discharge at a specified rate.

Ampere-Hour Efficiency. The ratio between the output quantity in ampere-hours from a battery during a test discharge and the input Ah required to charge the battery.

Ampere-Hour Meter. A direct-current electricity-supply meter which measures the current-time summation, or ampere-hours.

Ampere-Turn. The practical and M.K.S. unit for magnetomotive force. Abbreviation: AT. The magnetomotive force in AT is the product of the number of turns of a coil and the current in amperes passing through the turns.

Ampere's Law. The summation of the magnetic force round a closed path is equal to the total current flowing across the surface bounded by the path. See MAXWELL LAWS.

Amphoteric Substance. A substance which can dissociate electrolytically in two different ways, so as to give either hydrogen (positive) or hydroxyl (negative) ions, according to conditions.

Amplidyne. Trade name for a range of ROTATING AMPLIFIERS of the cross-field excited type. It is a quick-response d.c. generator, the output of which is controlled by a very small field power. Because of this property, the Amplidyne is particularly suitable for use as an exciter in a closed loop control system.

The method of excitation can be explained by reference to Fig. A–3. If current is supplied to the control-field winding, a

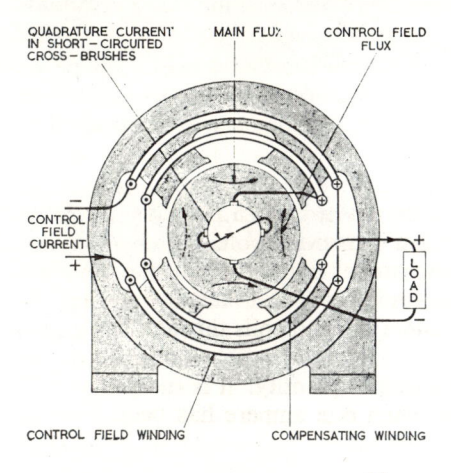

FIG. A–3. INTERNAL CONNECTIONS OF AMPLIDYNE.

Amplifier

flux (indicated by two dotted arrows) is set up on the vertical axis and induces an electromotive force in the armature conductors between the cross brushes. As these brushes are joined together, the e.m.f. causes a current to flow, and the passage of the quadrature current through the armature windings sets up an armature-reaction field (indicated by the two solid arrows). This field is used to induce the main ouptut voltage of the machine between the other pair of brushes shown on the the vertical centre-line.

The e.m.f. required to pass the quadrature current through the armature winding and the short-circuited cross brushes is small compared with the full output voltage of the machine, and the power needed by the control field winding is therefore only a small fraction of that which would be required to excite the machine in the normal way. In a fully compensated machine the power amplification ratio can be very large (20,000 or more).

Amplifier. Apparatus for producing a magnified version of an input signal. The term is commonly applied to an electronic circuit for amplifying a voltage applied to its input terminals. See also MAGNETIC AMPLIFIER, ROTATING AMPLIFIER.

Amplitude. The maximum instantaneous positive or negative value of an alternating, oscillating or vibrating quantity. It is preferably known as the *peak value*.

Amplitude Distortion. A change in waveform, when deviations in the ratio of the r.m.s. value of the output to that of the input occur for different input amplitudes due to a non-linear response in the system.

Amplitude Modulation. The imposition of a signal shape on to an alternating carrier wave by variation of the carrier amplitude. The maximum signal frequency must be considerably lower than that of the carrier.

Analogue. A system whose behaviour can be expressed in the same form as that of another distinct system. In particular an electric circuit can be arranged so that its voltages, currents and charges correspond completely to the forces, velocities and displacements in a mechanical system, with the advantage that the measurement of the electrical quantities, whether in the steady or the transient state, are normally very much more direct and simple than the comparable quantities in the actual system.

Analogue Computer. Computing device in which the variables are represented by physical quantities such as lengths, pressures, electric charges, etc., the calculations consisting in the manipulation and measurement of these quantities, the values of which may change continuously. The device of greatest interest in engineering is the electronic differential analyzer in which magnitudes such as electric charge or magnetic strength represent the variables. Cf. DIGITAL COMPUTER.

Anchor Clamp. A fitting for attaching an overhead-line conductor to a tension insulator or support. A grooved plate clamped down by bolts may be satisfactory for small fittings but heavier conductors require a SNAIL CLAMP.

Anchor Tower. A TOWER, placed at intervals along an overhead line to provide longitudinal rigidity.

Anderson Bridge. A six-arm A.C. BRIDGE, convenient for measuring inductance in terms of a fixed capacitance.

Andreau Generator. A form of WIND-POWER GENERATOR. It is shown diagramatically in Fig. A-4. The tubular blades

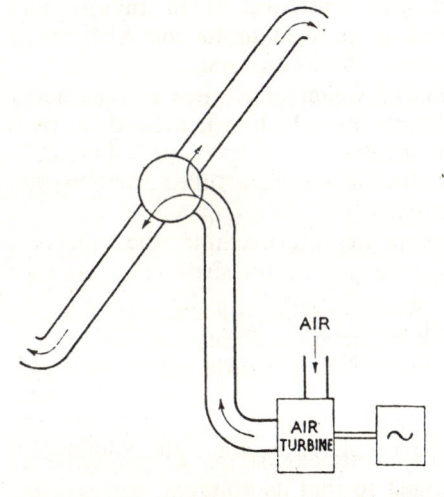

FIG. A-4. ANDREAU WIND-POWER GENERATOR.

As the wind drives the blades round, air is thrown out centrifugally from the blade tips.

throw air out at their extremities, thus permitting air to enter at the hub from a duct in which is situated an air turbine driving a generator. This scheme is theoretically less efficient than the more common propellor type, but has the advantage that the generating equipment is fixed and located at ground level.

Ångström Unit. A unit of length equal to 10^{-10} m or 10^{-8} cm or $10^{-4}\,\mu$, employed in measuring the wavelengths of short electromagnetic waves, particularly in the X-ray spectrum (abbreviation: Å).

Angular Frequency. Measurement of frequency of a regular recurring phenomenon. It is expressed in radians per second, i.e. $\omega = 2\pi f$ rad/sec, for a frequency f in cycles per second. It corresponds to the angular velocity of rotation in radians per second of a two-pole synchronous machine, and to the time angle in radians per second of a time-varying sine wave of frequency f.

Anion. A negatively charged Ion, existing in an electrolyte or gaseous discharge. It will travel towards the positive electrode or anode under the influence of an applied difference of potential.

Anisotropic Conductivity. Property possessed by a body which has different conductivity for various directions of current flow.

Anisotropic Magnetism. Property exhibited by certain alloys when heat-treated in a strong magnetic field. They develop extremely strong magnetic properties in the direction of the field at the expense of those in other directions. This leads to very powerful and stable magnets. Typical materials are Alcomax II and Hycomax.

Anode. The electrode out of which an electric current flows, with the convention that the current direction is defined in terms of the movement of positive charges. The term is applied typically to the appropriate electrodes of electron tubes, mercury-arc rectifiers and electrolytic baths.

Anode Drop. In an electric welding machine, the difference in voltage between the positive end of the arc-stream and the adjoining conductor.

Anodic. (1) Having a more negative electrode potential.

(2) Indicating an element above hydrogen in the electrochemical scale.

Anodic Etching. A preparatory process for deposition by electroplating. The metal to be plated is made the anode in a suitable electrolyte.

Anodizing. The production, on the surface of parts made of aluminium or its alloys, of a protective film of hydroxide. The part is made the anode of an electrolytic bath, and a current is passed.

Anolyte. The portion of an electrolyte surrounding the anode, which is affected chemically by reactions taking place at the anode.

Arc-Control Device

Antiferromagnetism. Phenomenon exhibited by materials in which the magnetic moments of adjacent ions are in opposition. Their susceptibility passes through a maximum as the temperature increases. Examples of such materials are manganese oxide, chromic oxide, manganese selenide and zinc ferrite.

Aperiodic. Incapable of sustained oscillation.

Apparent Resistance. An alternative term for IMPEDANCE (q.v.).

Arc. A column of ionized gas, appearing as a luminous discharge. Any gas or vapour will become an electrical conductor if it is heated to a sufficiently high temperature, and when cooled again will become quite a good insulator.

An arc, Fig. A–5, is usually divided into three parts in series: the cathode, the column or plasma, and the anode.

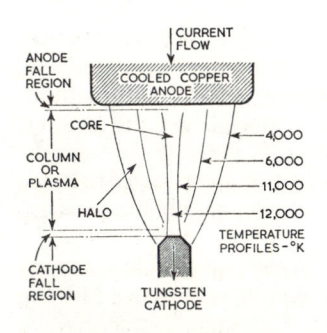

FIG. A–5. A CHARACTERISTIC ELECTRIC ARC.

The arc shown is of 200 A in nitrogen at atmospheric pressure.

Arc-Back. An alternative term for BACKFIRE (q.v.) in mercury-arc rectifiers.

Arc Chute. Device employed in an air-break circuit-breaker to confine the arc and prevent it from spreading to adjacent metalwork. A typical form is shown in Fig. A–6. Several slotted plates of insulating material as shown in (*a*) are arranged between two side plates so that a sectional plan on the plane PQ forms the labyrinth (*b*). The contacts are located in the lower portion of the slot so that, after striking, the arc rises up in the slots, eventually taking up the zig-zag form shown in (*b*) and becoming lengthened and cooled as required.

Arc-Control Device. Equipment for controlling the arc which inevitably occurs between the contacts of the interrupting device when a circuit is interrupted. Almost all circuit-breakers must be fitted with arc-control devices designed to keep the duration

Arc-Drop

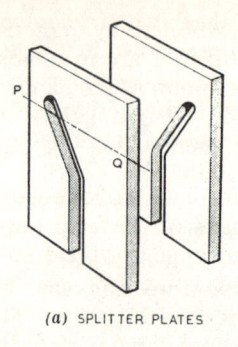

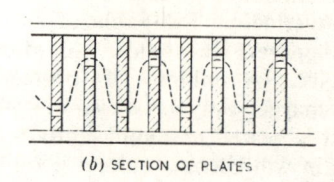

(a) SPLITTER PLATES

(b) SECTION OF PLATES

FIG. A-6. ARC CHUTE.

of the arc to a value between about 10 and 100 ms. See ARC CHUTE, ARC SPLITTER.

Arc-Drop. The total voltage between anode and cathode during the passage between them of an arc. Near the cathode there is a steep voltage gradient and a total *cathode fall*. At the anode surface there is an *anode fall*. In the arc plasma between anode and cathode there is a voltage gradient roughly independent of its length.

Arc Furnace. An electric melting furnace in which an arc is struck between electrodes, the metallic charge sometimes forming one of the electrodes (see DIRECT-ARC FURNACE, INDIRECT-ARC FURNACE).

Arc Rectifier. A rectifier containing a cathode and an anode between which an arc is maintained (see MERCURY-ARC RECTIFIER).

Arc Splitter. An arc-control device comprising an ARC CHUTE in which the side plates are separated by metal plates or pins. As the arc travels up it is split into a number of smaller arcs in series. At the anode and cathode of any arc there are significant voltage drops, so that by introducing a number of such drops, as well as by some lengthening and cooling, a high effective arc-resistance can be achieved.

Arc-Stream Voltage. In an electric welding machine, the difference in voltage between the two ends of the liquid or gaseous arc.

Arc-Suppression Coil. A neutral-earthing inductor in a three-phase, overhead-line, power-distribution system. It is also known as a *Petersen coil*.

Arc Voltage. In an electric welding machine, the sum of the CATHODE DROP, ARC-STREAM VOLTAGE and ANODE DROP.

Arc Welding. Fusion welding, the heat necessary to melt both the parent plate and filler material being derived from an electric arc. The process may involve either a consumable or a non-consumable electrode.

In consumable electrode processes, the arc is struck between the parent plate and an electrode which melts off to form the filler metal required in the joint. See, for example, SUBMERGED ARC WELDING. In non-consumable electrode processes, the arc is struck between a non-consumable electrode and the workpiece, with filler metal fed separately into the joint if required (see ARGON ARC WELDING, ATOMIC HYDROGEN WELDING, CARBON ARC WELDING).

Arcing Contacts. Contacts incorporated in switchgear to protect the main contacts from any deleterious effects due to arcing. They open after and close before the main contacts. They are made arc-resisting by suitable choice of material (e.g. carbon). They are also easily renewable.

Arcing Horn. Metal projections placed at ends of overhead-line insulator strings to provide an alternative path for an arc which might otherwise damage the insulator units.

Arcing Shield. See GRADING SHIELD.

Argon Arc Welding. The commonest form of non-consumable electrode ARC WELDING. It uses a tungsten electrode. A shield of inert gas, usually argon but occasionally helium, protects both the weld pool and the electrode from atmospheric contamination.

Armature. (1) Of a machine, one of the main members in which rotational electromotive forces are induced, and which is associated with the conversion of electrical to or from mechanical energy. In a d.c. machine and an induction machine the armature is almost invariably the rotating member; in a synchronous machine it is normally the stationary member (or stator).

(2) Of an electromagnetic circuit, a movable part which, by attraction, can be made to do useful mechanical work, as in a relay or contactor.

(3) Of a permanent-magnet, see KEEPER.

Armature Head. Alternative term for the end-plate of a laminated armature core.

Armature Reaction. The effect on the total working flux in an

Armouring

electrical machine of the armature magnetomotive force, produced by armature currents.

Armouring. The steel wires or tapes wrapped round a cable for mechanical protection.

Aroclor. A chlorinated diphenyl used as a dielectric in industrial capacitors.

Askarel. An American term for a class of synthetic liquids used as non-inflammable filling media in transformers. A typical example is Pyroclor.

Astatic System. An arrangement of magnets or coils so designed that no resultant force is exerted on it by an external uniform magnetic field.

Asymmetrical Breaking Capacity. The r.m.s. value of the combined a.c. and d.c. components of current that can be broken at a stated voltage by any one pole of a circuit-breaker.

Asynchronous Machine. An a.c. electrical machine whose normal operating speed (unlike that of a synchronous machine) is not restricted to the synchronous speed. The operating characteristics are, however, functions of the speed-difference, or slip, by which the operating speed departs from the synchronous.

Asynchronous Condenser. An Asynchronous Machine used for power-factor correction. See Synchronous Condenser.

Atmospheric Electricity. Atmospheric charges existing in the Earth's atmosphere. The voltage gradient near the ground is of the order of 150 V/m in fine weather, 1·5 kV/m in fog, 10 kV/m in thunderstorm conditions. The gradient decreases with height above ground, and is affected by the time of day. A positive gradient, with the potential increasing positively with height, implies a negative charge on the surface of the ground, of the order 1 μC/km^2. The "leakage" current from air to ground is of the order of 2 μA/km^2.

Atomic Energy. See Nuclear Fission.

Atomic Hydrogen Welding. A form of non-consumable electrode Arc Welding which employs two tungsten electrodes between which the arc is struck. A stream of hydrogen is directed across the arc and the molecules of hydrogen break down into atoms. As the gas stream passes out of the arc the atoms of hydrogen recombine and the exothermic reaction produces a very high temperature "flame" with which the welding is carried out. The hydrogen burns as it leaves the arc, thus protecting the weld

Attenuator

zone from atmospheric contamination. It is also called *atomic arc welding*.

Atomic Number. The number of PROTONS or positive charges in the nucleus of an atom, and hence the number of electrons in the neutral atom.

Atomic Structure. In simplified terms, an atom consists of a small dense nucleus of PROTONS and NEUTRONS, which is positively charged, surrounded at relatively great distances by orbital electrons having, in the neutral stage of the atom, an equal and opposite total charge.

Attenuation. A reduction in the magnitude of a quantity; particularly that reduction in the amplitude of a voltage or current due to its transmission over a line or through an ATTENUATOR. In each case the series resistance and shunt conductance losses reduce the signal energy during its passage. See also FREQUENCY DISTORTION.

Attenuation Coefficient. The ratio of input to output voltage or current in a transmission line or ATTENUATOR expressed in exponential terms. If the input voltage is v_1 and the output voltage is v_2, the ratio v_2/v_1 is expressed as $\exp(-\alpha)$ where α is the attenuation coefficient in nepers.

Attenuator. A device for introducing a specified loss between a signal source and a matched load without upsetting the impedance relationship necessary for matching. Resistors arranged in T or π formation, called attenuator pads, may be used for this purpose. Fig. A–7 shows at (*a*) a simple symmetrical-T pad. If the attenuation has to be adjustable in steps, the bridged-T

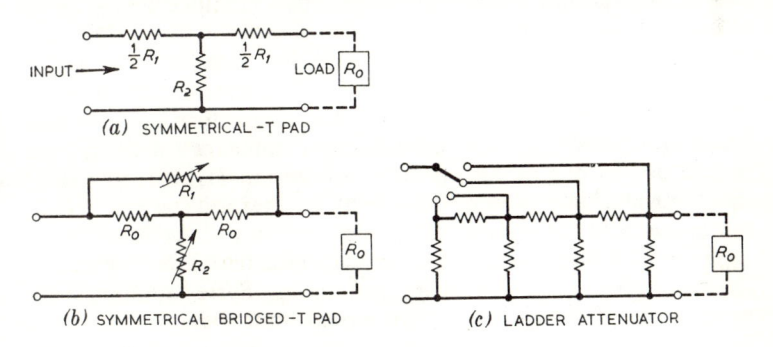

FIG. A–7. ATTENUATORS.

Attracted-Armature Relay

form shown in (*b*) is preferred, the variation being obtained by switching R_1 and R_2 simultaneously. More economical from the viewpoint of simple switching is the ladder attenuator (*c*); but on the minimum-attenuation position (at the right-hand end) the network still shunts the load, giving a loss of 3 dB.

Attracted-Armature Relay. A relay consisting of an electromagnet, the field of which influences an armature arranged to operate the contacts. When the coil is energized the force on the armature is proportional to (ampere-turns/gap-length)² and hence increases as the armature moves to reduce the gap.

Audiometer. Instrument used for diagnosis of hearing ability and otological research. It is an accurately calibrated variable-frequency audio oscillator producing pure sounds over the range 50 Hz–20 kHz. The amplitude of the signal can be varied, and in testing auditory acuity the patient is asked to indicate when the sound appears and disappears for various frequencies.

Automatic Control. The control of a device, system or manufacturing process with the minimum of human intervention. It is usual to exclude from this category the many forms of semi-automatic operation (as, for example, automatic passenger-operated lifts) that require some form of human initiation, even if it be only minor.

Automatic Reclosure. Feature incorporated in transmission and distribution systems to avoid unneceesary prolonged disconnection of lines by transient faults.

Automatic Voltage Regulator. A control device which automatically regulates the voltage at the exciter of an alternator, to hold the output voltage constant within specified limits, or to allow it to vary in a predetermined manner. There are four main types: rheostatic, pulse, normally inactive and static.

A *rheostatic-type regulator* consists of an automatically operated rheostat that usually controls the main exciter field current, and is generally used for the control of a small or medium-sized generator. Simple motor operation of the rheostat is rather slow, and the rheostat is usually in the form of a carbon stack or rolling sector (see CARBON-PILE REGULATOR). A PULSE-TYPE REGULATOR feeds the excitation current into the exciter field in a series of variable pulses. A *normally inactive regulator* combines the advantages of the two previous types. As the name implies, it functions only when some event demands a change of excitation. These electromechanical regulators described are not frictionless,

Auto-transformer

and most large generators are now equipped with some form of STATIC REGULATOR.

Automation. The automatic control of manufacturing and other processes through all their stages, involving self-examination and feedback of corrections. The development of automation opens up the prospect of operating factories and offices with the minimum of direct human labour.

Auto-transformer. A transformer possessing only one winding per phase, part of this winding being common to both primary and secondary sides. This results in an appreciable saving in first cost and higher operating efficiency compared with a double-wound transformer, provided that the ratio of transformation is not greater than about 2. A *single-phase* auto-transformer consists of a single winding. The most widely used *three-phase* auto-transformer is star-connected, as shown in Fig. A–8. A

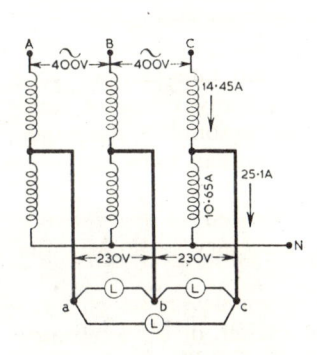

FIG. A–8. CONNECTIONS FOR 10 kVA 400/230 V THREE-PHASE AUTO-TRANSFORMER.

neutral point is available for giving a four-wire supply. The auto-transformer suffers from a number of disadvantages which restrict its use. As there is only a single winding, its inherent reactance is low, and it has therefore restricted self-protection against damage from short-circuits. Owing to the electrical continuity of the windings the higher voltage is liable in the event of a fault to become impressed on the low-voltage system, which makes added insulation and additional precautions necessary.

Auto-transformers find their greatest applications as voltage-regulating devices, for processes involving a variable control, such as the loading on furnace transformers, and for supplying a reduced voltage when starting up induction and synchronous motors. See also VARIAC.

Auto-transformer Starter

Auto-transformer Starter. A tapped AUTO-TRANSFORMER by means of which reduced voltage can be applied to a motor to limit the current taken during starting. The motor is started by connecting its primary to tappings on the starting transformer. After a time delay, it is re-connected direct to the supply. The basic diagram is shown in Fig. A-9. The simple auto-transformer

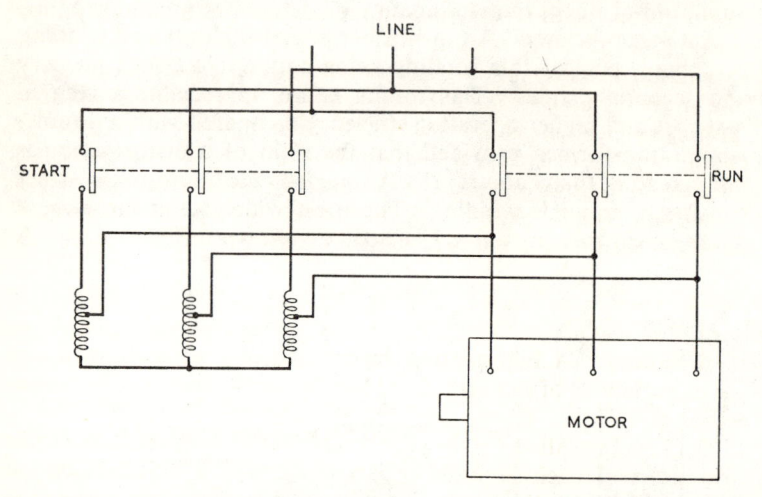

FIG. A-9. BASIC DIAGRAM OF AUTO-TRANSFORMER STARTER.

starter has the disadvantage that at the instant of transition from *start* to *run*, the supply to the motor is interrupted. This means that the insulation may be stressed by high transient voltages. This is avoided in KORNDORFER STARTING.

Auxiliary Contacts. Contacts in a CONTACTOR for operating auxiliary devices or for carrying the circuit current during make and break. See also ARCING CONTACTS.

Auxiliary Plant. Plant supplied to a generating station, such as feed pumps, oil pumps, circulating-water pumps, coal and ash-handling gear, switch tripping and lighting, etc., the requirements depending on the type of station.

Avalanche Breakdown. A component of the ZENER EFFECT in semiconductors, occurring as the result of the presence of holes and electrons. During acceleration across a *p–n* junction the charge carriers collide with electrons in the valence band to create electron-hole pairs. These additional charge carriers are, in turn, accelerated and give rise to further ionization. The effect is similar to the electron avalanche which occurs in a TOWNSEND DISCHARGE.

Avalanche Transistor. A JUNCTION TRANSISTOR in which the collector is operated into the current multiplication breakdown region. This gives a current gain greater than unity and produces a negative resistance region.

Average Value. With reference to an alternating quantity, the mean value over an integral number of repetitions. With a pure alternating quantity there is an average value of zero over a whole number of cycles; in this case the term is taken to describe the average value over one half cycle taken between successive zeros.

Ayrton Shunt. A form of SHUNT designed for open-circuit work or for use in circuits of high resistance.

B

B-Battery. A high-tension battery, usually consisting of one or more blocks of dry cells, used for the anode circuit of a thermionic valve.

B-Class Insulation. One of seven classes of insulating materials for electrical machinery and apparatus defined in B.S. 2757 on the basis of thermal stability in service. Class B insulation is assigned a temperature of 130°C. It consists of materials such as mica, glass fibre or asbestos suitably impregnated or coated.

B/H Curve. The graph of the relation between the magnetic flux density B and the magnetizing force H for a ferromagnetic material. The ratio of B to H is the absolute permeability, which varies strongly with flux density. Fig. B–1 shows the B/H curve

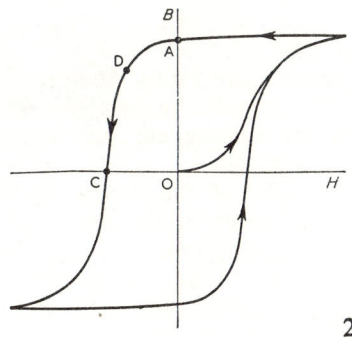

FIG. B–1. HYSTERESIS LOOP OF A PERMANENT-MAGNET MATERIAL.

B.O.T. Ohm

of a permanent-magnet material in which $B = OA$ is the *retentivity*, $H = OC$ is the *coercivity* and ADC is the *demagnetization curve*.

B.O.T. Ohm. See INTERNATIONAL OHM.

Back E.M.F. Voltage set up in a circuit opposing the applied voltage. For example, a back e.m.f. is generated in the armature conductors of a d.c. motor owing to the rotation of the armature in its magnetic field.

Back-Pressure Plant. Steam-driven generating sets delivering exhaust steam at a positive pressure suitable for industrial process work.

Back-To-Back Test. A method of testing motors or generators under full-load conditions with only a small consumption of power. Two exactly similar machines must be available, and they are coupled together so that one tends to drive the other, the driving current being supplied by the second machine. It is only necessary to supply from an external source the power to make up for the losses in the two machines.

Three methods of making back-to-back tests on series motors are shown in Fig. B–2. In method (*a*) a variable-voltage booster

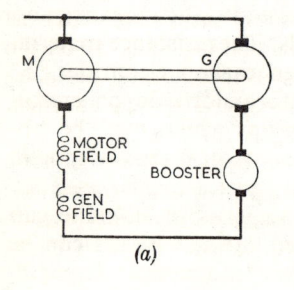

FIG. B–2. BACK-TO-BACK TESTS ON SERIES MOTORS.

The motor acting as a generator (G) must be similar to the supplied motor (M).

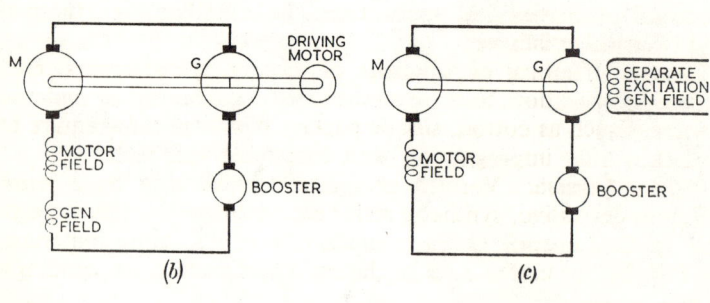

is added to the generator circuit to raise its voltage slightly above the motor voltage, so that the output of the generator can be returned to the supply. Method (*b*) employs an auxiliary driving motor, which supplies the iron, friction and windage losses, and a low-voltage booster which supplies the copper losses. Method (*c*) uses a separate source of excitation, although on small motors the field can be put in series with the generator and controlled by a diverter. See also HOPKINSON TEST.

Back-Up Protection. A scheme of equipment protection, the stability of which is dependent on the operation of protective systems of other apparatus, and the operation of which extends over a range of protected or unprotected apparatus.

Backfire. Of a mercury arc rectifier, its sudden failure due to over-heating of the anode and consequent excessive density of mercury vapour. It is also called *arc-back*.

Backward Lead, Backward Shift. Of the brushes on a motor commutator, their angular displacement from the neutral open-circuit position in a contrary direction to that of the rotation of the commutator.

Bacon Cell. A FUEL CELL employing hydrogen and oxygen as the reactants. It is shown in Fig. F–6.

Baily Furnace. A type of resistance furnace used for annealing and for heating bars, ingots, etc., in rolling mills. The resistance material consists of crushed coke packed between carbon electrodes.

Bakelite. Trade name for a phenolformaldehyde or cresolformaldehyde thermosetting synthetic resin, which may be obtained (before curing) as a semi-liquid or solid. It is readily soluble in methylated spirit or acetone, in which form it can be used for coating paper, fabric and wood. Some varieties of Bakelite can be cast by pouring into moulds and then curing. Fillers can be included in the cast to give the required mechanical strength, electrical properties and appearance. Bakelite has high thermal and electrical resistance.

Baking. Heating of windings of motors, transformers, etc., before impregnation, to drive off the moisture absorbed by material coverings such as cotton, silk or paper. Windings also require to be baked after impregnation with insulating varnishes.

Baking Varnish. Varnish or enamel, which may be a resin, cellulose derivative, synthetic resin, etc., used to provide a tough and resilient coating in the manufacture of insulating materials, or for the protective and insulating impregnation of windings.

Balance

The evaporation of solvents and the hardening of the material is effected by baking at, say, 100°C for several hours.

Balance. Term used in connection with bridge measurements. A balance is obtained when the impedances forming the arms of the bridge have been adjusted so that no current flows through the galvanometer (see A. C. BRIDGE, D. C. BRIDGE).

Balanced Current Protection. A form of system protection arranged to detect phase faults and earth faults or, more simply, earth faults only. In both cases current transformers are mounted on each circuit connected to the section being protected. These have their secondary windings connected in parallel through wires across which a relay is connected. Under healthy conditions the balance of primary current is reproduced in the secondary circuit and the relay remains unenergized. During internal faults, the current-balance condition fails and the relay operates. By suitable switching of the secondary circuits complete discrimination between faults on the various sections is achieved for all normal operating conditions.

Balanced Load. Symmetrical arrangement of loads on a system. A d.c. three-wire system is said to be balanced when the load is equally shared between the middle wire and the outers. An a.c. polyphase system is balanced when the loads taken from each phase are equal and at the same power factor.

Balancer. A device used to maintain, automatically or otherwise a BALANCED LOAD in an a.c. or d.c. system. It may also be used to convert a single-phase supply from two- to three-wire, or a three-phase supply from three- to four-wire.

Balancing Battery. A FLOATING BATTERY which helps a generator to supply current on peak loads. It is connected in parallel with the generator busbars, discharging during heavy load and charging during light load. At normal load, the current is supplied entirely by the generator, the battery being neither charged nor discharged.

Balancing Machine. A machine for determining and measuring by electrical means the amount and position of the unbalance forces in bodies such as armatures and rotors while rotating (see DYNAMIC BALANCING, STATIC BALANCING).

Balancing Ring. Ring conductors attached to an armature, to which equipotential or equalizing connections are made, to allow an interchange of current so that currents in different parts of a lap armature winding are kept symmetrical.

Ballast. A device, such as a resistor or choke, inserted in series in a circuit to limit the current and provide stable operation under changing conditions (see also BARRETTER).

Ballistic Galvanometer. A GALVANOMETER with a moving system which has a considerable moment of inertia and a low damping factor, so that the time of swing is long compared to the duration of the current impulse it measures. The swing is then proportional to the current-time integral of the impulse.

Bar Suspension. A method of mounting a traction motor on an electric truck. One side of the motor is supported on the axle by means of suspension bearings and the other by lugs projecting from the frame and bolted to a spring-suspended bar lying transversely across the vehicle. This is also known as *yoke suspension*.

Bar Winding. An armature winding which consists of bars of rectangular section. The winding is made of straight copper strips, insulated on the slot portion, forming a half-coil. One end is bent to shape and the half-coil is pushed through the slot. It is then bent at the other end and joined at both ends to form the winding.

Barium Titanate. A material possessing piezoelectric properties. A polycrystalline ceramic of barium titanate and various binding materials can be made to act as a single crystal by the application of an electrostatic field, the direction of the field determining the direction of piezoelectric activity.

Barkhausen Effect. Effect observed when a ferromagnetic material is subjected to a change in an applied magnetic field. When the magnetizing force is increased, small discontinuities appear in the magnetization curve of the substance, owing to reorientation of the domains.

Barkhausen Oscillator. A high-frequency generator depending for its action on the transit time of the electrons from cathode to anode in a three-electrode valve.

Barn. A unit of area applied to the concept of nuclear cross-section. 1 barn = 10^{-24} cm^2.

Barrel Plating. An electroplating process in which the articles to be plated are placed in a rotating "barrel" or container.

Barrel Winding. An armature winding in which the end connections lie flat upon a cylindrical surface.

Barretter. A resistive element having a high positive temperature coefficient of resistance. It is employed to maintain the current in a circuit approximately constant in spite of variations

Base Load

of voltage. The barretter operates by varying the resistance of the circuit proportionately with the voltage. Iron and nickel are suitable materials and one of these may be employed as a filament in an atmosphere of hydrogen or helium.

Base Load. The minimum demand on a power supply system. It is met as far as possible by the more efficient generating stations, the rest of a PEAK LOAD being supplied intermittently by the less efficient stations.

Base-Plate. See BED-PLATE.

Basket Winding. A DISTRIBUTED WINDING for a.c. machines. The name is derived from the appearance of the linked coils. It is also known as a *chain winding*.

Batch Furnace. See RESISTANCE FURNACE.

Battery. An assembly of similar units; in particular a series connection of voltaic PRIMARY CELLS or secondary cells. Batteries of secondary cells are also known as storage batteries or ACCUMULATORS.

Battery Vehicle. A road or rail vehicle, self-propelled by means of electric motors supplied from an accumulator.

Bayonet Cap. A lamp cap in which the cylindrical outer wall carries pins (usually two or three in number) for engaging in slots in the lampholder. It normally carries two end contacts insulated from the outer wall. These caps may be referred to as B.C. (bayonet cap) or S.B.C. (small bayonet cap) depending on size. See also CENTRE-CONTACT CAP.

Bearing Current. A stray current that flows between the shaft and bearings of an electrical machine.

Bed-Plate. Base on which machine frame and bearings are mounted.

Bedding. Of an armoured cable, a fibrous layer on which the armouring is applied. The bedding may be permeated with a waterproof compound.

Bel. The common logarithm of a power ratio. The numeric is very rarely used; but in the form 10 N, i.e. in DECIBEL units, it is commonly employed in communication technology.

Bell. A sounding apparatus in which a hammer for striking a bell or gong is operated by an electromagnetic mechanism. This mechanism may be arranged to give either a single-stroke or a trembling action (see also TREMBLING BELL).

Bell Transformer. A double-wound transformer which may be connected to the mains supply to supply a bell circuit. A choice

Bimetal

of outputs, 3, 5, or 8 V, or in larger sizes, 12 or 24 V, is usually provided from the terminals on the secondary side. The transformer is permanently connected to the mains, as the magnetizing current taken when idle is very small.

Belted Cable. A form of power cable in which three insulated cores are laid up together (with wormings to fill the interstices), and the whole assembly is then wrapped with a "belt" of paper tapes (see Fig. B-3). The design gives excellent service up to 11 kV, but above this voltage breakdown can occur because of tangential electrical stresses.

Fig. B-3. Belted Cable.

Bentley-Galloway Discriminator. A crane LOAD DISCRIMINATOR consisting of an electric motor driving a centrifugal brake connected to the load motor.

Beta Particle. A high-velocity electron, emitted from a radioactive nucleus.

Beta Ray. A stream of BETA PARTICLES.

Betatron. An ORBITAL ACCELERATOR in which electrons are both directed into circular orbits and also accelerated by means of a magnetic field. See SYNCHROTRON.

Bifilar Suspension. Suspension of the moving part of an instrument by two vertical wires or threads which give a considerable controlling torque. The period of torsional vibration, T, is given by $T = 4\pi (Il/mgd^2)$ where I is the moment of inertia of a suspended body of mass m, and the threads are of length l and distance d apart.

Bifurcating Box. A closed box in which the two cores of a twin cable can be connected to external conductors.

Bimetal. A temperature-sensitive device produced by firmly bonding together two or more metals or alloys having sensibly different coefficients of expansion. The composite material is rolled into a flat strip, from which a variety of shapes may be cut. A change of temperature produces a change of curvature,

Bimetallic Instrument

utilizable in several forms such as the deflection of a straight strip, the opening or closing of U-shaped pieces, the rotation of spirals and helices, or the "dishing" of washers and discs.

Bimetallic Instrument. Indicating instrument incorporating a bimetallic strip which is fixed at one end while the other end operates a pointer. Current to be measured may be passed through the strip, but usually there is a separate heater, with or without a thermal barrier to control the heating rate. The instrument is sometimes used as a *maximum-demand ammeter*.

Binary Scale. A method of writing a number in terms of powers of 2. A number expressed in the binary scale requires about 10 digits for every 3 decimal digits, but has a particular advantage in requiring only two numerals (0 and 1). It can therefore be represented in a DIGITAL COMPUTER by a row of devices needing only two "states" each—*on* and *off*.

Binding Energy. Of a molecule, the energy which must be supplied in order that the molecule may be dissociated. Conversely, when a molecule is being formed this energy is released. It is also known as *dissociation energy*.

Biot-Savart Law. The magnetizing force H, at a point P (Fig. B–4), due to a current I flowing in a network, is the same as

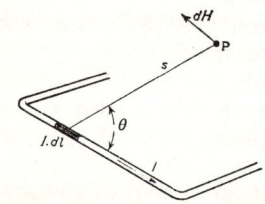

FIG. B–4. THE MAGNETIZING FORCE DUE TO THE CURRENT IN A CONDUCTOR.

if each element dl of the network contributed a vector

$$dH = I\, dl \sin \theta / 4\pi s^2$$

The direction of this contributory vector is at right angles to dl and to s, in such a sense that dH forms a right-handed screw about the direction of the current.

Bi-polar Electrode. In an electrolytic cell, an additional electrode, insulated externally from the anode and cathode, which is in the path of the current. Various parts therefore function as anodes, and parts as cathodes.

Blow-Out Coil

Bi-polar Machine. A machine in which the field magnet has only two poles. It is also known as a *two-pole machine*.

Birmingham Wire Gauge. A method of designating the diameters of wires and rods by numbers. It is commonly used for steel, and is an alternative to the STANDARD WIRE GAUGE.

Bismuth Spiral. A flat spiral of bismuth wire used for measuring the intensity of a magnetic field. The resistivity of bismuth increases in the presence of a magnetic field, and so the change in resistance of the spiral can be used to measure field intensity (e.g. in the short air-gaps of electrical machines) after calibration in known fields.

Bitter Figure. A method of making magnetic domains visible. Where domains meet in the surface of a crystal there is a surface distribution of magnetic polarity across the boundary (*Bloch wall*) between them. This, in turn, produces a stray magnetic field above the surface. If finely divided, permanently magnetized particles are placed over the surface, they settle in the neighbourhood of the domain boundaries, and can be viewed under a low-power microscope.

Bitumen. A natural product consisting of a complex mixture of hydrocarbons with inorganic materials, and melting at about 90–100°C. *Refined bitumen* is used in the manufacture of insulating varnishes, impregnating compounds and joint-box compounds. *Vulcanized bitumen* is employed as a cable insulant.

Black-Body Emitter. An ideal perfect emitter of radiation. The ideal is approached in practice by a cavity with a small aperture, the inner surface being non-reflecting and of a uniform temperature.

Block-Rate Tariff. A TARIFF in which a fixed number of units of energy supplied in a given period are charged at a high rate, and successive quantities of units are charged at diminishing rates. The quantities are variable in the case of the VARIABLE-BLOCK TARIFF.

Blow-Out Coil. A device for lengthening an arc in air by magnetic action, particularly in d.c. contactors. The coil is mounted on a steel core between steel pole-pieces, and is connected in series with the circuit to be broken. The current in the coil produces a magnetic field across the contact tips. When an arc is formed, a mechanical force is produced on it, urging it upwards and away from the contacts. The arc is consequently lengthened and rapidly extinguished, reducing contact wear and

Bobbin Insulator

burning. A blow-out coil is commonly associated with an ARC CHUTE to prevent the arc from wandering.

Bobbin Insulator. Alternative name for SHACKLE INSULATOR (q.v.).

Bobbin Winding. Transformer winding, used mainly for the high-voltage windings of small transformers, in which all the turns are arranged on a bobbin (cf. DISC WINDING).

Bohr Magneton. The unit of magnetic moment in atomic phenomena. Electron spin causes an electron-charge rotation on an axis, equivalent to a current loop magnetizing along this axis having a magnetic moment. The value of the Bohr magneton is $1 \cdot 165 \times 10^{-29}$ Wb-m.

Bolometer. Instrument for measuring radiant energy. The resistance of a fine wire or strip varies when it is exposed to the radiation.

Bond. In a lightning protection system, a short length of conductor providing electrical connection between the system and other conductive material, or linking conducting parts.

Booster. A device to give a small increase to a normal-voltage supply, being connected in series with one line. D.C. boosters are normally low-voltage d.c. generators employed for adjusting a supply voltage, in line-drop compensation and as an aid in controlling the charging of large accumulator batteries. See also BOOSTER TRANSFORMER.

Booster Transformer. A device for regulating the voltage on a power supply network. It is used to add or subtract a voltage and is used at the sending end of a long h.v. radial feeder. It is connected in series with the feeder as shown for one phase in Fig. B–5,

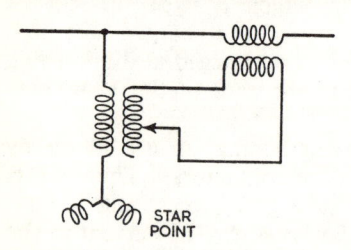

FIG. B–5. BOOSTER TRANSFORMER. The connections are shown for one phase of a feeder.

this transformer being supplied from another transformer connected between one phase and neutral. Similar transformers are connected in each of the three phases. The added voltage is in

Bridge

phase with the phase-to-neutral voltage of the system, i.e. it adds to it (or subtracts from it) arithmetically. The magnitude of this added voltage can be varied in steps by varying the tappings on the secondary of the supply transformer.

Bow Collector. A sliding current-collector used in electric traction with overhead wires. It consists of a bow-shaped contact strip mounted on a hinged frame. See also PANTOGRAPH.

Boy's Camera. Camera having two lenses moving in opposite directions. It facilitates the study of time characteristics of lightning and similar phenomena.

Braiding. Of a cable, a plaited protective covering.

Branch Joint. A form of cable joint where a connection to one continuous main cable is made by another and, often, smaller cable forming a branch connection. It may be a *breeches joint* in which the branch cable is taken parallel to the main cable and out at one end of a lead sleeve where a double plumb is made to the cables, or a *tee joint* in which the branch cable is taken away at right angles to the main cable and needs a pressed lead or copper box of suitable shape.

Break. The total minimum gap introduced in one pole by the contacts of a switch or circuit-breaker when this is in its fully open position.

Break Time. Of a circuit-breaker, see TOTAL BREAK TIME.

Breakdown. See DIELECTRIC BREAKDOWN, IONIC BREAKDOWN, THERMAL BREAKDOWN.

Breaking Capacity. Of a circuit-breaker, see ASYMMETRICAL BREAKING CAPACITY, SYMMETRICAL BREAKING CAPACITY.

Breather. Desiccating device fitted to conservator tank of a transformer to ensure that no moisture is drawn in with inspired air.

Breeches Joint. A form of BRANCH JOINT in which the branch cable is taken away parallel to the main cable. It is usually confined in a lead sleeve. It is also known as a *trousers joint*.

Breeder Reactor. A NUCLEAR REACTOR in which the weight of new fissile material produced exceeds that of the original fissile material put into the reactor.

Bridge. An arrangement of resistors, capacitors, inductors, etc., usually in a four-sided circuit, for the purpose of measuring an electrical quantity by a null method. One of the diagonals is connected to the current source and the other to a measuring instrument. See A.C. BRIDGE, D.C. BRIDGE, etc.

Bridge Transition

Bridge Transition. Of d.c. motors, a method of changing the connection from series to parallel without breaking the main circuit and while maintaining a similar current in all the motors.

Bridged-T Network. See QUADRIPOLE, ATTENUATOR.

Brightness. A term used to describe either luminance or luminosity.

Britannia Joint. A joint used for overhead transmission lines. The two lines to be joined are bound together over a length of several inches.

British Thermal Unit. A unit of heat (abbreviation: Btu). It is the quantity of heat required to raise the temperature of 1 lb of water from 60° to 61°F, equivalent approximately to 1,054 J. A *mean Btu* is 1/180 of the quantity of heat required to raise 1 lb of water from 32° to 212°F. This is approximately equivalent to 1,055 J.

Broad-Base Tower. A LATTICE TOWER, each leg of which is anchored separately.

Brown and Sharp Gauge. A system, frequently used in the U.S.A., of designating the diameter of wires by numbers. It is alternatively known as the *American Wire Gauge* (B. & S. or A.W.G.). See WIRE GAUGE.

Brush. A conductor used to make contact between a stationary and a moving surface. Operating conditions vary over a wide range, and numerous types and grades of brush have been developed over the years to match progress in machine design (see CARBON BRUSH, ELECTROGRAPHITIC BRUSH, METAL-GRAPHITE BRUSH, NATURAL GRAPHITE BRUSH).

Brush Discharge. Discharge from a conductor when the voltage is not quite high enough to cause a true spark or arc. It has a feathery appearance and is usually accompanied by a hissing or crackling noise.

Brush Shift. The movement of the brushes on a commutator away from the neutral position. In a d.c. machine without commutating poles a shift of the brush position may be made in order to improve commutation; in a repulsion or Schrage machine to vary the speed. It is also known as *brush lead*.

Brushless Generator. A generator employing silicon rectifiers as static commutation devices. It is of particular value as an aircraft generator, difficulties having been experienced with sliding contacts under conditions of high running speed, dry rarefied air and wide temperature range.

Buffer Battery

Buchholz Relay. A protective relay on a power transformer. An electrical fault will always produce gas, and the relay is situated so that all gas released must pass into it, e.g. in the pipe connection to the conservator vessel. The relay consists of a chamber, normally full of oil, in which a hinged float carries a mercury switch which is open when the float is in its highest position. If gas is formed it will collect in the chamber, the float will fall about its hinge and at a predetermined point the switch will close, and, in common practice, produce an alarm signal. A second similar float and switch is mounted on a hinge in the opening of the pipe connection from the transformer. It operates on the surge of oil up the pipe which will result from the rapid formation of gas due to a severe fault. The switch on the surge float is invariably set to trip the circuit-breakers. (See Fig. B-6.)

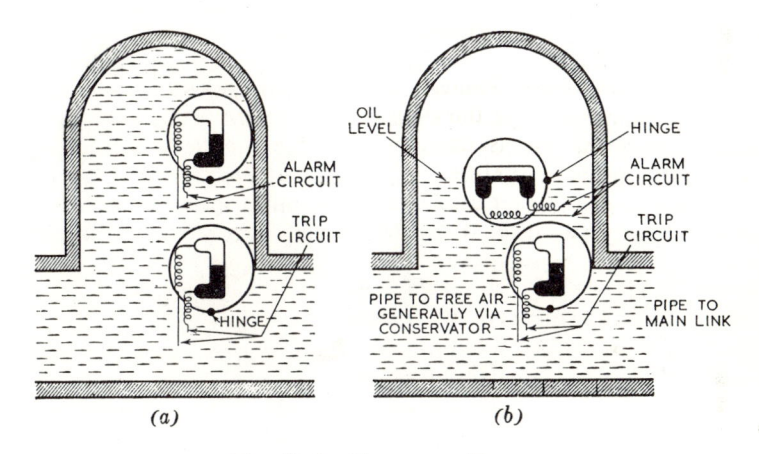

FIG. B-6. BUCHHOLZ RELAY.
The normal condition is shown in (a) with the relay full of oil. In (b) gas has formed and collected in the relay, the oil level has fallen and the first float has turned about its hinge allowing the mercury switch to close an alarm circuit.

Buck. Effect of a BOOSTER connected to reduce rather than increase the resultant voltage.

Buffer Battery. A battery included in a d.c. generation system, connected across the lines, to smooth out variations in supply voltage and current.

Bulk-Oil Circuit-Breaker

Bulk-Oil Circuit-Breaker. A CIRCUIT-BREAKER in which arcing takes place in oil, the contacts being enclosed in an oil-filled earthed tank. It is sometimes referred to as a *dead-tank circuit-breaker*. Circuit-breakers of this type have been developed for a.c. systems for all voltages and rupturing capacities up to 330 kV.

Bull Ring. In overhead construction, a metal ring employed at the junction of three or more straining wires.

Bunch Filament. A lamp filament having two levels of support with the filament suspended between. It is bunched for compactness for use in applications involving lens systems, where the focal point is critical.

Bunched Cables. Two or more cables enclosed within a single conduit or way.

Bunched Conductor. A CONDUCTOR composed of several wires twisted together in the same direction, and with the same lay throughout (cf. STRANDED CONDUCTOR).

Bundle Conductor. A conductor in which each phase conductor comprises two, three or more small stranded cables with a proper spacing maintained between them and a total section equal to that needed for current conduction. The bundle construction, although raising mechanical difficulties in erection and operation, makes possible a significant reduction in the surface electric field intensity (see Fig. B–7). Moreover, it confines the higher

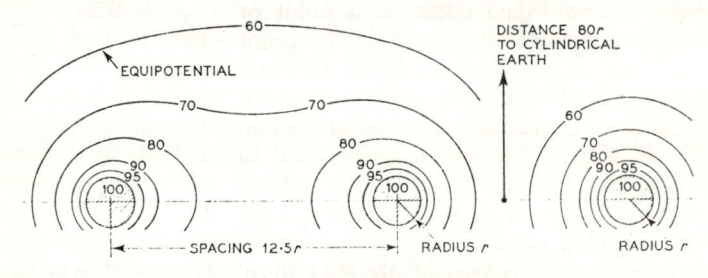

FIG. B–7. BUNDLE CONDUCTOR.
The equipotential field is compared with that of single conductor (*right*) of the same radius.

intensities to a comparatively small surface area, so reducing corona loss on two counts.

Burden. The load in volt–amperes on the secondary of an instrument transformer.

Bus Line. In a railway train, a cable (or TRAIN LINE) extending throughout the length of the train, used to interconnect collector shoes of similar polarity.

Busbar. A substantially equipotential conductor forming a terminal or junction point in a power system, for the connection of supplies and feeders.

Busbar Protection. A form of protection which automatically disconnects a busbar system from its source of supply in the event of an insulation failure. Two basic methods are FRAME FAULT PROTECTION and BALANCED CURRENT PROTECTION. The choice of scheme for a given application is governed to some extent by economics. Schemes may have varying complexity, either affording earth-fault protection only, or providing phase-to-phase fault protection as well, and covering either the individual busbar sections or the complete installation.

Bushing. An insulator, generally cylindrical in shape, for insulating and fixing a conductor where it passes through earthed structures such as the walls separating parts of an X-ray equipment, or the tanks of power switches and transformers. It is also known as a *lead-in insulator*.

Butt Contacts. Contacts brought together with a minimum of relative movement between fixed and moving components, relying mainly on the applied pressure for satisfactory conduction. Contact is established either at a point or along a line. The current is generally interrupted at the point where normal conduction occurs (cf. ROLLING-BUTT CONTACTS).

Butt Seam Welding. A variety of BUTT WELDING in which roller electrodes are used to conduct the current to the two sides of the joint and separate devices are used to force the two edges together, thus forming a continuous weld as in the case of tube formed from strip.

Butt Welding. A form of RESISTANCE WELDING in which the parts being joined are butted together end to end or edge to edge. Varieties of butt welding include UPSET BUTT WELDING, FLASH BUTT WELDING and BUTT SEAM WELDING.

Buzzer. Signalling device like an electric bell without hammer or gong. The sound is produced by the electromagnetic vibration of an armature.

C

C-Class Insulation. One of seven classes of insulating materials

C-Core

for electrical machinery and apparatus defined in B.S. 2757 on the basis of thermal stability in service. Class C insulation is assigned temperatures over 180°C. It consists of materials such as mica, porcelain, glass, quartz or asbestos, with or without an inorganic binder.

C-Core. See CUT-WOUND CORE.

C.G.S. System. A system of physical units based on the centimetre, gramme and second. In particular the old-established electromagnetic C.G.S. and electrostatic C.G.S. systems of units are based separately on the C.G.S. system. Both systems are obsolescent, and were replaced by the metre-kilogramme-second (M.K.S.) system, with the additional recognition of the interrelation between the electric and magnetic space constants, and the use of a rationalized approach. See also SYSTÈME INTERNATIONAL.

Cable. A length of one or more conductors laid up together, each conductor having its own insulation. The three principal components of a cable are the CONDUCTOR, the *dielectric* and the *protection*.

The two principal dielectric materials most commonly used are oil-impregnated paper and natural or synthetic vulcanized-rubber compounds, but in recent years plastic compounds such as polyvinyl chloride and polyethylene have been extensively used in place of rubber. The protection for an impregnated-paper cable is, firstly, a lead or aluminium sheath to prevent ingress of moisture and, secondly, layers of bituminized-fabric tapes with or without a metal armour to protect the sheath against corrosion and mechanical damage. For rubber- and plastic-insulated cables, the protection is governed by actual service requirements.

Cable Bond. Electrical connection for the metallic covering of a cable.

Cable Coupler. A device consisting of two engaging parts for connecting two sections of flexible cable. An interlock may be incorporated to prevent operation of the coupler when the circuit is live.

Cable Jointing. The connection of one piece of cable to another in such a way that the conductivity and insulation at the joint are equal in quality to that of the body of the cables concerned. The main types of joint used on normal distribution cables are the STRAIGHT-THROUGH JOINT and the BRANCH JOINT.

Cable Laying. Placing cables in position for the passage of electric power. Cables are *laid direct*, i.e. buried directly in the earth, or *drawn in*, i.e. drawn into conduits or ducts already laid.

Cadmium Cell. A standard cell alternatively known as a WESTON CELL.

Cadmium Electrode. A rod of metallic cadmium within a perforated insulating tube, and provided with a flexible connection. By connection through a suitable voltmeter to either the positive or the negative terminals of a lead-acid cell, the cadmium electrode when immersed in the electrolyte gives readings that indicate the condition of the cell plates.

Cadmium-Nickel Cell. See NICKEL-CADMIUM CELL.

Cadmium Test. A method of testing the condition of the plates in an accumulator cell. It involves the insertion of an electrode made of cadmium into the electrolyte near the middle of the cell and the measurement of voltage between the electrode and a terminal.

Cage Motor, Rotor, Winding. See SQUIRREL-CAGE MOTOR, SQUIRREL-CAGE ROTOR, SQUIRREL-CAGE WINDING.

Calculated Effective Area. Of the conductor of a cable, the cross-sectional area of a solid conductor of similar resistivity and having the same resistance as an equal length of the cable.

Calibration. Determination, by experimental observation, of the values (in appropriate units) of the readings of an instrument having an arbitrary or inaccurate scale.

Calomel-Electrode System. A mercury electrode in a potassium-chloride solution saturated with calomel (mercurous chloride).

Calorie. The quantity of heat required to raise 1 g of water by 1°C from 14·5 to 15·5°C. This is the *International Calorie*. It is equated to 1/860 Wh as being more directly realizable. The calorie has been replaced in S.I. units by the joule. It is approximately equivalent to 4·18 J.

Cam. In an electric-lift system, a wedge-shaped component fixed in the well or on the lift car or counterweight which operates the control apparatus when the lift moves.

Candela. The S.I. unit of luminous intensity. It is of a magnitude such that the luminance of a full radiator (black body) at the temperature of solidification of platinum is 60 cd/cm^2.

Cap. (1) A metal fastening on an insulator by which it may be suspended and connected to another insulator in a string.

(2) The terminal base of a lamp.

Capacitance

Capacitance. Property of a body by virtue of which a quantity of electricity must be imparted to it to create a difference of potential between it and neighbouring bodies. It is also the ratio between the quantity of electricity imparted and the potential difference produced. The unit is the farad (abbreviation: F). 1F = 1 C/1 V.

Capacitive Load. See LEADING LOAD.

Capacitor. A circuit element, the predominant characteristic of which is CAPACITANCE, and which stores energy in its electric field. A practical capacitor will have energy loss in dielectric hysteresis and conductivity, and may be represented by a pure capacitor with a series or a shunt loss-resistor. See ELECTROLYTIC CAPACITOR, INDUSTRIAL CAPACITOR.

Capacitor Motor. A single-phase machine with a split stator winding, with an external capacitor in the auxiliary circuit to increase starting torque and decrease starting current. The capacitor provides phase displacement so that the current in the auxiliary winding leads that in the main winding. It may be used during the starting period only (*capacitor-start motor*), or be permanently connected. Alternatively, a motor may start with one value of capacitor and run with a different value (*capacitor-start-run motor*).

Capacitor Transformer. Voltage transformer for transmission lines operating at 100 kV and upwards. It comprises two capacitors connected in series across line and earth, so as to give a voltage-divider effect. A burden is connected through a step-down magnetic transformer with an appropriate leakage reactance (see Fig. C-1). As an alternative to the magnetic voltage trans-

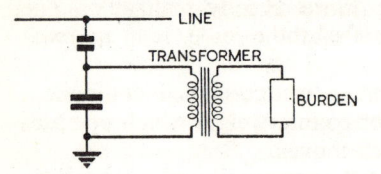

FIG. C-1. CAPACITOR VOLTAGE TRANSFORMER.

former it has lower cost and greater service reliability, and can also be used as a connection element in carrier-frequency protection.

Capacity. (1) The quantity of electricity which may be taken from an electric cell or battery at a given rate of discharge. It is usually expressed in ampere-hours.

Carbon-Pile Regulator

(2) The ratio of the charge to the potential between two bodies (see CAPACITANCE).

Carbon. Non-metallic element of adjustable conductivity. It is highly refractory, and as its oxides are gaseous, no surface films are produced. It is chemically resistant and self-lubricant. Its contact resistance varies inversely with the pressure.

Carbon is used in various forms for brushes (see CARBON BRUSH, ELECTROGRAPHITIC BRUSH, etc.), for arc electrodes (see CARBON ARC LAMP, CARBON ARC WELDING) for cell electrodes (see LECLANCHÉ CELL), for contacts, for current collectors in electric traction, for pressure-sensitive resistors in microphones, CARBON-PILE REGULATORS, etc., and as a furnace material.

Carbon Arc Lamp. Lamp consisting basically of two carbon electrodes between which an electric arc is maintained. The carbon arc is obsolete as a source of light for general illumination, but its high brightness, compact source and colour find application in film projection and searchlights.

Carbon Arc Welding. A form of non-consumable electrode ARC WELDING in which the arc is struck between a carbon rod and the workpiece. The quality of the weld is low owing to inadequate protection of the weld pool from atmospheric contamination.

Carbon Brush. One of the first types of BRUSH to be developed to supersede the metal wire or foil brushes used on the earlier electric machines. The raw material, amorphous carbon in various forms such as lamp black, gas carbon or coal, is first reduced to a very fine powder. It is then mixed with a suitable binding material such as pitch or tar and compressed into blocks which are baked under controlled conditions at a fairly high temperature to carbonize the whole mass of the material and to impart the required mechanical and electrical characteristics.

Carbon-Pile Regulator. In the simplest form, an assembly of carbon plates with means for compressing the assembly, usually a simple handwheel and screw mechanism. The resistance of the stack of plates varies with the pressure, giving a convenient method of controlling currents up to the order of 100 A.

Some types of AUTOMATIC VOLTAGE REGULATOR employ a carbon-pile rheostat. A stack of carbon plates is connected in the main exciter-field circuit, and is controlled by a suitable torque motor energized by the voltage under control. As this voltage rises, the pressure on the carbon-stack is reduced, which increases

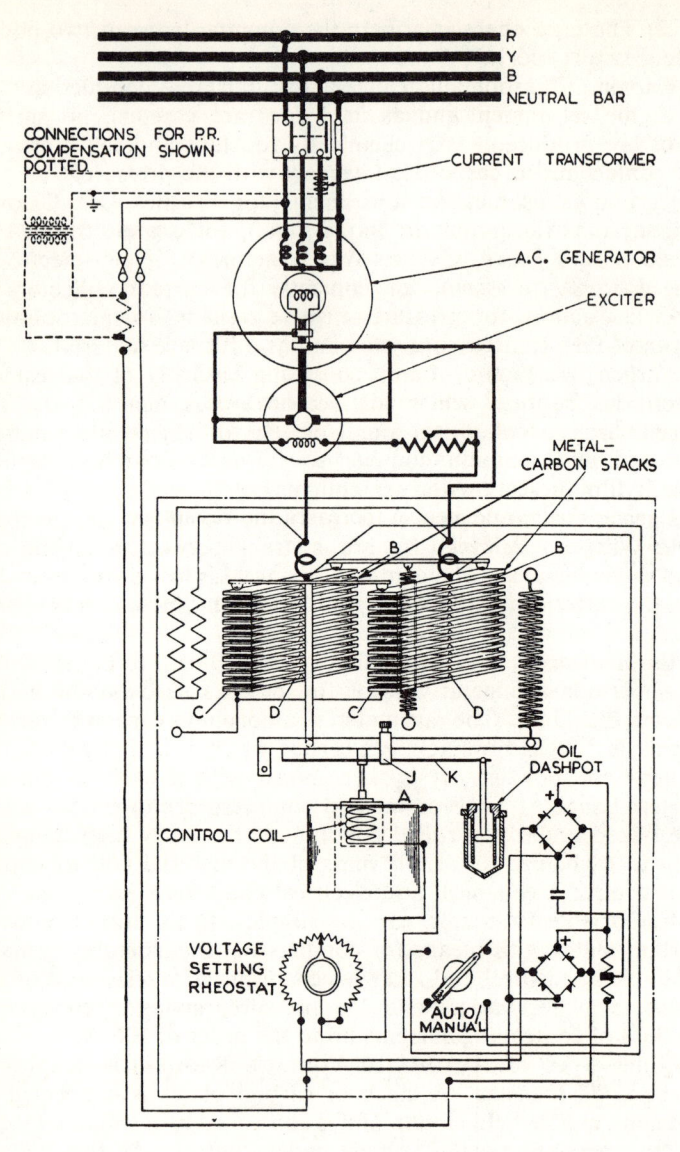

Fig. C-2. Carbon-Pile Rheostatic Regulator.

the resistance of the stack, reduces the excitation current, and hence reduces the generator voltage. As this voltage falls, the pressure on the carbon-stack is increased, the resistance decreases, the excitation current increases, and the generator voltage is raised.

Another type of carbon-pile rheostatic regulator is illustrated in Fig. C-2. The measuring element A is a permanent-magnet moving-coil device, which is mechanically connected to the rheostatic element by a lever system. The rheostatic element is made up of two stacks, each consisting of a series of rectangular carbon plates; near the right-hand side of each carbon plate is a silver contact B. At the left-hand side of each plate is a high-resistance spacer C. Under the centre of each plate is fixed a second resistance spacer D, about which the plates can rock. When the voltage under control is low, all the silver contacts are closed, the overall resistance being a minimum. As the voltage rises the measuring element tilts the carbon stacks and causes the silver contacts to open successively, and hence, increases the resistance.

Carey-Foster Bridge. An adaption of the WHEATSTONE BRIDGE giving a greater accuracy with resistances of low and medium value. It is especially useful for obtaining the divergence between a nominal-valued resistor and a supposedly identical standard. A feature is the use of a low-resistance slide-wire of length l and resistance r per unit length. The ratio arms P and Q are nominally equal, and the unknown R is nominally equal to S. Balance is obtained with the conditions of Fig. C-3 so that

$$\frac{P}{Q} = \frac{R+ar}{S+(l-a)r} = \frac{R+S+lr}{S+(l-a)r} - 1.$$

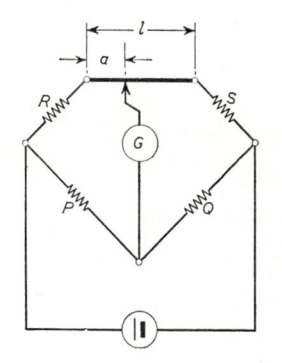

FIG. C-3. CAREY-FOSTER BRIDGE.

Carrier

R and S are now interchanged, and balance obtained at a setting b. Then

$$\frac{P}{Q} = \frac{R+S+lr}{R+(l-b)r} - 1$$

whence

$$S - R = (a-b)r.$$

Carrier. A sinusoidal voltage or current used for signalling by line or radio. Intelligence is impressed on it by a process of MODULATION.

Carrier Current Protection. A form of protection for power transmission lines which employs a high-frequency signal transmitted over the line itself. Fig. C–4 shows the essential components

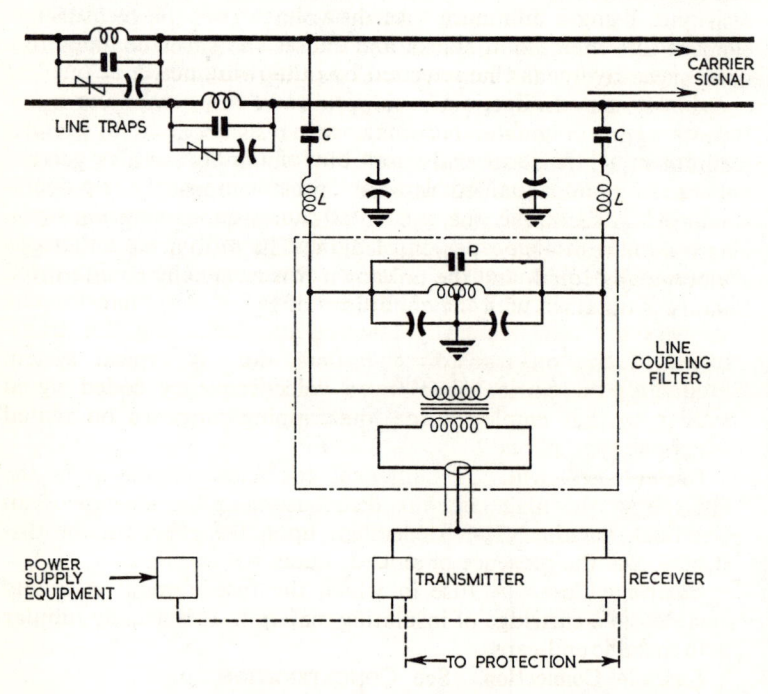

FIG. C–4. MAIN COMPONENTS OF A CARRIER CURRENT PROTECTION SYSTEM.

at one end of a line. The transmitter and receiver are connected in parallel and feed the line-coupling equipment via a low-loss cable. Connection to the line is through two high-voltage capacitors CC; they are series tuned by h.f. inductors LL to accept the carrier frequency. A path for the 50 Hz current of the capacitors is provided by a parallel-tuned circuit having a high impedance at the carrier frequency. The line traps are parallel-tuned circuits which reject the carrier frequency; the signal is thereby confined to the protected line.

There are three principal types of carrier system in common use, also a scheme of carrier intertripping. All employ fault detecting or "starting" equipment to avoid the necessity of continuous high-level carrier transmission.

Phase comparison systems compare the phase angle of the current entering and leaving the protected section, and trip if there is a phase difference exceeding, say, $\pm 60°$. *Direction comparison* systems use the criterion that a through-fault condition on a healthy line is characterized by outflowing power at one line end. Polyphase directional relays at each end permit the transmission of an unmodulated carrier signal to the line only so long as power at the local end flows out of the section. A carrier/distance system is the "acceleration" scheme in which a trip operation occurring at one end of the line causes an accelerating signal to be sent to the opposite end of the line permitting instantaneous tripping at that end also. This is convenient where high-speed auto-reclosing circuit-breakers are employed. *Carrier intertripping* schemes are used primarily for non-line faults, e.g. for transformer faults on transformer-feeders, etc. A typical system modulates the carrier signal by a voice-frequency coded signal similar to that employed for intertripping purposes on rented telephone-type pilots.

Carter Coefficient. A numerical coefficient employed in the analysis of the magnetic flux distribution in the air-gap of an electrical machine. It is dependent upon the effect on the distribution of the presence of slots or ducts.

Cartridge Fuse. A fuse in which the fuse element is totally enclosed in a cartridge of insulating material. It is usually tubular with metallic end-caps.

Cascade Connection. See CONCATENATION.

Castell Interlocking. A method of ensuring that switchgear and other equipment is operated in the correct sequence or under

Cataphoresis

the right conditions. The lock itself comprises, within a brass body, a centre spindle bearing a block which carries raised letters, figures, or other characters up to three in number. These characters will mate only with their engraved counterparts in the end of the appropriate key. The lock may be designed so that the key is trapped in either the open or the locked position as required.

Cataphoresis. In electrochemistry, the migration of colloidal particles when a voltage gradient exists through a substance. It is also known as *electrophoresis*.

Catenary. The curve taken up by a chain (or a cable) of uniform weight per unit length when freely supported from two points. Consider a cable of length $2l$. The catenary equation is given by the expansion $y = c \cosh(x/c)$. The sag is $s = y_1 - c$, and the half-length l of the cable is $l = c \sinh(d/c)$. See Fig. C–5.

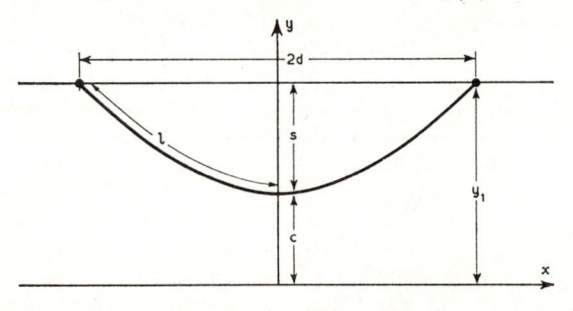

FIG. C–5. A CATENARY.

Catenary Suspension. In electric traction, a type of construction for overhead conductors including a support (*catenary*) wire, strung as a catenary between structures, from which a *contact* wire is suspended by means of droppers (see SIMPLE CATENARY SUSPENSION, DOUBLE CATENARY SUSPENSION). An *auxiliary catenary* wire is frequently interposed between the catenary and contact wires (see COMPOUND CATENARY SUSPENSION, STITCHED CATENARY SUSPENSION).

Cathode. That electrode (e.g. in an electron tube, mercury-arc rectifier, electrolytic bath, etc.) to which the current flows from the ANODE. The current direction is taken conventionally as that of the movement of positive charges. In an electronic vacuum tube the current is, in fact, solely a transfer of electrons emitted from the cathode.

Cathode Drop. The difference in voltage, in an electric welding machine, between that at the negative end of the arc-stream and that of the adjoining conductor.

Cathode-Ray Tube. Electron tube containing a thermionic cathode and electron gun for the production of a stream of electrons which are directed in a beam onto a flattened surface coated with a phosphor. Electric or magnetic fields are used to focus the electron stream to form a spot of light on the screen. The beam may be deflected by the application of electric or magnetic fields. The tube is employed in a *cathode-ray oscilloscope* for examining varying electrical quantities.

Cathodic. (1) Having a more positive electrode potential.

(2) Indicating an element below hydrogen in the electrochemical scale.

Cathodic Protection. A defence against corrosion which can be applied to cables drawn into ducts or buried direct. It depends on the fact that electrochemical corrosion results from the existence of electrical potentials between metals and the electrolyte surrounding them. The corrosion tends to take place at the anode.

The technique involves making the potential of the sheath and armour more negative to the electrolyte (water or soil) than a limiting value (about -550 mV for lead). All parts of the metal then become cathodic. This result can be achieved by connecting to the cable sheath reactive anodes composed of an alloy of base metals such as magnesium or zinc. If a more powerful effect is wanted transformer-rectifiers can be used, with anodes or "groundbeds", designed to have a long life, and consisting either of a large system of interconnected metal such as steel rails, or of a non-metallic conducting material such as graphite. In either case, the anodes are buried in the soil near the cables to be protected, and connected to them by substantial, well-insulated conductors.

Cathodoluminescence. The excitation of luminescence in a phosphor by the impact of electrons from an external source. Certain phosphors can be excited with an electron (or cathode ray) beam of 100 V or less; this is *low-voltage cathodoluminescence*. On the other hand, the accelerating voltage in a conventional television picture tube is about 15 kV, producing an image by *high-voltage cathodoluminescence*.

Cation. A positively charged Ion, existing in an electrolyte or gaseous discharge. It will travel towards the negative electrode or cathode under the influence of an applied difference of potential.

Catolyte

Catolyte (or *catholyte*). The portion of an electrolyte surrounding the cathode, which is affected chemically by cathodic reactions.

Caustic Potash. Potassium hydroxide, used as an electrolyte in a STEEL-ALKALINE CELL.

Cavity Resonator. Device formed by closing a length of WAVEGUIDE at both ends, so that a wave of appropriate frequency is excited in it. Standing waves will be set up due to reflections from the ends, and the device becomes a resonator.

Cell. (1) A chemical source of electrical energy consisting of two electrodes in an electrolyte.

(2) Compartment of cellular switchgear.

Cell Constant. Of a CONDUCTIVITY CELL, the ratio of the specific conductance of the electrolyte to the actual conductance. It is numerically equal to the separation of the electrodes divided by the average cross-sectional area of the current path.

Cellular Switchgear. A form of switchgear in which the components are housed in individual cells. Each circuit has a separate cell divided into compartments for circuit-breakers, busbars, isolators, voltage transformers and cable sealing boxes. Each compartment has its metal access door, which may be interlocked with other equipment to ensure the safety of maintenance personnel. The cells are often made of moulded stone, carried in a steel framework.

Cellulose Acetate. A thermoplastic, available as sheet, film, moulding powder, extruded sections or lacquer solutions, which is used for a number of applications in the electrical field. Typical applications of film are wire and coil inter-layer insulation, while mouldings are used for such components as terminal blocks and switch dollies. In sheet form it is used for instrument dials.

Centre-Contact Cap. A BAYONET CAP in which the shell forms one contact, the other consisting of a central projection. It may be referred to as S.C.C. (small centre contact).

Ceramic Insulators. Insulators made of CERAMICS. They are used in high-voltage power transmission, switchgear, transformer bushings and fuse-holders because of their good mechanical strength, weathering resistance, high electrical strength and resistance to arcing. In aerials, electron tubes, resistor bodies and coil formers they are useful because of high electrical resistivity, and dimensional and thermal stability. The table summarizes their properties.

Ceramics. Articles formed from substances containing clay and converted into hard brittle bodies by a high-temperature treatment, or other products such as silica refractories that are made by a similar process. They are used to a large extent in the manufacture of insulators (see above).

Ceramic Material	Relative Permittivity ϵ_r	Loss Power-Factor $\tan \sigma \times 10^4$	Properties
Porcelain	5–7	50–200	Plastic for forming
Steatite	6	<10	Low loss
Forsterite	6	4	High thermal expansion
Cordierite	4·5	40–100	Low thermal expansion
Alumina	8·5	3–9	Gas-tight

Chain Reaction. Reaction in which the products assist in sustaining the process, such as in nuclear fission.

Chain Winding. Alternative name for BASKET WINDING (q.v.).

Change-Of-Linkage Law. A descriptive term for FARADAY'S LAW OF ELECTROMAGNETIC INDUCTION. If there is any closed linear path in space, and the magnetic flux ϕ surrounded by the path varies with time, then an electric field appears in the path, summing in the closed path to

$$e = -\partial \phi / \partial t.$$

The negative sign denotes that the direction of the electromotive force is such as to produce a current opposing the change of flux.

Change-Pole Motor. An INDUCTION MOTOR in which the speed is altered by changing the number of primary (stator) poles.

Characteristic Impedance. A function of a transmission line. The propagation of a voltage disturbance v along a line is accompanied inevitably by a current i, because the travel of the voltage implies the charging of the line capacitance C, for which i is necessary. Similarly the travel of the current sets up magnetic flux in accordance with the line inductance L. The ratio $v/i = Z_0$ is the characteristic impedance, and in a loss-free line $Z_0 = \sqrt{(L/C)}$, where L and C are expressed per unit length of line.

Charge. (1) An accumulation or deficiency of electrons, an ELECTRON being a subatomic particle endowed with a negative

Charge Indicator

charge of $e = 1 \cdot 602 \times 10^{-19}$ C. This is the most fundamental manifestation of electricity. The unit is the coulomb (abbreviation: C). A *conductor* is positively charged if a small proportion of its free electrons is removed, and negatively charged if it is supplied with excess electrons.

(2) The quantity of electricity on the positive plate of a *capacitor*. This is equal to the product of the applied voltage difference and the capacitance.

(3) The electricity passed into an *accumulator* to restore its plates to a state in which they can supply electrical energy. This is usually measured in ampere-hours.

Charge Indicator. A visual indicator of the state of charge of a LEAD-ACID CELL in terms of its specific gravity. A direct measure of the specific gravity is given by a hydrometer, a hollow-glass bottom-weighted tube which floats in the electrolyte and is marked with an appropriate scale. Pilot-ball indicators are wax or hollow glass spheres that will sink or float in accordance with the density of the electrolyte.

Charging Current. The current taken by a capacitor when its potential difference is being changed.

Charging Resistor. A resistor connected in circuit with a circuit-breaker or switch to prevent the current rising at an excessive rate when the breaker or switch is closed.

Choke. A colloquial term for an INDUCTOR or inductive reactor having inductance as its primary parameter.

Chopped Wave. A wave in which the voltage decreases abruptly from near maximum to zero. Used to determine the ability of a transformer to withstand the effects of high unidirectional voltages resembling lightning surges, it may be obtained by paralleling an IMPULSE GENERATOR with a rod-type spark gap, which has a delayed breakdown. The chop simulates the sudden disappearance of lightning surge consequent upon spill-over or the action of a lightning arrester in service.

Chopping. The action of a circuit-breaker in interrupting an alternating current prior to a zero-pause.

Chorded Coil. A DISTRIBUTED WINDING in which the coil span is less than the pole-pitch. It is also known as a *short-pitch coil*.

Chording. Reduction (or, occasionally, the increase) of the span of a coil in an electrical machine compared with full pitch (equivalent to a pole pitch).

Chromel. Nickel-chromium alloy, sometimes containing iron.

Closed-Circuit Alarm System

It is used in thermocouples, for heating elements and in heat-resistance applications. See NICKEL ALLOYS.

Circle Diagram. A term usually understood to mean the polar diagram of the stator current of an induction motor, but actually descriptive of the current in several other kinds of a.c. machine (see also LOCUS DIAGRAM).

Circuit. The closed loop of a solenoidal function; the endless path around which either in reality or by analogy, a physical quantity flows; in particular the path of an electric current or of a magnetic flux.

Circuit Analysis. See NETWORK ANALYSIS.

Circuit-Breaker. A device for automatically making and breaking the normal current in a circuit or the current which may flow under fault conditions such as short-circuit. Circuit-breaking is effected by the mechanical separation of contacts connected in series with the circuit, in a dielectric medium which assists in extinction of the arc so formed and which restores the dielectric strength of the break. Dielectrics in common use are air and oil, although other media are possible and water has been used successfully in continental designs. *Miniature circuit-breakers* are replacing fuses in domestic installations.

Air circuit-breakers include AIR-BREAK CIRCUIT-BREAKERS and AIR-BLAST CIRCUIT-BREAKERS. *Oil circuit-breakers* include BULK-OIL CIRCUIT-BREAKERS, SMALL-OIL-VOLUME CIRCUIT-BREAKERS and IMPULSE CIRCUIT-BREAKERS. See also Fig. O–1.

Circular Mil. Unit of area, equal to the area of a circle whose diameter is one MIL ($= 2 \cdot 5\pi \times 10^{-7}$ in^2).

Clamping Die. Of a resistance welding machine, alternatively known as a CONTACT JAW (q.v.).

Clark Cell. An early form of standard cell developing 1·433 V at 15°C. One electrode is of zinc in zinc sulphate, the other of mercury in mercurous sulphate.

Class A, B and C Amplification. Conditions under which a thermionic valve may be operated as an amplifier or oscillator. The conditions are determined by the amount of negative grid-bias applied, in accordance with which the valve introduces varying amounts of distortion.

Cleat. An insulated incombustible cable support. It is usually employed for vulcanized-rubber-insulated, taped and braided or p.v.c. insulated cable.

Closed-Circuit Alarm System. Burglar alarm system in which

Closed-Circuit Television

the alarm is sounded if wires are broken, in distinction from an open-circuit system in which the alarm is given when the circuit is closed. The closed alarm-contacts are wired in series with a battery and the coil of a relay, the relay controlling a separate alarm-bell circuit. Thus, with all contacts closed, the relay coil is excited and keeps the alarm-bell circuit open.

Closed-Circuit Television. Television system in which the signal is fed from the camera to the receiver by cable and not by radio transmission.

Closed-Loop Control. A system of control of a machine or process in which the result of an *input* control setting is compared with the input, and any difference (*error*) is used to correct the result (*output*) to bring the error towards zero.

Coaxial Cable. Cable suitable for transmission of radio-frequency signals with frequencies about the v.h.f. range. It occupies a place intermediate between an open-wire feeder and a WAVEGUIDE. A central conductor is separated from a surrounding coaxial conductor by a solid dielectric such as polythene, or by air. In the latter case, the separation is maintained by a series of discs.

Cockroft–Walton Multiplier. A voltage multiplier having two banks of series capacitors alternately connected by rectifiers acting as changeover switches according to the instantaneous output polarity of the energizing transformer. A machine of this type is commonly used to inject protons into a larger PARTICLE ACCELERATOR such as a proton linear accelerator. See Fig. C–6.

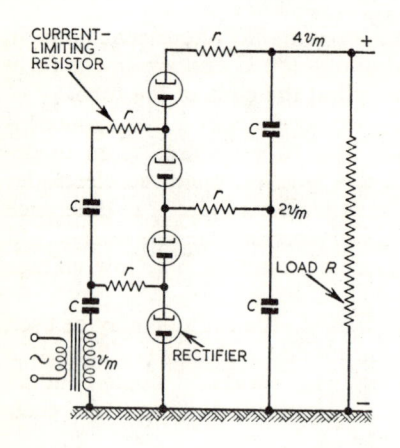

FIG. C–6. TWO-STAGE FOUR-VALVE COCKCROFT-WALTON HIGH-VOLTAGE D.C. GENERATOR.

v_m = transformer peak output voltage.

Coefficient of Coupling. Ratio of the mutual impedance components (inductive, capacitive or resistive) of two circuits to the square root of the products of their total impedances of the same kind, i.e. inductive, capacitive or resistive. That is, $k = x_m/\sqrt{(x_1 x_2)}$ where x_m is the common mutual reactance; and $x_1 x_2$ are the total similar reactances of the primary and secondary respectively taken separately. The coefficient expresses the degree of closeness with which the primary and secondary are coupled.

Coefficient of Mutual Induction. See MUTUAL INDUCTANCE.

Coefficient of Self-Induction. See SELF-INDUCTANCE.

Coefficient of Utilization. The proportion of light emitted by a lamp which actually reaches a working plane.

Coercivity. The demagnetizing force required to reduce the magnetic flux density in a material from the value of the retentivity to zero (see B/H CURVE and Fig. B–1).

Coil Ignition. A system of h.t. supply in automobiles. The interruption of the current from a battery causes a high secondary electromotive force to be induced in a coil winding for distribution to the sparking plugs.

Coil Span. The distance between the two sides of an armature coil. It is measured round the periphery of the armature and stated, usually, in electrical degrees or slots.

Cold-Cathode Emission. Emission occurring from an unheated cathode. When an electric field intensity of the order of 10^9–10^{10} V/m is applied to a metal surface and is so directed as to accelerate electrons from the surface, emission will occur even at low temperature. This is also know as *field emission*.

Cold-Cathode Lamp. A form of long-life fluorescent lamp in which the electrodes operate at about 200°C and are in the form of shells of high-purity steel housed at the ends of the tube.

Cold-Cathode Rectifier. A device dependent on the asymmetrical current/voltage characteristic of a tube containing an oxide-coated electrode of large area and a smaller, uncoated electrode: the former is the cathode and the latter the anode. The difference in work-function of the two kinds of electrode surface makes field emission possible from the cathode at comparatively low voltage gradients.

In practice an auxiliary anode close to the cathode may be used to initiate discharge to the main anode. A rare-gas filling at a relatively high pressure (e.g. 40 mmHg) makes adequate conduction possible. Advantages of cold-cathode rectifiers are their

Collective Control

ability to operate without cathode heaters, and their immediate availability.

Collective Control. A method of automatic lift control. Calls from the lift-car and landings are stored, and the lift-car stops in floor sequence at each landing from which a call has been made. In *directional collective control*, there are provided *up* and *down* buttons, and the lift-car deals with all calls in one direction then reverses for the remainder.

Collector Ring. Alternative term for SLIP-RING (q.v.).

Collector Shoe. In electric traction, a metal device which is kept in sliding contact with the conductor rail.

Colour-Matching Tube. Fluorescent tube coated with a phosphor to provide a colour temperature of 6,500°K. This approximates to the light from an overcast north sky.

Commissioning. Testing an electrical machine before it is connected to a supply system to ensure as far as possible that it is in a sound condition and will cause no disturbance when switched on. These tests should be designed to prove both the mechanical and the electrical conditions of the machine. Their number and complexity depend on the size and importance of the plant under consideration.

Commutating Pole. See INTERPOLE.

Commutation. The passage of current between the commutator of an armature and the brushes without appreciable sparking. As the sectors of an armature coil pass a brush, the coil is short-circuited and the current in the coil has to undergo an abrupt reversal or change in magnitude which facilitates sparking, unless specific remedies are applied.

Commutator. An essential part of all d.c. machines, consisting of a number of copper bars insulated from each other and from earth and suitably supported to withstand the stresses due to rotation. Each bar is connected to one end of at least two armature coils. Stationary carbon brushes rub on the surface of the commutator and pass current between the winding and the terminals of the machine.

Commutator Motor. A motor having an armature incorporating a COMMUTATOR. Single-phase commutator motors may be *series* or *repulsion*. A three-phase commutator motor is a variable-speed machine in which an auxiliary voltage supplements the current flow in the armature. A commutator machine may be cascaded with an induction motor to control the speed of the

Complex Permeability

latter. With motors of moderate or large outputs (above about 150 hp) the commutator machine is a separate unit (SCHERBIUS SYSTEM) but for smaller outputs it is integral with the induction motor (SCHRAGE MOTOR). See also STATOR-FED SHUNT MOTOR.

Compensated Induction Motor. INDUCTION MOTOR in which the excitation is supplied to the secondary circuits at slip frequency by means of either an external exciter or an additional winding, with commutator and brushes, incorporated in the motor. By suitable adjustment of the phase of the exciting current, unity power factor may be obtained in the primary circuits.

Compensated Voltmeter. An instrument designed to indicate the voltage at a remote point. It is compensated for the voltage drop in the circuit between the remote point and its point of connection.

Compensating Winding. A winding on a machine designed to reduce some of the effects of armature reaction.

Compensation Theorem. A theorem employed in NETWORK ANALYSIS. For any given set of circuit conditions, any impedance Z in a network, carrying current I, can be replaced in the network by a generator of zero internal impedance and electromotive force $E = IZ$; that is, an e.m.f. directed against the current. An extension of the theorem states that if an impedance Z be changed by $+\Delta Z$, the effect on all other branch currents is that which would be produced by an e.m.f. of $-I \cdot \Delta Z$ in series with the changed branch. The theorem is valid for linear systems (see SUPERPOSITION).

Compensator. A large shunt capacitor inserted at major high-voltage substations to improve power factor. Such compensation can be by banks of static capacitors or synchronous rotating condensers.

Complex Algebra. An algebra for analytical handling of complex quantities, in particular those representing sine-wave voltages and currents as directed r.m.s. quantities, and those representing sine-wave impedance and admittance operators. It is sometimes called *vector algebra*.

Complex Operator. A quantity such as impedance or admittance which is not sine-varying. These are the ratios of COMPLEXORS, and algebraically are treated similarly.

Complex Permeability. An expression of the static permeability and core loss of transformers or reactors, used in the design of these equipments for communication and control circuits. If

Complex Permittivity

B and *H* are complexors representing sine-varying flux density and excitation, the complex permeability $\mu = \mu \angle \theta$. For under these conditions, **B** lags behind **H** by the angle θ as a consequence of hysteresis and eddy currents. Thus

$$\mu = B/H = (B/H)(\cos \theta - j \sin \theta).$$

The "real" term represents pure magnetizing effect corresponding to the more usual definition of absolute permeability. The quadrature term is proportional to the effect of core loss.

Complex Permittivity. A consequence of the hysteresis and residual conductivity effects in an insulant. Under alternating stress there is an effective conductivity σ which includes both ohmic and hysteretic effects. The total current density *J* in a material in an electric field of intensity *E* at angular frequency ω is consequently

$$J = \sigma E + j\omega\epsilon E = E(\sigma + j\omega\epsilon)$$
$$= j\omega\epsilon E$$

where ϵ is called the complex permittivity, given by

$$\epsilon = \epsilon - j(\sigma/\omega).$$

The "real" part is the absolute permittivity, and the quadrature term is associated with the loss component of the current density. The complex permittivity enables the straightforward relation

$$J = j\omega\epsilon E$$

to be given, where the complex permittivity includes the angle $\arctan (\sigma/\omega\epsilon)$ called the Loss Angle.

Complex Quantity. The sum of a real quantity and an imaginary quantity, e.g. $x + jy$, where x may be real or zero and y is real. $j = \sqrt{-1}$.

Complex Waveform. A sustained periodic waveform that cannot be completely described in terms of a single sine function. Such a wave is generally treated as the superposed sum of a series (infinite if necessary) of sine functions having various interrelated periodic times.

Complexor. Pseudo-vector used to represent the time phase of a scalar quantity. Sustained alternating quantities whose magnitudes vary sinusoidally with times are commonly represented on a time-phase diagram as *vectors*. Such representations are not actually vectorial, for a vector must represent a directed quantity in space, not a changing scalar value. These pseudo-vectors are called complexors, because in this form they are manipulated as though they were complex numbers on an Argand diagram. Alternative names for complexor quantities are *phasors* and *sinors*.

Compole. See INTERPOLE.

Compound Catenary Suspension. In electric traction, a form of construction for overhead conductors designed to eliminate the "hard spots" which occur in SIMPLE CATENARY SUSPENSION where the contact wire is attached to the supporting structure. The contact wire is suspended from an auxiliary catenary by solid-wire loop droppers which are free to slide vertically over the auxiliary catenary, which is itself suspended by rigid droppers from the main catenary (see Fig. C–7).

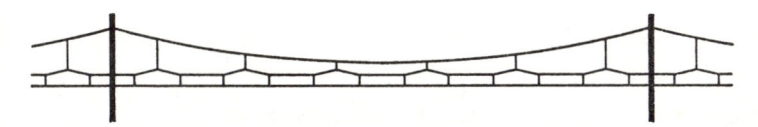

FIG. C–7. COMPOUND CATENARY SUSPENSION.

Compound-Filled Switchgear. Switchgear having the space between the metal casing and the conducting parts filled with an insulating compound.

Compound-Wound Motor. A D.C. MOTOR in which the load characteristic of a SHUNT-WOUND MOTOR is tailored to requirements by a winding on each main pole connected in series with the armature circuit. If this series winding is of such a polarity as to assist the main poles it is known as *cumulatively compound wound* and if in opposition to the main poles as *differentially compound wound* or *counter compound wound* or *reverse compound wound*. The effect on the speed characteristic is shown in Fig. C–8. It will be seen that the cumulative compound characteristic is similar to that of a SERIES-WOUND MOTOR. The degree of departure from the characteristic of the plain shunt motor depends on the number of turns in the compound winding.

Compounding. A method of adjusting the characteristics, usually of an electrical machine, in accordance with its loading.

Compression Cable. An impregnated lead- or polyethylene-sheathed single- or three-core cable, in which the conductor section

Compton Electrometer

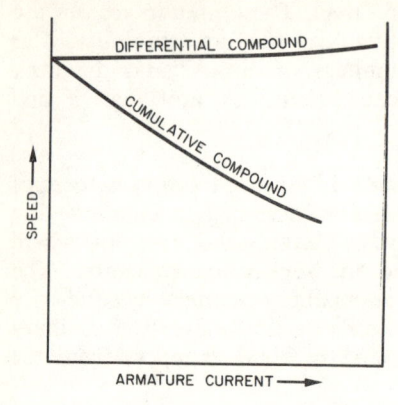

FIG. C–8. LOAD CHARACTERISTICS OF DIFFERENTIALLY COMPOUND-WOUND AND CUMULATIVELY COMPOUND-WOUND MOTORS.

is oval, and the sheath is thinner than normal (see Fig. C–9). The cable is included in a pressure container that is charged with nitrogen gas at about 200 lb/in^2. The pressure container is either another circular lead sheath, or a mild-steel pipe, in which case a three-core cable is always used. On load the cable section increases in size, and on cooling the gas pressure forces the high-viscosity impregnating compound back into the cable. It is also known as an *external gas-pressure cable*.

FIG. C–9. COMPRESSION CABLE.

Compton Electrometer. A sensitive quadrant ELECTROMETER, with a mirror on the moving system used with a fixed lamp and scale.

Computer. A mechanism for automatically solving arithmetic and other problems. See ANALOGUE COMPUTER, DIGITAL COMPUTER.

Concatenation. A connection in sequence of four-terminal networks; in particular the connection of two induction machines whereby the slip power of one machine is supplied to another as input for the purpose of speed control. Three speeds are normally possible: those of each motor alone and one determined by the sum of the number of poles of each machine. Cascade connection suffers from the disadvantages of lower power factor, efficiency and pull-out torque. See also DIFFERENTIAL CONCATENATION.

Concentric Winding. (1) Armature winding in which the coils of a group are arranged concentrically, as shown in Fig. C–10.

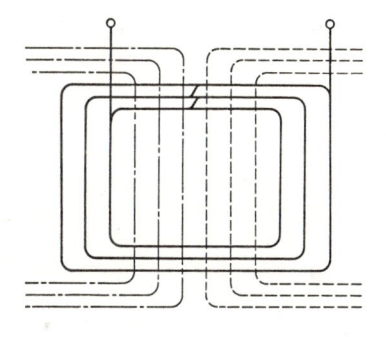

FIG. C–10. SINGLE-LAYER CONCENTRIC WINDING.

(2) Transformer winding in which the primary winding is arranged concentrically with the secondary winding.

Condenser. An old term for CAPACITOR (q.v.). See also SYNCHRONOUS CONDENSER.

Conductance. The dissipative part of ADMITTANCE. In the case of steady-state d.c. circuits, the conductance $G = 1/R$. The unit is the mho ($\mho$) or **reciprocal-ohm (or siemens).**

Conduction. The process by which an electric field is enabled to maintain an electric current. In a *metal* or a *vacuum*, the current comprises the transfer of electrons (negative elemental charges) in a direction opposite to that in which the electric field is conventionally directed; in an *electrolyte*, it is the movement with and

Conduction Pump

against the field respectively of positive and negative ions; in a *semiconductor*, it is the like transfer of electrons and "holes"; and in a *gas*, it is the movement of electrons and positive ions in opposite directions and at very different velocities.

Conduction Pump. A type of ELECTROMAGNETIC PUMP having an external current source. In the d.c. pump shown in Fig. C–11,

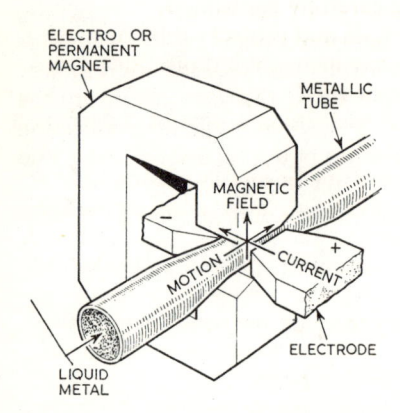

FIG. C–11. D.C. CONDUCTION PUMP.

This pump is used mainly with heavy metals such as lead, bismuth or mercury.

the current is supplied to the liquid via busbars brazed to the tube-wall; a few turns of busbar provide the magnetomotive force for the magnetic field. Alternatively, a permanent magnet may be used, though this is usually economical only in small pumps. The d.c. conduction pump is most versatile, covering a wide range of flow and pressure with any liquid metal. The inconvenient supply (thousands of amperes at 0·5–2·5 V) is the chief disadvantage and restricts this pump mainly to large sizes (500 l/min, 700 kN/m^2 and over) and to heavy metals.

The a.c. conduction pump is similar. Skin effect limits its application to smaller-size pumps of up to 500 l/min at 50 Hz or 2,500 l/min with a special low-frequency supply of 5 Hz.

Conductivity. A specific property of a material describing its ability to conduct an electric current. It is the conductance in mhos of a conductor of unit length and unit cross-section, under specified conditions (e.g. of temperature).

Conductivity Cell. In electrochemistry, a cell used for measuring the resistance of an electrolyte.

Conductor. (1) Material that offers a low resistance to the passage of electric current.

(2) Of a cable, the electrically continuous wire or wires which form the conducting portion. They are usually composed of high-conductivity annealed copper, and with the exception of very small sizes, are made up of a number of copper wires stranded together for flexibility. Aluminium of high purity is sometimes used, the other components being carefully controlled.

Conduit. Trough or pipe to contain and protect cables or wires. It may be of plain or screwed steel, earthenware, bituminized paper, fibre, plastics, etc., and be flexible or rigid.

Consequent Pole. (1) A pole occurring on a part of a permanent magnet other than a free end.

(2) A magnetic pole in an electromagnetic circuit occurring at a point between two magnetizing coils when their magnetomotive forces are in opposition (see below).

Consequent-Pole Winding. A winding in which the number of coil groups per phase is equal to half the number of poles. The current passes in a similar direction through the coils forming two poles of the same polarity in each phase. Poles of the opposite polarity are induced between the wound poles.

Conservator. A separate tank connected by a small diameter pipe to the main tank of an oil-immersed transformer and mounted above it. It has a capacity of about 10% of the main tank, and maintains a small head of oil above it. Its purpose is to ensure that the oil is kept in good condition, by excluding moisture and reducing the area of oil in contact with the atmosphere.

Constant-Flux-Linkage Theorem. The magnetic flux linkages associated with a closed electric circuit of zero resistance will not change. Where such a lossless closed circuit carries a current I and is in consequence associated with a magnetic flux, it possesses stored magnetic energy amounting to $\frac{1}{2}LI^2$, where L is the effective inductance. If the circuit has no dissipative element, the stored energy must remain associated with the circuit: i.e. the current must remain constant.

Constantan. Copper–nickel alloy with a low temperature/resistance coefficient. It is used in thermocouples and as the resistance wire in resistance boxes. See NICKEL ALLOYS.

Contact Bar. Of a resistance seam-welding machine, a device to exert pressure on the components being welded, and conduct the current to them. It is also termed an *electrode bar*.

Contact E.M.F.

Contact E.M.F. The electromotive force occurring when two dissimilar metals are brought into contact. Free electrons diffuse in each direction across the boundary, and initially more electrons will pass in one direction than in the other, so that one metal will become positive to the other by a small *contact potential*. A condition of equilibrium is reached in which the net electron transfer is zero. Two or more contact e.m.f.s tending to produce a direct current, with energy derived from a chemical source, are the basis of primary and secondary cells.

Contact Jaw. Of a resistance welding machine, a device that holds the components being welded, and conducts the current to them. It is also termed a *clamping die*.

Contact Rectifier. A device employing mechanically operated switches to convert alternating to direct current. The process of rectification is similar to that of any other polyphase rectifier, and similar transformer connections can be used as for single-anode or ignitron valves (see Fig. C–12).

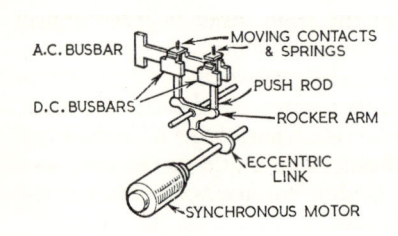

Fig. C–12. Typical Mechanism for Contact Rectifier.

Contact Wheel. Of a resistance seam-welding machine, a rotatable wheel that exerts pressure on the components being welded, and conducts the current to them. It is also termed an *electrode wheel*.

Contactor. A device for repeatedly opening and closing electrical circuits. Electrically, relays and circuit-breakers also fulfil the same function. The difference between these and contactors is one of detail design. Relays are of light construction, generally without any arc-control devices, and designed to operate frequently in circuits carrying only low currents. Circuit-breakers are built to interrupt heavy fault currents. They have heavy contacts and powerful arc-control devices, and operate infrequently. Contactors fall between these two groups; they are heavier than relays, possess effective arc-control means, but are not as heavy and powerful as circuit-breakers. They are employed for controlling currents from a few amperes up to a few thousand amperes at voltages up to 600 V; special designs are available for higher voltages. Depending on duty and installation conditions, the contacts are of the oil-immersed, mercury-in-glass, or air-break type. The last is the most usual.

Continuity Bond. In railway track, a bond connecting rails at a joint.

Continuous Discharge Current. In a lightning flash, the relatively low current which flows continuously and on which much larger currents are superimposed.

Continuous Maximum Rating. The load at which a machine may be operated for an unlimited period under the specified conditions. It is sometimes known as *c.m.r.*

Control Board. A switchboard incorporating indicating and operating devices and accessories for the remote control of switchgear.

Control Line. In a railway train, a cable or TRAIN LINE extending throughout the length of the train, used to interconnect master controllers or contactor control circuits.

Control Ratio. Of a thyratron, the ratio between the critical grid voltage and the related positive anode voltage.

Controlled Rectifier. Rectifier incorporating a means of regulating its output voltage. This will be a grid in the case of a MERCURY-ARC RECTIFIER or an additional layer of material in a semiconductor rectifier or THYRISTOR.

Convection Current. An electric current resulting from a transport of charges by means other than the application of an electric field.

Convection Heating. Heating which functions by warming the air and setting up air movements. It is more suitable for continuous than intermittent heating. Cf. RADIATION HEATING.

Converter. A device for the conversion of electrical energy from the a.c. to the d.c. form. Its inverse is an INVERTER. The common types of machine include the MOTOR GENERATOR, the MOTOR CONVERTER and the SYNCHRONOUS CONVERTER. A modern converter is rarely a rotating machine; it is generally a form of RECTIFIER.

Converting Station. A substation for conversion between alternating and direct current, or for changing phase or frequency of alternating current. The conversion is achieved with motor

Coolant

generators, rectifiers, etc. Rectifier converting stations include glass-bulb installations serving small public distribution areas, traction substations equipped with steel-tank units, and converter stations associated with high-voltage d.c. transmission schemes in which grid-controlled mercury-arc rectifiers can be inverted as required to enable power to flow in either direction in accordance with operational requirements. The greatest use now being made of converter substations is in connection with traction.

Coolant. Material employed to limit the temperature rise of a machine. Four common substances are air, hydrogen, oil and water (see table). Air and oil are by far the most common coolants because of their cheapness and insulating properties respectively; hydrogen has the merit of a low density which results in a low windage loss for rotating machines, and also a high specific heat which gives it a heat-removal capacity similar to that of air.

Coolant Material	*Air*	*Hydrogen*	*Water*	*Oil*
Density, g/cm^3	0·0013	0·00009	1·0	0·83
Specific heat, J/g°C	0·84	14	4·2	2·0
Volume required to remove 1 kW at 150°C, m^3/min.	3·4	3·4	0·00095	0·0025

Copper. Metal having high electrical and thermal conductivity, good all-round mechanical properties, good resistance to corrosion, easy and efficient jointing by a variety of methods, ready availability in almost any shape or form, and high scrap value. Copper of the highest purity is a more efficient conductor of electricity than any other known material except silver. *Annealed copper* that has a resistance equal to that specified by the International Electrotechnical Commission is said to have a conductivity of 100% i.a.c.s. (International Annealed Copper Standard), although in practice high-conductivity copper, which is both wrought and annealed, can in fact have a conductivity of 101% even 102% i.a.c.s. The resistivity of copper is $1·56 \times 10^{-6}$ Ω cm at 0°C.

Copper Alloys

Copper Alloys. Mixtures of copper with small quantities of added elements. The impurities which may be found in high-conductivity copper are controlled to within very close limits (see Fig. C–13). Phosphorus in particular has a drastic effect upon

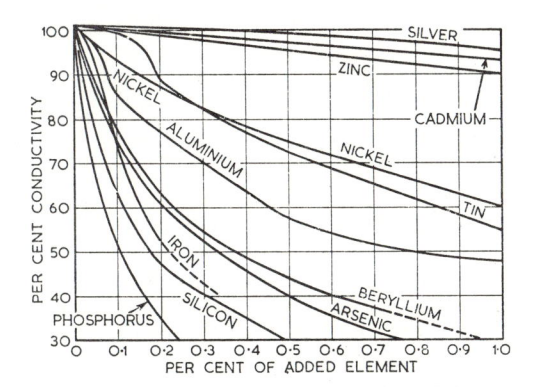

FIG. C–13. EFFECT OF ADDED ELEMENTS ON THE CONDUCTIVITY OF COPPER.

conductivity, as little as 0·05% being sufficient to reduce the value to about 70% i.a.c.s. Of the other impurities that copper may possibly contain, all but silver, cadmium and zinc cause appreciable reduction in the electrical conductivity when present in even the smallest quantities.

Ordinary high-conductivity copper will anneal fairly rapidly if subjected to a temperature of 200°C, but the addition of 0·1% of silver to the copper raises this figure to between 300°C and 350°C. Since such an amount of silver has a negligible effect upon the conductivity, the silver-bearing coppers are very suitable for certain electrical applications (such as the windings of turbo-alternator rotors) where an increased resistance to softening is required.

When hardness and strength at elevated temperatures are required, two of the most important copper alloys are cadmium–copper and chromium–copper. The former has good mechanical properties and retains its strength at moderately elevated temperatures. It is often used for the manufacture of welding electrodes. Chromium–copper is one of the few heat-treatable alloys and can be used at temperatures up to about 400°C. The addition of

Copper Loss

0·5% tellurium or selenium to copper produces a material that can be machined at high speeds.

It is usually the high conductivity of copper and its alloys that merits their consideration, but there are also several useful groups of copper-based resistance materials. Copper–manganese–aluminium and copper–manganese–nickel (manganin) alloys possess high resistivities and very small resistance-temperature coefficients and low thermal e.m.f.s relative to copper, and so are very suitable for the manufacture of instrument shunts and multipliers.

Copper Loss. A term loosely applied to the conduction, Joulian or I^2R Loss that occurs in conductors not intended specifically to produce heat.

Copper-Oxide Rectifier. A metal rectifier consisting of discs or plates of copper oxidized in a furnace at a temperature around 1·050°C and then annealed at a lower temperature. A layer of cupric oxide which forms on the outer surface is removed by controlled acid treatment, leaving a thin layer of cuprous oxide in contact with the copper matrix. This forms the semiconductor portion of the rectifier and good electrical contact must be made over the whole of its surface.

Core. (1) The part of an electromagnetic circuit situated within the winding.

(2) A single conductor with its insulation forming part of a cable.

Core Loss. The loss developed as heat in the ferromagnetic core of a magnetic circuit subjected to alternating magnetization. It comprises two components quite different in their origin, namely HYSTERESIS LOSS and EDDY-CURRENT LOSS.

Core Plate. Of a machine or transformer, each of the thin, metallic sheets which form part of the core. They are also referred to as *laminations*, *punchings*, or *stampings*.

Core-Type Transformer. A transformer in which the windings surround and generally enclose the iron core (cf. SHELL-TYPE TRANSFORMER).

Cored Carbon. A carbon rod for use as an electrode in a carbon arc lamp. It contains longitudinal channels filled with mixtures to modify the effect of the arc.

Coreless Induction Furnace. See HIGH-FREQUENCY INDUCTION FURNACE.

Corona. A discharge of electricity round a conductor caused by ionization of the surrounding air. It gives, in darkness, the visual appearance of a luminous sheath and appears when the potential gradient at the surface of the conductor exceeds a certain value. It results in a continuous rate of energy loss and may cause radio interference.

Corrosion. The chemical change of a metal from the elemental state into compounds—either solids such as oxides or sulphides, or salts such as sulphates or chlorides in aqueous solution. Most corrosion is due to chemical attack by substances in water, soil or air (especially near industrial areas or the coast), to electrolytic action, or to the action of sulphate-reducing bacteria in soil under anaerobic conditions.

Electrochemical corrosion results from the existence of electrical potentials between metals and electrolytes surrounding them (see CATHODIC PROTECTION).

Cosine Law. The illumination received on a surface is proportional to the cosine of the angle between the direction of the incident light rays and the normal to the surface at the point of incidence.

Coulomb. The practical unit of electrical quantity or charge; also the absolute unit in the M.K.S. system (abbreviation: C; symbol: Q, q). The coulomb is perhaps the most fundamental of all the electrical units because it can be related to the electron charge e, which is $1 \cdot 602 \times 10^{-19}$ C. As the ampere is defined in terms of length and magnetic space properties, the coulomb is derived from the ampere and second: 1 C = 1 As. Its value is unchanged in S.I. units.

Coulomb's Law. The force of attraction or repulsion between two electrostatically charged bodies is proportional to the magnitude of their charges and inversely proportional to the square of the distance separating them.

Coulometer. An electrolytic cell, also known as a *voltameter*, which measures quantity of electricity by the amount of a substance liberated as one of the products of electrolysis. It can be either a weight coulometer or a volume coulometer. Special types include a *silver coulometer* and *gas coulometer*.

Counter Compound-Wound Motor. Alternative term for differentially compound-wound motor (see COMPOUND-WOUND MOTOR).

Counter-Current Braking. A method of braking an electric motor by reversal of the supply connections. It is frequently known as *plugging*.

Counter E.M.F.

Counter E.M.F. The electromotive force developed by certain forms of circuit element, by virtue of which they absorb electrical energy and convert it into some other form such as mechanical or chemical energy (cf. BACK E.M.F.).

Counter-E.M.F. Acceleration. A method of automatic acceleration which can be employed when starting small d.c. motors. Use is made of the fact that the counter e.m.f. across the armature of a d.c. motor increases in proportion to the speed as the motor accelerates. The operating coil of the accelerating contactor is connected across the armature. When the supply is connected at starting, the voltage across the contactor coil increases gradually until a value is reached sufficient to close the contactor, and thus short-circuit a step of starting resistance (see Fig. C–14). The duration of starting time is proportional to the load on the motor.

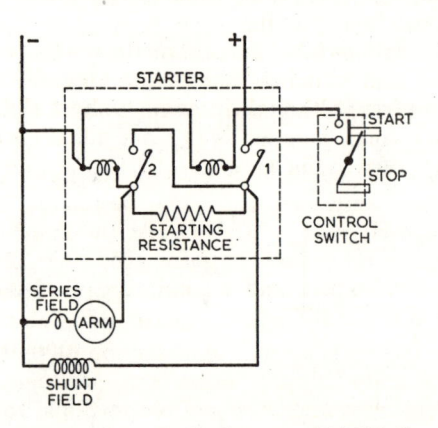

FIG. C–14. DIAGRAM OF D.C. COUNTER-E.M.F. STARTER.

Counterpoise Earthing. A method of earthing in which steel wires or tapes are buried in the ground, the length and number depending upon the resistivity of the ground and the value of earth resistance required. This method may be used for protecting transmission and high-voltage distribution lines, and where ordinary rod or plate earthing is unsuitable because of the nature of the soil.

Coupled Circuits. Circuits in which electrical energy can be passed from one to the other. This means that a simple two-mesh circuit is coupled, because the two meshes have one or more impedances in common. However, the term is generally restricted to circuits that are separable, and particularly to those in which the coupling is by means of a common magnetic field.

Coupled Surge. A surge induced in a conductor by a surge in another adjacent conductor.

Coupling Coefficient. See COEFFICIENT OF COUPLING.

Covalent Bond. A molecular binding force due to two electrons of opposite spin being shared mutually by adjacent atoms.

Cradle. In overhead-line construction, an assembly of GUARD WIRES to form a net.

Crater-Lamp Oscillograph. Instrument used for recording the slow portion of a lightning discharge.

Crest Factor, Value. See PEAK FACTOR, PEAK VALUE.

Critical Damping. The least value of DAMPING necessary to prevent oscillation in a system.

Crompton Potentiometer. A precise form of potentiometer. Resistance coils and switches replace the long slide-wire of the basic POTENTIOMETER, although in some a short slide-wire, commonly in circular form, is retained as a fine adjustment. A simplified diagram is shown in Fig. C–15.

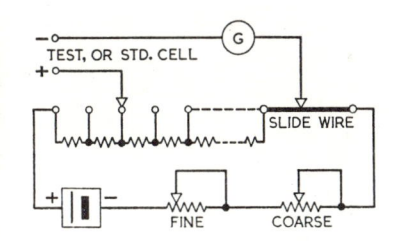

FIG. C–15. SIMPLIFIED DIAGRAM OF CROMPTON POTENTIOMETER.

Cross Axis. Alternative name for QUADRATURE AXIS (q.v.).

Cross Bond. In railway track, a bond connecting two rails of a track, or the rails of adjacent tracks.

Cross-Compound Turbo-alternator. A TURBO-ALTERNATOR unit, generally restricted to large outputs, which consists of two lines, each with its own alternator (cf. TANDEM-COMPOUND TURBO-ALTERNATOR). The two lines may have the same or dissimilar speeds.

Cross-Field Machine. A d.c. machine in which an essential feature is the magnetic flux which is produced on the armature

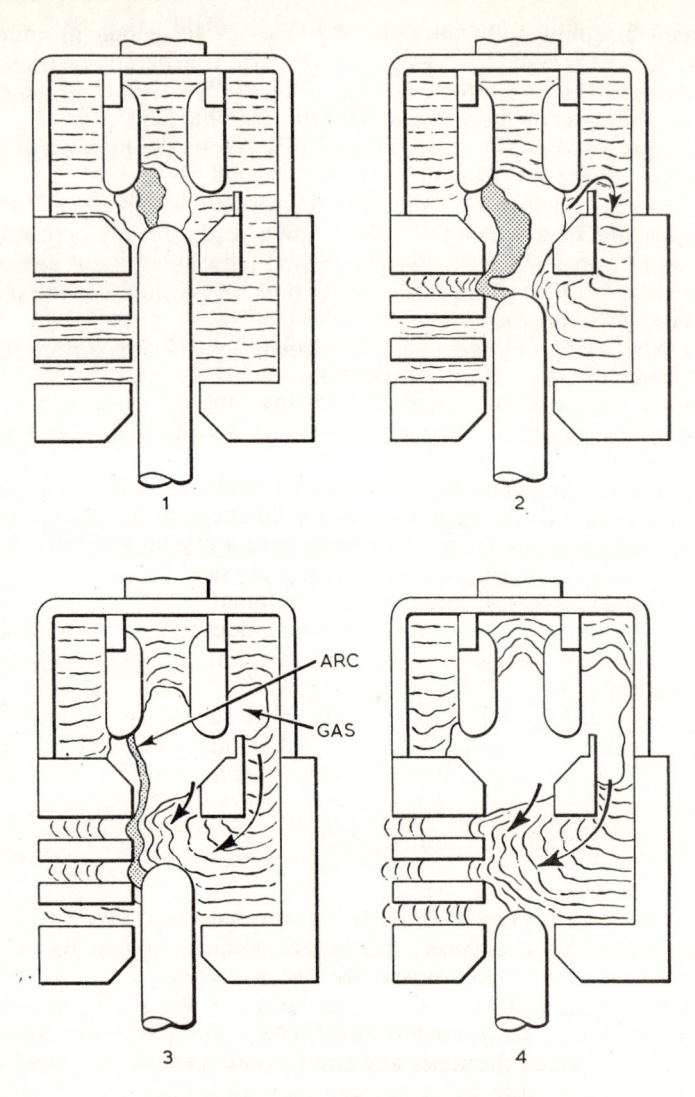

Fig. C-16. Cross-Jet Explosion Pot.
Four stages of operation are shown.
(*Associated Electrical Industries Limited*).

axis, in addition to that produced on the field axis (see AMPLI-DYNE).

Cross-Jet Pot. An EXPLOSION POT in which high pressures may be generated at a low current without exceeding a safe pressure at high currents. The gas formed in the early stage of contact movement generates pressure causing the oil to flow across the path of the arc, which is driven on to insulating barriers acting as arc splitters (see Fig. C–16).

Cryotron. A superconductive device comprising a fine superconducting wire surrounded by a coil of fine wire having normal resistance characteristics. Superconductance in the first wire can be destroyed by the magnetic field generated by passing a current through the coil. The circuit also contains means for detecting whether, at a given instant, the wire is superconducting or not. The complete device can be employed in a computer as a binary memory element.

Crystal. An assembly of molecules having a definite internal structure and the external form of a solid enclosed by a number of symmetrical plane faces. All metals have a crystal structure, and the nature of this may determine their physical properties.

Crystal Diode. In general, the combination of a crystal and a point contact to produce unilateral conductivity. Recent crystal diodes consist of germanium or silicon with a tungsten wire as point contact.

Crystal Lamp. A semiconductor light source consisting of a chip of material, such as gallium phosphide, encapsulated in a block of transparent resin. Its efficiency is low, but its switching-on speed is extremely rapid.

Crystal Rectifier. Combination of a crystal and a point contact, or of two crystals, the junction possessing unilateral, or at least asymmetrical, conductivity.

Cubicle Switchgear. Switchgear divided into compartments constructed of steel sheet. The circuit-breakers, which are either air-break or oil-break types, are not withdrawable as in truck gear (see Fig. C–17).

Cumulatively Compound-Wound Motor. A COMPOUND-WOUND MOTOR in which the series and shunt windings assist one another electromagnetically.

Curie Point. That temperature, reached from cold, at which a given material ceases to be ferromagnetic. Such a material will regain its ferromagnetism on cooling, but the Curie point for

Current

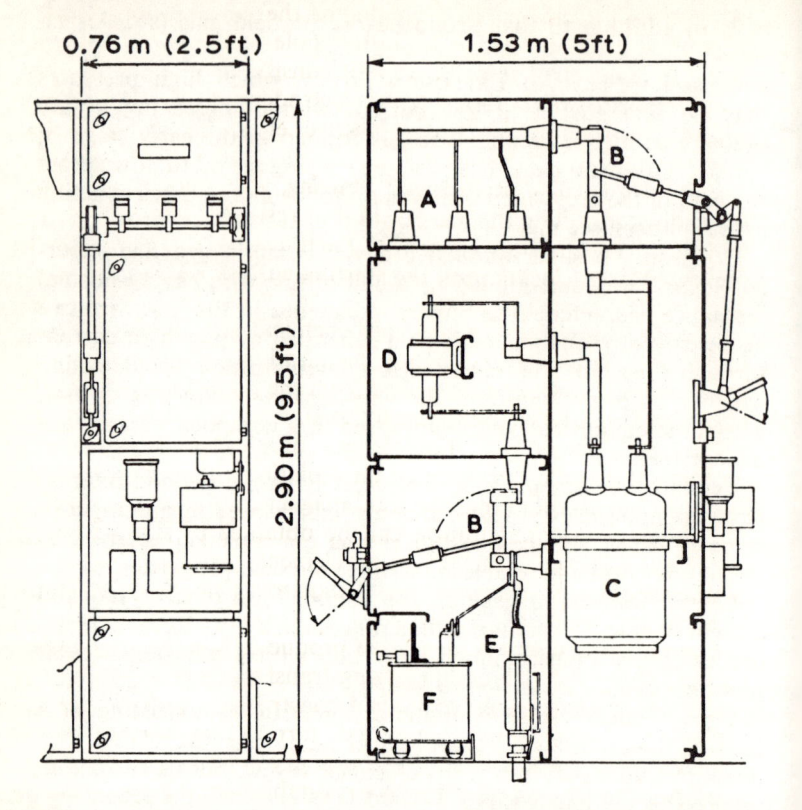

FIG. C-17. 6·6 KV SINGLE-BUSBAR CUBICLE-TYPE SWITCHGEAR.

A. Busbars.
B. Isolator.
C. Dead-tank oil circuit-breaker.
D. Current transformer.
E. Cable sealing end.
F. Voltage transformer.

(*Associated Electrical Industries Limited*)

falling temperature is not always the same as that for rising temperature.

Current. A movement of electric charge. The direction of a current is taken arbitrarily as that of the movement of positive charges and opposite to that of negative ones. A current may also result from the combined motion of positive and negative charges

Current Transformer

in opposite directions. In a conductor the current consists of a drift of electrons towards the positive pole of the applied electric field. In an electrolyte or in a gas it consists of the migration of positive ions towards the negative electrode and of negative ions and/or electrons towards the positive electrode. Unit is the *ampere*.

Current-Control Acceleration, Starting. A method of automatically controlling the acceleration of resistance-started electric motors during the starting period. Each accelerating contactor is prevented from closing until the starting current falls to the minimum value.

Current Efficiency. In an electrochemical process, the ratio of the mass of a substance affected chemically or electrolytically to the mass which would be expected from theoretical considerations (see FARADAY'S LAWS OF ELECTROLYSIS).

Current Element. A short length of conductor carrying a current. It is often considered in isolation, a concept that is fictitious but useful in that the total magnetic effect of a current in any conducting configuration can be obtained by summation of the effects of its constituent current elements taken separately.

Current-Element Starter. See CURRENT-CONTROL ACCELERATION, STARTING.

Current Impulsing. A method of producing large unidirectional current pulses. A current-impulsing transformer is employed. It resembles a transformer of normal construction, except that the secondary winding comprises a single turn of large section. With the secondary open-circuited, the primary is connected to a d.c. supply. After the primary current is established, the secondary is closed. The primary current is then broken as rapidly as possible: a corresponding current is developed momentarily in the secondary circuit. Thus if a current of 10 A in a primary of 2,000 turns is suddenly broken, a current pulse approaching 20,000 A will flow in the one-turn secondary. The underlying principle is that of the CONSTANT-FLUX-LINKAGE THEOREM. An essential part of a current-impulsing equipment is a high-speed circuit-breaker capable of interrupting currents of the order of 10 A in about 2 ms.

Current-Limiting Reactor. An inductor, in a power-supply system, arranged for series connection in order to limit short-circuit currents at various points on the network.

Current Transformer. An INSTRUMENT TRANSFORMER for the transformation of current. The primary winding is connected in

Cut-Out

series with the load to be measured or controlled, and this load decides the current flowing through it. The secondary winding is loaded with a constant impedance, for any given set of conditions. The core flux and the current in the secondary circuit depend upon the primary current. It may be known alternatively as a *series transformer*.

Cut-Out. A means of interrupting a circuit, usually of low power. A fusible cut-out is a FUSE, but the term is more usually applied to a magnetic device, such as that used in automobile electric systems to close and open the circuit between the generator and the battery in accordance with the effect of engine speed on the generator electromotive force.

Cut-Wound Core. A magnetic-core construction using continuous grain-oriented steel strip. The strip is wound into a closed core of suitable shape, such as circular or rectangular, and impregnated.

Cyclic Rating. A form of cable rating. Tables of permissible current ratings (maximum permissible currents) for transmission and distribution cables are based on the continued application of a steady alternating or direct current. In practice the maximum load is usually applied only for a limited time, and the current may then be increased without the maximum permitted conductor temperature rise being exceeded. Such increases in current rating, where the applied load is cyclic (usually with a period of once a day), are given by the application of cyclic rating factors.

Cyclone Furnace. Boiler employing crushed coal. The ash is reduced to a fluid state so that it may be tapped and subsequently quenched to a solid state for disposal.

Cyclotron. An ORBITAL ACCELERATOR consisting essentially of a vacuum chamber between the poles of a fixed field magnet. Inside this chamber are two hollow *Dee* electrodes which load the

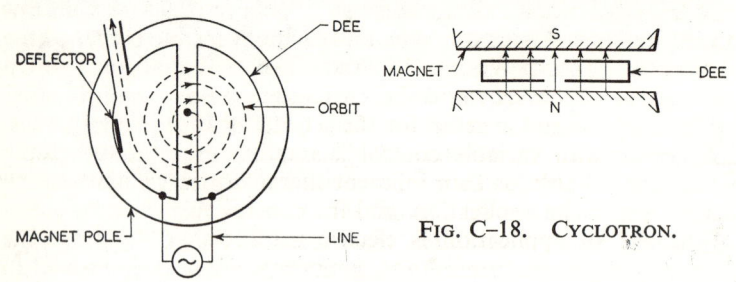

FIG. C-18. CYCLOTRON.

end of a quarter-wave resonant line so that **10–20 MHz voltage** appears across the accelerating gap (see Fig. C–18). Protons or other positive ions are introduced from an ion source near the centre of the magnet and are accelerated twice per revolution as they spiral out from the centre of the machine (see SYNCHRO-CYCLOTRON).

Cylindrical Winding. A transformer winding consisting of a single coil of one or more layers, its axial length usually being considerably greater than its diameter.

D

D.C. Amplifier. An electronic amplifier arranged to deal with d.c. signals, or a.c. signals of very low frequency.

D.C. Balancer. A d.c. motor–generator or rotary transformer used in a multiple-wire system to distribute the voltage equally between the wires.

D.C. Bridge. A BRIDGE circuit employing direct current for the measurement of circuit parameters. The basic form is the WHEATSTONE BRIDGE. See also CAREY–FOSTER BRIDGE, KELVIN DOUBLE BRIDGE.

D.C. Current Transformer. A term used to describe a magnetic amplifier employed to provide an output current proportional to a d.c. input.

D.C. Injection Braking. A method of braking an electric motor by disconnecting it from the supply and feeding direct current to the stator winding. The effect is to build up a static magnetic field in the rotor space, and the current thereby generated in the rotor winding produces a powerful braking torque.

D.C. Motor. A machine which produces mechanical power when supplied with a direct current. There are three principal types: the SERIES-WOUND MOTOR, the SHUNT-WOUND MOTOR and the COMPOUND-WOUND MOTOR. The field and armature windings are connected in parallel, with respect to the supply, for the shunt motor and in series for the series motor. Where a wide speed range with variable control is required, d.c. motors are particularly suitable, as their inherent characteristics can be easily tailored to suit the application, and the control gear is simple. A typical field of application is electric transport vehicles which require a speed range from zero to maximum speed. See Fig. D–1.

D.C. Resistance

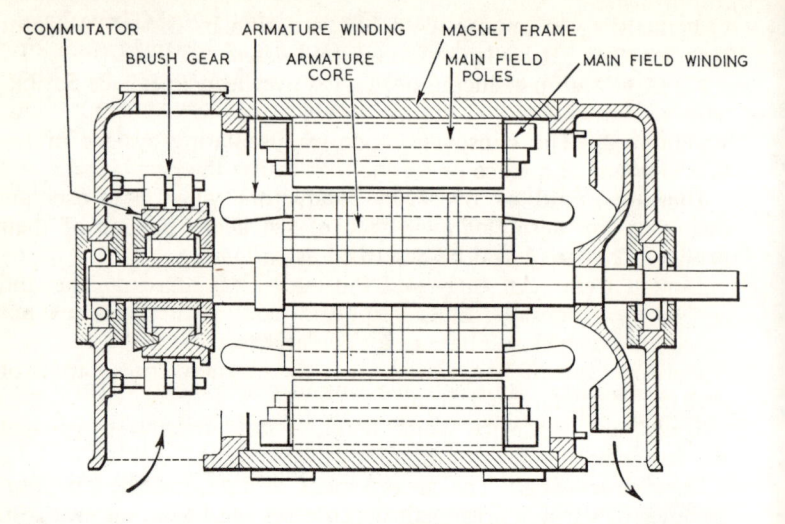

FIG. D–1. LONGITUDINAL SECTION SHOWING MAIN COMPONENTS OF A D.C. MOTOR.

D.C. Resistance. The resistance presented by a conductor, or conducting device, to a constant direct current. The term is used in contradistinction to *a.c. resistance*, or EFFECTIVE RESISTANCE, in which either (*a*) the current/voltage characteristic is such that the resistance $r_a = v/i$ for a given voltage and current differs from $r_a = \Delta v/\Delta i$ for small variations Δv and Δi, or (*b*) the I^2R loss is increased, e.g. by eddy-current effects, which do not occur with constant direct currents.

D.C. Voltage Transformer. Equipment analogous to the D.C. CURRENT TRANSFORMER, but including a resistor in the primary circuit of the magnetic amplifier. The output current depends on the value of the direct input voltage.

Damper. (1) A winding, usually consisting of solid copper bars connected at the ends by partial or completed rings, inserted in the pole-face of a low-speed synchronous machine. Its purpose is to damp oscillations of angular speed about the mean (synchronous) speed. Under steady conditions the bars lie in, and rotate with, a constant magnetic field. If the speed is subject to cyclic disturbance, the bars cut the field and have e.m.f.s induced in them.

Currents flow to oppose the speed variation by induction-motor torque action. It is also called an *amortisseur* winding.

(2) A vibration-reducing device for overhead lines (see STOCKBRIDGE DAMPER).

Damping. The consequence on an oscillating system of the development of a force or torque that resists the oscillation.

Dancing. Said of overhead conductors when they oscillate owing to the aerodynamic effect of ice accretions or of their profile. They may also be described as *galloping*.

Daniell Cell. An early wet PRIMARY CELL, having zinc and copper electrodes in dilute sulphuric acid, with a porous pot containing copper sulphate as a depolarizer.

Daraf. Unit of ELASTANCE. It is the elastance of a capacitor that has a capacity of 1 farad.

Dash-Pot. A device which employs fluid friction to prevent sudden or oscillatory motion of a moving part.

Data Processing. The arrangement of data and the carrying-through of various arithmetical processes (addition, subtraction, etc.) on these data in a digital computer.

De Sauty Bridge. An A.C. BRIDGE for the comparison of capacitors. In Fig. D-2, if C_1 is the unknown and C_2 is a standard

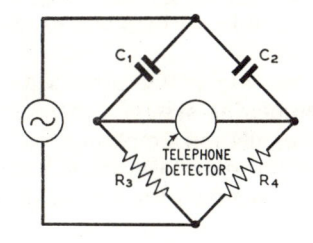

FIG. D-2. DE SAUTY BRIDGE.

capacitor, then on balance

$$\frac{C_1}{C_2} = \frac{R_4}{R_3}$$

The simple bridge shown is inadequate in most practical cases as, in general, the power factor of the capacitor under test is significant on account of the internal dielectric losses.

Dead. At or near earth potential and disconnected from any live system.

Dead Beat Mechanism. One in which the oscillatory motion of its moving parts dies rapidly owing to the damping present.

Dead Earth

Dead Earth. A connection between a conductor and earth by means of a low-resistance path.

Dead-Ended Feeder. Alternative name for INDEPENDENT FEEDER (q.v.).

Dead-Front Panel. A switchboard in which all the live parts of the switches, fuses, etc., are mounted behind the panel.

Dead-Man's Handle, Pedal. In electric traction, an attachment to the handle of a controller or to a brake valve, which is normally held by the driver's hand or pressed by his foot. If the pressure of the hand or foot is released, the device acts to cut off the current and apply the brakes. A more comprehensive safety device being fitted in locomotives is the *driver's safety device* (D.S.D.).

Decay Constant. A measure of the rate of breakdown of a radioactive nucleus. It is also known as the DISINTEGRATION CONSTANT (q.v.).

Decibel. One-tenth of a *bel*. It is equal to ten times the common logarithm of a power ratio. It is a method of expressing gain or loss in signal level in telecommunication circuits.

Decimal Prefix. An internationally agreed code to indicate the denary order of magnitude of a unit in any metrically based system. A list of prefixes is given in the Appendix.

Decomposition Voltage. For an electrolyte, the minimum voltage between two electrodes immersed in it which will produce continuous electrolysis.

Decrement. A measure of the decrease in the amplitude of oscillation of a system. A freely oscillating system undergoes a more or less gradual reduction of oscillation amplitude as the system losses dissipate the stored energy responsible for oscillation. The decrement is the rate of decay expressed as the ratio of the amplitudes of successive swings, the oscillation being assumed to decay exponentially.

Deep-Bar Winding. A winding, particularly a SQUIRREL-CAGE WINDING, on the rotor of an induction motor, designed to increase production of eddy-current loss and so increase the effective resistance. The additional loss is proportional approximately to the fourth power of the conductor depth, and a significant increase in starting torque is obtained without much effect on the slip and efficiency at normal load. The power factor at full load, is, however, somewhat reduced because of the additional rotor slot reactance to which the generation of eddy currents is due. Fig. D–3 shows typical conductor shapes.

Delay Line

FIG. D–3. DEEP-BAR CONDUCTORS FOR A SQUIRREL-CAGE WINDING.

Deflection Potentiometer. A variation of the simple d.c. POTENTIOMETER in which a small deviation from a nominal voltage is read by balancing the potentiometer in the ordinary way for the nominal voltage, but using a calibrated galvanometer in order to read the actual departure from the nominal. The accuracy is somewhat lower than that of an ordinary potentiometer but better than that of a voltmeter.

Degaussing. The process of arranging current-carrying coils in a ship to counteract its effect on the earth's field, so as to avoid setting off the detector mechanism of a magnetic mine.

Deion Grid. An ARC-CONTROL DEVICE used in certain air and oil circuit-breakers. The contacts separate within a stack of U-shaped insulating plates. Some of the insulating plates have an iron insert so that the arc current sets up a considerable magnetic field which encircles the arc and forces it towards the closed end of the U. This cools the arc and breaks it up into a series of shorter arcs, facilitating its rapid extinction.

Delay. See PHASE-SHIFT DISTORTION.

Delay Angle. A term used in connection with grid control of mercury-arc rectifiers. Transfer of the arc from anode to anode in such a rectifier normally occurs at the instant in each cycle when the potential of an idle anode begins to exceed that of the anode carrying the current. The transfer occurs quite naturally at such an instant, but by means of grid control, it can be delayed by a fraction of a cycle, normally expressed in angular form as the delay angle. By delaying the transfer in this way, i.e. increasing the delay angle from zero, the mean value of the rectifier output voltage can be reduced.

Delay Cable. A concentric cable connected to an object under test to delay the arrival of a surge from an impulse generator.

Delay Line. Any passive network capable of delaying a signal of any waveform without introducing appreciable distortion, the parameters being chosen to decrease the propagational velocity.

Delta Connection. A three-phase connection in which the three windings are connected in series, the supply being applied to, or taken from, the junctions. It is a special case of MESH CONNECTION, and is sometimes written as a Δ *connection*.

Delta Voltage. (1) The voltage between any two lines of a symmetrical three-phase system (also known as *line voltage*, or *mesh voltage*).

(2) The voltage between alternate lines of a symmetrical six-phase system.

Demagnetization Curve. The plot of magnetic flux density in a material against demagnetizing force, as sufficient force is applied to reduce the magnetic flux density from the value of the retentivity to zero. It is the characteristic that determines the quality of the material for permanent-magnet purposes. See B/H CURVE and Fig. B–1.

Demodulation. The inverse process to MODULATION, known also as *detection*. It involves the recovery of the modulating signal (or intelligence) from the modulated carrier wave after passage through the transmission medium.

Derived Unit. A unit which is defined in terms of two or more FUNDAMENTAL UNITS.

Diamagnetism. Phenomenon exhibited by materials having a permeability less than that of a vacuum, i.e. less than unity.

Diametral Connection. A form of connection to brushes on a commutator, or to tappings on a closed double-layer winding. The circuit is taken through brushes or tappings which, on a two-pole machine, are diametrically opposite to each other; or on a multipolar machine are displaced from each other by 180 electrical degrees.

Diametral Voltage. The voltage between opposite lines of a symmetrical six-phase system.

Diaphragm. (1) A sheet of finely perforated or porous material used in a cell of an ACCUMULATOR as a separator between the plates of opposite polarity.

(2) A partition used in electrochemistry which is permeable to ions (see ELECTROLYTIC DIAPHRAGM).

Diathermic Coagulation. The use of current at high frequencies in electromedicine for destroying parts of the human body by heating to the point where albumen coagulates.

Diathermy. Electromedical treatment employing sustained high-frequency current. A heating effect is obtained throughout the tissues. See SHORT-WAVE DIATHERMY, SURGICAL DIATHERMY.

Dielectric Heating

Dielectric. A material that will support the application of an electric field without conduction; more loosely, an insulating material, which may be a solid, a liquid or a gas.

Dielectric Absorption. The property of a dielectric material whereby a charging or discharging current remains present, at an appreciable level, for a period longer than would be expected if the decay were exponential.

Dielectric Breakdown. The failure of an insulating material to withstand an applied electric stress. It may be due to IONIC BREAKDOWN or THERMAL BREAKDOWN, or *thermo-ionic breakdown* which involves both the other processes.

Dielectric Constant. The ratio of the capacitance of a dielectric to the capacitance between identical electrodes at the same spacing in a vacuum (see also RELATIVE PERMITTIVITY).

Dielectric Heating. A method of HIGH-FREQUENCY HEATING applied to insulating materials. When an alternating voltage is applied across the plates of a perfect capacitor, an alternating current flows which leads the applied voltage by 90°, the power factor is zero, and no power is lost. A practical capacitor contains a dielectric material with losses, which appear as a shunt resistance. A component of the current now flows in phase with the applied voltage, and results in a power loss, which appears as heat throughout the dielectric. Fig. D-4 (*a*) shows the simplest form of dielectric heating, in which the workpiece to be heated is placed between two flat sheets of conducting material across which the high-frequency voltage is applied, and between which exists an electric field of uniform voltage gradient. If the non-conducting material forming the dielectric is itself homogeneous, it will be heated up uniformly throughout. Fig. D-4 (*b*) indicates that

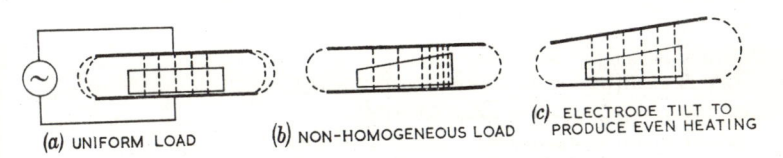

FIG. D-4. DIELECTRIC HEATING.

uneven work placed between parallel electrodes will be heated non-uniformly, and (*c*) shows how a graded air-gap can overcome this trouble.

Dielectric Hysteresis

Dielectric Hysteresis. A phenomenon by which energy is expended, and heat produced, as the result of the reversal of electrostatic stress in a dielectric material subjected to alternating electric stress.

Dielectric Loss. The dissipation of energy within a dielectric material when it is subjected to an alternating electric stress.

Dielectric Polarization. The change of physical state in a dielectric material under electric stress, such that each small element becomes a DIPOLE.

Dielectric Strength. A measure of the ability of insulating material to withstand an electric stress. It can vary from values as low as 50 V/mil up to around 4,000 V/mil for a synthetic material such as polyester film. The maximum possible strength attainable from any given material is the INTRINSIC STRENGTH.

Diesel-Electric Plant. Equipment employing a generator driven by a Diesel engine, i.e. a compression-ignition engine in which fuel oil is injected into a heated compressed-air charge. In the case of a *Diesel-electric locomotive*, the generator supplies direct current to electric motors connected to the driving axles.

Differential Booster. A BOOSTER having differentially connected field coils.

Differential Concatenation. A method of cascade connection or CONCATENATION, in which two motors may be adjusted to oppose one another, thus giving four possible speeds.

Differential Follow-Up Unit. Protective device sometimes fitted to an automatic voltage regulator. One can be seen in Fig. S–20. A resistor, R_4 is connected in series with a rheostat R_3, which is mechanically ganged to the main-field rheos at R_2, the whole circuit being energized by the pilot exciter. R_4 can be regarded as a phantom field; the relay measures the voltage across R_4, and compares it with the voltage across the main exciter field. When a difference occurs between the two, the relay operates the rheostat motor and moves R_3 to a position where the voltages are equal. Since R_3 is ganged to the main-exciter-field rheostat, the main-field rheostat, although not in use when the machine is on automatic control, will be positioned to provide the correct resistance to meet the generator-load conditions. This ensures that the rheostat automatically follows the loading conditions, and should it be desirable to change over from automatic to manual control at any time, this relay provides the correct position for the main-field

Differential Protection. One of the most common and effective means available for the protection of a.c. systems and generators. It acts by comparing the current at the two ends of a feeder, or a machine or transformer winding, so that any divergence is detected and relay action initiated. See also MERZ-PRICE PROTECTION.

Differential Selsyn. A unit interposed in the electrical tie between a conventional pair of SELSYN units without direct connection to the supply. The differential Selsyn records angular displacement or rotates at a speed which is either the sum or difference of that of the other machines. Conversely it is possible to make the shafts of the transmitter and receiver Selsyns rotate with a predetermined speed difference by driving the differential machine at that difference speed.

Differential Windings. Two windings on a piece of apparatus, so arranged that, when excited by direct current, their electromagnetic effects are opposed. Such an apparatus is known as *differentially wound* or *differential*.

Differentially Compound-Wound Motor. A COMPOUND-WOUND MOTOR in which the series and shunt windings are opposed electromagnetically.

Diffused-Junction Transistor. A TRANSISTOR in which the width of the base is very small, rendering the device suitable for high-frequency operation. It is fabricated by the diffusion of impurities into a solid crystal.

Diffusion Pump. A mechanical pump used in the production of evacuated valves, etc. A stream of mercury or oil vapour carries away atmospheric molecules from a partially evacuated container. It is capable of producing pressures of the order of 10^{-7} mmHg.

Digital Computer. A calculating machine that handles numbers in digital form. In modern use the term implies that class properly called an electronic automatic digital computer.

Dimensional Analysis. The use of the dimensions of a physical quantity, i.e. its statement in terms of fundamentally defined physical concepts, to determine the units in which it must be measured and to verify the physical validity of expressions and equations in which it appears. See table on page 84.

Diode. A device with two electrodes, in particular an anode and a cathode, and a non-linear current/voltage characteristic (see also VALVE).

Dipole

DIMENSIONS OF ELECTRICAL QUANTITIES

Quantity	Defining Equation	Dimensional Formula
Energy	w	L^2MT^{-2}
Power	$p = dw/dt$	L^2MT^{-3}
Current	i	I
Current density	$j = di/da$	$L^{-2}I$
Voltage	$v = p/i$	$L^2MT^{-3}I^{-1}$
Electric field intensity	$E = -dv/ds$	$LMT^{-3}I^{-1}$
Resistance	$R = v/i$	$L^2MT^{-3}I^{-2}$
Resistivity	$\rho = R(a/s)$	$L^3MT^{-3}I^{-2}$
Magnetic flux	$\Phi = -\int v.dt$	$L^2MT^{-2}I^{-1}$
Magnetic flux density	$B = d\Phi/da$	$MT^{-2}I^{-1}$
Magnetomotive force	$F = Ni$	I
Magnetizing force	$H = dF/ds$	$L^{-1}I$
Permeance	$\Lambda = \Phi/F$	$L^2MT^{-2}I^{-2}$
Inductance	$L = N\Phi/i$	$L^2MT^{-2}I^{-2}$
Permeability (abs.)	$\mu = \Lambda(s/a)$	$LMT^{-2}I^{-2}$
Charge, electric flux	$q = \int i.dt$	TI
Electric flux density	$D = dq/da$	$L^{-2}TI$
Capacitance	$C = q/v$	$L^{-2}M^{-1}T^4I^2$
Permittivity (abs.)	$\epsilon = C(s/a)$	$L^{-3}M^{-1}T^4I^2$

Dipole. An arrangement of two separated quantities (e.g. two electric charges) of opposite polarity. If point charges $+q$ and $-q$ are separated by a small distance l, such a system is an electric dipole of dipole moment equal to ql. The term *dipole* is applied to double-ended aerials, which in their simplest form are a whole number of half-wavelengths long.

A magnetic dipole comprises two fictitious sources, $+m$ and $-m$, of magnetic flux, separated by a small distance l, and having a moment ml, so that it tends to turn into alignment with a magnetic field. It is also known as a *doublet*.

Direct-Arc Furnace. An electric melting furnace in which an arc is formed between three equally spaced graphite electrodes and the metallic charge. Compensating reactors are placed in the circuit to take care of surges on the supply caused by solid metal coming into contact with the electrodes. Direct arc furnaces are used as steel-making units, smelting furnaces and copper-melting furnaces.

Direct Axis. The axis of the main field system in an electromagnetic machine, electrically at right-angles to the *cross* or *quadrature axis*. See Fig. D–5.

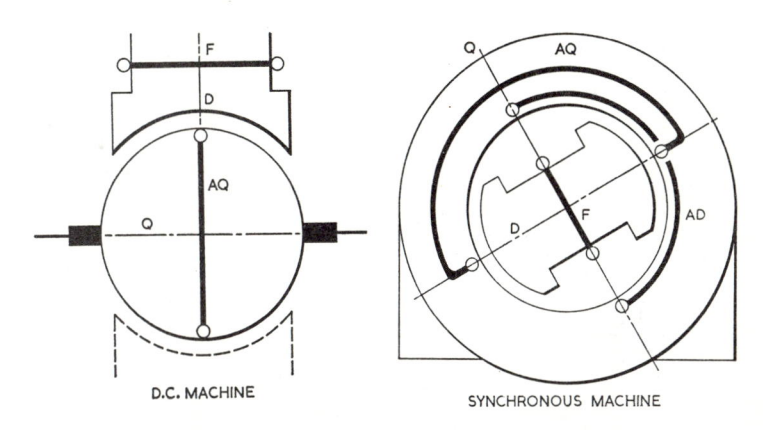

FIG. D–5. DIRECT AND QUADRATURE AXES.
A. Armature; F. Field; D. Direct Axis; Q. Quadrature Axis.

Direct Cooling. The application of a coollant direct to the conducting metal of a winding in a machine or transformer, by means of hollow conductors.

Direct Coupling. Coupling of COUPLED CIRCUITS in which the two circuits share a common impedance, or in which one circuit is fed from the other through an impedance link.

Direct Current. A current having continuously the same direction of flow: in particular a current that has a rigidly constant magnitude and direction. A unidirectional current of magnitude varying with time may be called a *pulsating current* if its variations are rhythmic.

Direct Load Loss. The I^2R Loss in an alternator stator based on the measured ohmic resistance corrected to 75°C.

Direct-On-Line Starting. Direct connection of a motor to the supply for starting. With single-phase motors it is possible only in the case of self-starting machines such as the series or repulsion types of commutator motor; with two- and three-phase squirrel-cage motors it is limited to small machines. Direct-on-line

Direct Suspension

contactor starters are designed round the basic circuit of Fig. D–6. Overload protection is by magnetic or thermal relay. An isolating switch may be incorporated in the starter. If reversing is required, two contactors, one for each rotation, are required and are interlocked so that only one can close at a time. A hand-operated oil switch with under-voltage trip coil may be used with larger motors.

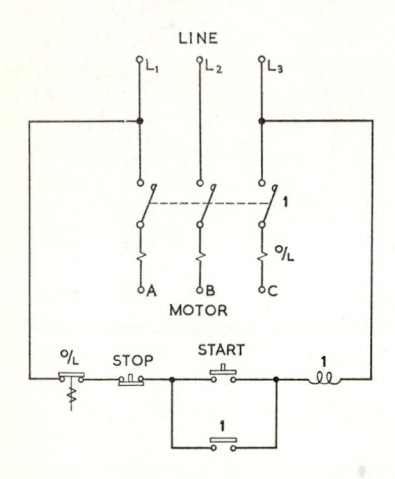

FIG. D–6. BASIC CIRCUIT OF DIRECT-ON-LINE CONTACTOR STARTER FOR SQUIRREL-CAGE MOTOR.

Direct Suspension. In electric traction, a form of construction for overhead conductors in which the contact wire is suspended directly from the supporting structures. Compare the various forms of CATENARY SUSPENSION.

Direct-Trip Circuit-Breaker, Starter. A circuit-breaker or starter which is tripped by the current flowing in the circuit in which it is connected, without the aid of an auxiliary electricity supply (cf. INDEPENDENT-TRIP CIRCUIT-BREAKER).

Direct Voltage. A voltage having constant polarity. It is frequently referred to loosely as *d.c. voltage*.

Disc Insulator. An insulator, in an overhead-line system, which consists of a number of separate units cemented to metal fittings which interlock to form a flexible string. It may be used as a SUSPENSION INSULATOR or a TENSION INSULATOR.

Disc Winding. Transformer winding, used mainly for the high-voltage windings of medium and large transformers, in which the turns are made up into a number of annular discs wound singly and then jointed (*single-disc winding*), or in pairs (*double-disc winding*), or from a continuous length of conductor (*continuous-disc winding*). Cf. BOBBIN WINDING.

Discharge Detector. Instrument used in the non-destructive testing of insulating material. An a.c. discharge detector isolates and amplifies the discharge current in the earthy end of a specimen or the voltage appearing at the high-voltage terminal. In the latter case the voltage is reduced to a suitable level by a discharge-free capacitor divider. The signal is displayed on a cathode-ray tube and can be compared with a calibrating pulse to obtain its magnitude and hence the discharge energy.

Discharge Lamp. A DISCHARGE TUBE containing gas and/or metal which is vaporized during operation, the discharge causing the emission of useful light with a colour characteristic of the gas. Sodium and mercury vapours are used for lighting; neon is used for sign work, giving a strong red.

Two factors affecting light output are the quantity of electrons in the discharge (i.e. the current density) and the gas pressure, both affecting the number of electron-atom collisions. With increase of gas pressure, increased voltage is required to operate the lamp and the spectral emission also changes. To increase the flow of electrons, heated electrodes are introduced in all except the cold-cathode types, which are run at higher voltages as a substitute for heating; all hot-cathode filaments have chemical treatment to increase electron emission still more. Mercury- and sodium-vapour lamps run hot and rely on the power in the lamp to produce sufficient heat to vaporize the metal. Mercury lamps operate off mains voltages of 200 V–260 V, but sodium lamps require a step-up transformer.

Discharge Tube. A tube of insulating material into which electrodes are fitted so that a discharge passes between them when raised to a sufficiently high voltage, the tube being evacuated to a sufficiently low gas pressure.

Discharge-Tube Rectifier. A rectifier consisting of a DISCHARGE TUBE having its electrodes arranged to permit current flow in only one direction.

Discharger. Alternative name for SPARK GAP (q.v.).

Discrimination. Means by which protective equipment separates faulty apparatus from sound (which may, nevertheless, be carrying the fault current). Discrimination tripping signals are sent only to those circuit-breakers whose operation will disconnect the fault from the system. Two basic systems of discrimination are available: *graded-time-delays*, and the *unit system*.

In the former, tripping of circuit-breakers is delayed by a time-lag. The time-lags are made longer for locations near the supply point so that breakers near the fault will disconnect it before others have begun to operate. Circuits beyond the fault will also be disconnected so that the system is only partially effective when the supply point is at one end only.

The unit system of discrimination necessitates measurement of the magnitude and/or direction of the current entering and that leaving each section of a system. If these are the same the section is sound; if they differ it is faulty, and the difference is made to initiate tripping signals to cause the circuit-breakers at each end of the faulty section to open.

Disintegration Constant. A measurement of the rate of breakdown of a radioactive nucleus, also known as the *decay constant*. This rate is characteristic of each nuclear species, so that if N_0 is the population of the species in a sample at the present, the population at some future time t will be given by

$$N = N_0 \exp(-\lambda t)$$

where λ is the disintegration constant. Another measure of the disintegration rate is the HALF-LIFE.

Dispersion. The variation of capacitance with respect to frequency or time. In a *dispersion test* of an insulating material a range of **0·5–50 Hz** for a sinusoidal supply or 3–300 ms for a square-wave supply is suitable.

Displacement Current. A rate of change of dielectric flux, which has a magnetic effect equal to a corresponding conduction current. Displacement current is present in any material, conducting or insulating, whenever there is an applied electric field which changes with time. There is a displacement current in a copper wire carrying alternating current, but the conduction current is so vastly greater, even at very high frequencies, that the presence of the displacement current is ignored. In poor conductors and insulators the displacement current may exceed the conduction current if the frequency is high enough; and in free space and perfect dielectrics the current is displacement current alone.

Disruptive Strength. See DIELECTRIC STRENGTH.

Distribution Fuse-Board

Distorted Waveform. Generally, any waveform that is not sinusoidal.

Distortion. The change in waveform that occurs when the scaled output response of an electrical (or mechanical) signal transfer-system is an imperfect reproduction of the input stimulus; for example, the change that occurs between input and output in an amplifier or transmission network. See AMPLITUDE DISTORTION, FREQUENCY DISTORTION, HARMONIC DISTORTION, INTERMODULATION DISTORTION, PHASE-SHIFT DISTORTION.

Distributed Winding. Winding that is arranged uniformly over the surface of a stator or rotor, each coil having the same dimensions. In Fig. D-7, a slot contains one side of a coil, the other

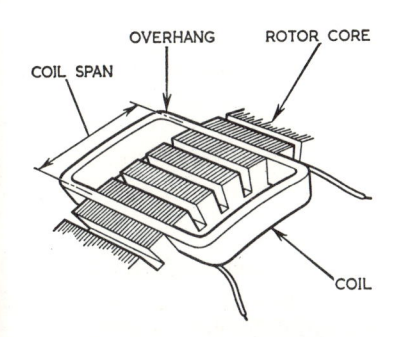

FIG. D-7. COIL OF A DISTRIBUTED WINDING.

side being about one pole-pitch away, the actual distance being the *coil span*. If the span is exactly a pole-pitch it is a *full-pitch coil*, but commonly it is less than this by one, two, three or more slot pitches, being then known as *short-pitched* or *chorded*.

Distribution Board. Equipment consisting of busbars and possibly switches, fuses, links, etc., for connecting, controlling or protecting a number of branch circuits fed from one main circuit of a wiring installation.

Distribution Factor. Factor used in the calculation of the electromotive force generated in the winding of an a.c. machine. Its use is necessary as the total e.m.f. is less than the arithmetic sum of the e.m.f.s in each coil, these not being in phase with one another.

Distribution Fuse-Board. A DISTRIBUTION BOARD having fuses in each of the branch circuits.

Distribution Pillar

Distribution Pillar. A pillar containing switches, links or fuses for interconnecting distributing mains.

Distribution Switchboard. A DISTRIBUTION BOARD having switches in each of the branch circuits. These circuits usually also include fuses, and a main incoming switch may be incorporated in the board.

Distributor. (1) Device used in automobile electrical engineering to ensure that the voltage is applied to the SPARKING PLUGS of an internal combustion engine in the correct sequence.

(2) Part of power distribution system to which consumers' circuits are connected. It is also known as a *distributing main*.

District Heating. A system whereby heat is supplied to a housing estate or commercial premises from the exhaust steam of back-pressure turbines in a generating station.

Diversity Factor. The ratio of the sum of the individual maximum demands of a group of consumers to their actual simultaneous demands.

Diverter. A low-value resistor connected in parallel with the series field winding of a d.c. machine in order to divert some of the current from it, either for control purposes or to correct some fault in design.

Dividing Box. A closed box in which the cores of multicore cable can be connected to external conductors.

Doctor. In electroplating, an anode covered with a permeable material saturated with the plating solution. It is applied to parts of the object to be plated, the latter being made the cathode.

Dolly. Of a tumbler switch, the operating member consisting of a lever pivoted on the face of the switch. It normally projects through an outer cover.

Domain. In magnetic theory, an elementary magnet, many of which compose the body of a ferromagnetic substance. Each domain consists of a fairly large group of atoms, magnetically aligned, which are spontaneously magnetized to saturation by atomic interaction, owing to the spinning electrons. In any unmagnetized specimen the domains are randomly oriented with a zero resultant magnetic moment.

Donor Impurity. An element introduced into a semiconductor material having a higher valency than the semiconductor. It liberates free electrons which become available as negative charge carriers. This gives rise to n-type conductivity. Cf. ACCEPTOR IMPURITY.

Doppler Effect. Change in observed wavelength or frequency of radiation due to relative motion between the wave source and the observer.

Dose Meter. An instrument for measuring quantity of radiation, particularly of X-rays in X-ray therapy.

Double Amplitude. The overall range of a wave, equal to the sum of the positive and negative peak values. It is sometimes referred to simply as the amplitude, though this may cause ambiguity (see PEAK VALUE).

Double-Break Circuit-Breaker, Switch. A CIRCUIT-BREAKER or SWITCH in which there are two breaks in series in each pole or phase.

Double Bridge. See KELVIN DOUBLE BRIDGE.

Double-Cage Winding. A form of SQUIRREL-CAGE WINDING using two (generally separate) cages, one set near the surface of the rotor slots and of high resistance, the other deeper and of low resistance. As the depth of slot controls the cage reactance, the outer cage has high resistance with low reactance, while the inner cage has the reverse. At starting, the leakage reactance of the inner cage is large enough to cause the rotor current to flow chiefly in the outer cage, the resistance of which produces considerable I^2R loss and therefore starting torque. When the speed is normal the reactance of each cage is negligible so that the rotor current is shared between the two in accordance with their resistances. Typical rotor slottings are shown in Fig. D-8.

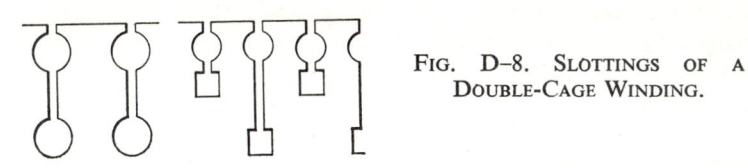

FIG. D-8. SLOTTINGS OF A DOUBLE-CAGE WINDING.

Double Catenary Suspension. In electric traction, a form of construction for overhead conductors employing two catenaries, supported at the same level from the same structure and having the same vertical sag. They form, with the contact wire, a triangular suspension formation.

Double-Delta Connection. A six-phase connection in which the six windings may be represented diagrammatically as two triangles or two DELTA CONNECTIONS.

Double-Element Wattmeter

Double-Element Wattmeter. A WATTMETER which automatically adds two measurements of power, two moving systems being mounted on a common shaft. The method of connecting such an instrument for the measurement of total power in a three-phase three-wire circuit with an unbalanced load, using voltage and current transformers, is shown in Fig. D-9.

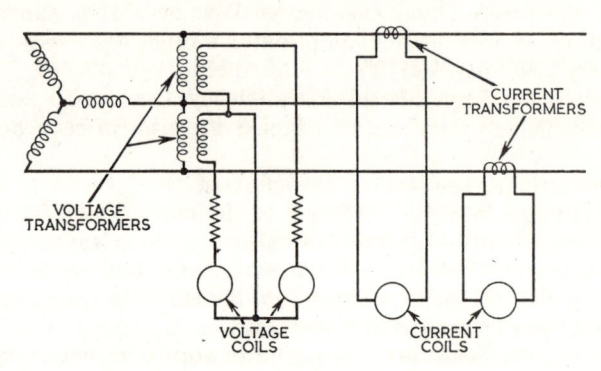

FIG. D-9. CONNECTIONS OF A DOUBLE-ELEMENT WATTMETER IN A THREE-PHASE CIRCUIT.

Double Insulation. Appliance insulation in which external metal parts are separated from live parts by two distinct layers of insulation—an *inner* or *functional* insulation which insulates live parts from each other and from non-exposed metal parts, and an *outer* or *protective* insulation which insulates non-exposed from exposed metal. In some cases a single *reinforced* insulation having characteristics at least equal to the two layers may be used.

Double-Layer Winding. A winding in which each slot contains two coil sides, one above the other.

Doublet. See DIPOLE.

Down Conductor. Of a lightning protective system, the CONDUCTOR connecting the AIR TERMINATIONS with the EARTH TERMINATIONS.

Drag-Cup Machine. A small machine (motor, generator or tachometer) having a *drag-cup rotor*, comprising essentially a hollow copper cylinder not attached to any rotor iron. The stator core and winding, and the "rotor" core, are both fixed, while the drag cup is free to rotate in the intervening air-gap.

Drained Cable. A mass-impregnated cable from which the free impregnating compound has been drained at a temperature above the maximum working temperature of the cable.

Draw-In Box. Box through which cables can be inserted into or withdrawn from a conduit, sometimes including facilities for cable jointing.

Drift Transistor. Form of TRANSISTOR in which a field is produced in the base region by arranging a resistivity gradient, so that carriers are swept across at high speed, reducing the transit time and increasing the maximum frequency.

Driving-Point Function. A type of SYSTEM FUNCTION, related to the input terminals of a network.

Driving-Point Impedance. The ratio of the voltage applied across a pair of network terminals to the current entering (and leaving) the network at those terminals.

Drooping-Characteristic Welding Set. A welding set in which the voltage droops automatically from the striking voltage to the ARC VOLTAGE as the current is increased.

Drum Starter, Controller. A starter or controller which has its moving contact parts arranged on a cylindrical surface (cf. FACEPLATE STARTER, CONTROLLER).

Drum Winding. A winding consisting of coils arranged either on the inside or on the outside of a cylindrical core, and situated either on the surface or in slots.

Dry Cell. A LEAD-ACID CELL in which the sulphuric acid electrolyte is completely absorbed and held by porous electrodes and separator material. The term may also apply to a dry PRIMARY CELL.

Dry Gas-Pressure Cable. An unimpregnated paper-insulated internal-pressure cable having the space inside the lead sheath raised to a high pressure by means of an inert gas in contact with the dielectric.

Drysdale Potentiometer. A polar type of A.C. POTENTIOMETER. It is basically of the Kelvin–Varley slide type for magnitude comparison, using first a direct current and a standard cell, then alternating current and an a.c. galvanometer; a precision milli-ammeter forms the transfer instrument from d.c. calibration to a.c. use. The a.c. source is a phase-shifting transformer, which is manually adjustable over 90°, so that a phase-angle measurement can be made by reading the phase-shifter angular position when the potentiometer is balanced.

Duality

Duality. A parallelism between the equations and theorems of physical systems (in particular of electric networks). The impedance of a series *RLC* circuit can be written

$$Z = R + j\omega L + 1/j\omega C,$$

and the admittance of a parallel connection of pure *RLC* branches is

$$Y = G + j\omega C + 1/j\omega L.$$

The two expressions have precisely the same form; they are, in fact, the duals of each other. Thus, the dual of circuit elements connected in series is a parallel connection of the duals of these elements.

Duddell Oscillograph. A mechanical oscillograph employing electromagnets. Its moving element is a very light bifilar loop to which is attached a tiny mirror. The waveform under investigation is traced out by a beam of light reflected from this mirror which is vibrated by the loop. By using a suitable arrangement of mirrors oscillating, or rotating, in synchronism with the wave, a stationary trace is obtained for visual observation (see Fig. D–10).

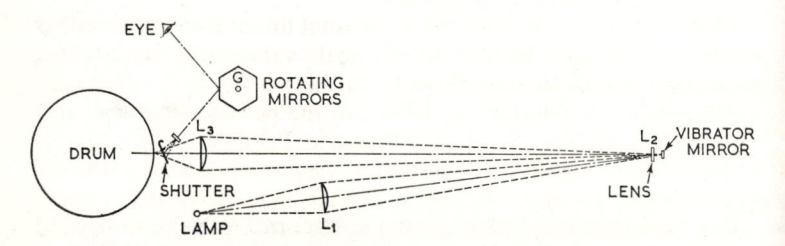

FIG. D–10. OPTICAL SYSTEM OF DUDDELL OSCILLOGRAPH. (*Cambridge Instrument Co. Ltd.*)

Duplex Winding. A winding comprising two distinct windings connected to alternate segments of the commutator and joined in parallel by the brushes, as shown in Fig. D–11 for a lap-connected arrangement. The number of parallel circuits through a duplex winding is thus twice that for an ordinary SIMPLEX WINDING. Duplex windings are used to increase the number of parallel paths through the commutator winding for high-current low-voltage machines.

Dynamic Balancing

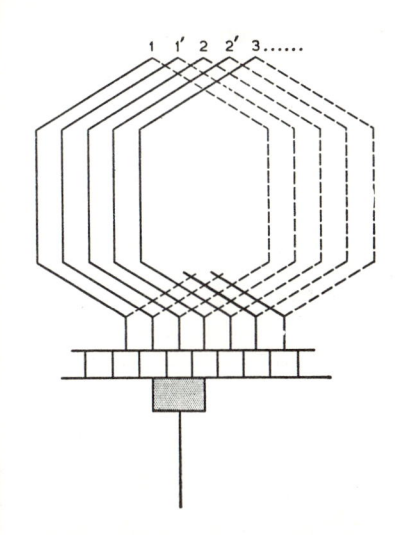

FIG. D–11. DUPLEX LAP WINDING.

Duralumin. Trade name for a high-tensile aluminium alloy containing copper, manganese, magnesium and silicon. It is used for the moving parts of switchgear.

Duration. Of a WAVE-FRONT, the total time taken by an impulse voltage or current to rise to its peak value from zero. It is nominally 5/4 times the time interval between points where the voltage or current is 10% and 90% of the peak value.

Dust Precipitator. See ELECTROSTATIC PRECIPITATOR.

Dynamic Balancing. A means of determining and correcting unbalance in rotating machinery. When the axial length is significant in relation to the diameter, STATIC BALANCING is not sufficient. Although the centre of gravity may coincide with the axis of rotation and thus show a state of static balance, the inertia axis may be at an angle to the rotational axis, and a dynamic couple of forces will be produced when the rotor revolves. Dynamic balancing machines are produced for the purpose of rapidly determining the state of unbalance whether it is static or dynamic or both.

A dynamic balancing machine in its essentials consists of flexibly supported bearings on which the rotor runs, suitable means for driving the rotor at a convenient speed (which may or may not be the working speed) and means for measuring the amount and angular position of the unbalance forces.

Dynamic Braking

Dynamic Braking. A means of braking an electric motor in which the machine is disconnected from the mains and a flux is established which causes e.m.f.s to be induced in the rotating windings. Discharge paths are arranged so that currents flow, resulting in the loss of energy and subsequent braking of the motor. See D.C. INJECTION BRAKING.

Dynamo. A direct-current generator.

Dynamometer. A device for measuring the mechanical output of a prime mover or electric motor. An ABSORPTION DYNAMOMETER absorbs the output as well as measuring it, while a TRANSMISSION DYNAMOMETER measures the output transmitted to some other equipment. A dynamometer is also an electrical instrument that measures the force exerted between fixed and moving coils.

Dynamometer Relay. A moving-coil relay in which the field flux is produced by one control current, the second control current being applied to the moving coil which operates the contacts. The operating force is proportional to the product of the fluxes, and by using spring control the contact movement is made proportional to the product of the two input currents. If the two coils carry the same current a square-law response is obtained.

Dynamometer Wattmeter. Instrument for precision and industrial measurement of power at power frequencies. As a precision instrument it can be constructed so that it is equally accurate when used with direct or alternating current and voltage.

Dynamotor. Alternative name for ROTARY TRANSFORMER (q.v.).

Dyne. The C.G.S. unit of force, being that force which endows the mass of 1 g with the acceleration of 1 cm/sec². The work done by a force of 1 dyne acting through a distance of 1 cm is 1 erg or 10^{-7} J. In S.I. units, 1 dyne = 10^{-5} N.

E

E-Class Insulation. One of seven classes of insulating materials for electrical machinery and apparatus defined in B.S. 2757 on the basis of thermal stability in service. Class E insulation is assigned a temperature of 120°C. It is particularly applicable to motors.

Ear. In overhead construction, a metal fitting that is attached to a contact wire to suspend the wire or hold it in position. Various types are the *straight-line ear*, *anchor ear*, *splicing ear*, etc.

Earth. The conducting mass of the earth or any conductor connected to it. The main body of the earth is considered to be at zero potential.

Earth Current. (1) A FAULT CURRENT which flows to earth. (2) A large current circulating in the body of the earth. Such currents may be induced by changes in the earth's magnetic field. Variations of the magnetic field by up to half a gauss for two or three days, due to magnetic storms related to sunspots, can induce currents of up to 5,000A.

Earth Detector. An instrument for measuring or indicating a leakage current to earth. It is also known as a *leakage indicator*.

Earth Electrode. A metal plate, rod, strip, etc., buried in the ground to make contact with the general mass of the earth, usually for protective purposes.

Earth Fault. An accidental connection, or low-resistance path, between a live conductor and earth.

Earth-Fault Protection. System of protection whereby, should an insulation defect occur causing earth-leakage current to flow, the faulty circuit is automatically disconnected by means of a fuse or circuit-breaker (see EARTH-LEAKAGE CIRCUIT-BREAKER).

Earth-Leakage Circuit-Breaker. A device which disconnects the supply if the voltage on non-current-carrying metalwork or the out-of-balance current in the supply due to leakage exceeds a predetermined value. Such a system may be made to operate more rapidly and at lower values of fault current than one depending on over-current protective devices. It does not require a low-resistance connection with earth.

Where the earth-leakage circuit-breaker is actuated by a trip coil collected between the metal to be protected and earth, it is said to be *voltage-operated*. If actuated by a differential transformer or other device in which the operating current is proportional to the total earth-fault or leakage current from the system, it is said to be *differential-current-operated*.

Earth Return. The return path of an electrical circuit having only one insulated conductor. The current returns to the source through the earth.

Earth Shield. Of a cable, the metal sheath, immediately inside and in contact with the lead sheath (cf. TEST SHIELD).

Earth Termination. Part of a lightning-conductor system designed to collect discharges from the general mass of the earth or to distribute charge into the earth. It includes all the parts below the TESTING JOINT in a DOWN CONDUCTOR.

Earth Testing. A process comprising: (*a*) the measurement of the resistance to earth of earth electrodes, such as plates, rods,

Earth Wire

strips, etc., buried in the ground to make contact with the main body of the earth, usually for protective purposes; (*b*) the measurement of the resistivity of the soil itself, for use in the design of an earth-electrode system, and in investigating the nature of the underlying strata.

The resistance of any earth electrode is made up of several components: those of the connecting lead, the contact between the electrode and the soil, and the body of earth surrounding the electrode. The last-named is the most important, for the lead resistance is usually low and the contact resistance is known to be negligible.

Earth Wire. A conductor which is connected to earth and which runs parallel to and in close proximity with the conductors of an overhead transmission line.

Earthed System. A system of distribution in which the neutral point or one conductor is maintained at earth potential (cf. UNEARTHED SYSTEM).

Earthing. Connection of part of a system or equipment to earth. The object of earthing is (*a*) to ensure the security of the system by limiting the potential of the live conductors, with respect to earth, to values that their insulation will withstand. This is achieved by adequate connections of the neutral or one pole of the system to earth (*system earthing*); (*b*) to safeguard life and property by ensuring that dangerous potentials with respect to earth are not maintained on parts of electrical equipment or on non-electrical metalwork normally not alive. This is achieved by connection of the parts concerned to the general mass of earth in such a manner as to discharge any dangerous potential and isolate the faulty parts of the system from the supply (*equipment earthing*).

Earthing Reactor. A reactor connected between the midpoint or neutral point of a system and earth to limit the earth-fault current which may flow.

Earthing Switch. Device used to connect to earth an isolated circuit to ensure safety of maintenance personnel. It is frequently combined with an isolator which has a blade arranged to engage fixed earth contacts. On high-voltage outdoor isolators it is usual to have a blade hinged to the base of the isolator to connect the live end of a post insulator to earth.

Earthing Transformer. Transformer intended primarily to provide a neutral point for earthing purposes.

Eddy Current. A parasitic current flowing in a closed conducting path, generated by a magnetically induced e.m.f. On occasion an eddy current may be encouraged for some useful purpose, such as high-frequency heating, or the braking of an energy meter, but most cases of eddy-current production are unwanted and cause waste of energy in I^2R LOSS.

Eddy-Current Brake. Braking system utilized mostly in energy meters or absorption dynamometers. In these cases the magnetic field is developed by a permanent magnet or d.c. electromagnet, and the eddy currents are generated by the movement of a conductive mass across the field. For convenience the mass is usually a rotatable disc.

Eddy-Current Heating. The form of heating involved in INDUCTION HEATING.

Eddy-Current Loss. An I^2R LOSS resulting from currents induced in a conductor. There is a loss in the iron of any magnetic circuit carrying an alternating flux.

Edison Screw Cap. A lamp cap in which one contact is the outer case, in the form of a screw thread, and the other contact is a central projection. These caps are known, in decreasing order of size, as Goliath Edison Screw (G.E.S.), Medium Edison Screw (E.S.), Small Edison Screw (S.E.S.) and Miniature Edison Screw (M.E.S.).

Effective Resistance. Of an element in an a.c. circuit, the component of the voltage across it which is in phase with the current, divided by the current. It is also known as the *a.c. resistance*. The value of the effective resistance may be obtained by dividing the total losses caused by the current (e.g. the power dissipated in heat) by the square of the r.m.s. value of the current. See also D.C. RESISTANCE.

Effective Value. Of an alternating current or voltage, the value of a constant direct current or voltage that, when applied to a given circuit for a given time, produces the same expenditure of energy as when the alternating quantity is applied to the same circuit for a similar time. It is sometimes called the *virtual value*, and is equal to the ROOT-MEAN-SQUARE VALUE.

Effectively Earthed System. A system in which during a line-to-earth fault the maximum voltage to earth of the other two lines does not exceed 80% of the normal line-to-line voltage. This is the case when, for all switching conditions and unlimited supply, the ratio of the ZERO PHASE SEQUENCE reactance to the POSITIVE

Elastance

PHASE SEQUENCE reactance is less than three, and the ratio of the zero-sequence resistance to the positive-sequence reactance is less than unity. A system in which all the transformers have star-connected windings with all neutrals directly earthed is regarded as being effectively earthed.

Elastance. The reciprocal of capacitance. The defining relation for capacitance is that between the charge Q and the potential difference, V, i.e. $Q = CV$ or $V = Q/C$. The same proportionality can also be expressed as

$$V = QS$$

where $S = 1/C$ is the elastance of the capacitor in darafs. For a series connection of elastances the overall equivalent elastance is

$$S = S_1 + S_2 + S_3 \ldots$$

Elbow. A conduit fitting, being a short radius length of tubing to connect two lengths of conduit at right-angles.

Electret. The electric counterpart of a permanent magnet; a piece of material which possesses a permanent bulk electric moment. It is produced by the solidification, in a strong electric field, of mixtures of certain organic waxes. As with a permanent magnet, subdivision produces smaller electrets, while permanency is preserved over long periods of time by the use of a keeper which effectively short-circuits faces of opposite polarity. A charge on an electret of the same sign as the forming electrode (usually of tinfoil) is known as a *homocharge*, while one of opposite polarity is a *heterocharge*.

Electric Fence. Fence consisting of a single wire, usually barbed, suspended on posts at a convenient height and nowhere in direct contact with the ground, which carries a high-frequency high-voltage charge from a small battery-operated charging set in a weatherproof box. If an animal touches the wire it receives a shock but is not harmed.

A screen to keep fish from power-station water consists of a row of aluminium electrodes that discharge a mild current into the surrounding water.

Electric Field. A concept devised to explain the phenomena associated with an electric CHARGE. An electric charge produces, in the surrounding space, an energy state such that mechanical forces are exerted on other charges. The charges are said to be in an electric field, and the concept can be used to explain electrical phenomena without always referring them back to the charges by virtue of which they occur. By an extension of the concept, the electric field appears as a constituent of ELECTROMAGNETIC RADIATION.

Electric Field Intensity. See VOLTAGE GRADIENT.

Electric Flux. The surface integral of the *electric field intensity* normal to the surface. It is considered to be in a direction outwards from a positive charge.

Electric Furnace. See RESISTANCE FURNACE, INDUCTION FURNACE, DIRECT-ARC FURNACE, INDIRECT-ARC FURNACE.

Electric Resistance Welded Tube. Tube manufactured by continuously BUTT SEAM WELDING two edges of steel strip. Linear speeds up to 50 m/min are in use, and the product is sometimes referred to as *ERW tube*.

Electric Shock. Sensation produced on the nervous system by the passage of an electric current through the body. The severity of the shock depends primarily upon the magnitude of the current, the path which the current takes through the body, and the duration of the contact. In extreme cases the shock causes failure of the normal action of the heart and lungs, resulting in unconsciousness or death. Prolonged exposure to a current of 10–30 mA will in most cases be lethal; a current of 100 mA can kill instantly if it passes through the heart.

Shocks experienced from direct and alternating current differ in their effects, alternating current at normal frequencies **(25–50 Hz)** being the most dangerous. With the increasing use of equipment operating at high frequencies, an added danger arises from the passage of h.f. current through the body. At a frequency of about **100 kHz,** the sensation of shock begins to disappear, although serious internal burns can result.

Electric Space Constant. An alternative (and technically preferable) term for the absolute permittivity of free space. Its value in M.K.S. terms is $\epsilon_0 = 8·854 \times 10^{-12}$ F/m.

Electrical Thread. A special form of thread used on screwed steel conduit.

Electrically Exposed. Term applied to an installation, usually one connected to an overhead line, in which the apparatus is subject to over-voltages originating in the atmosphere.

Electroacoustical Transducer. A TRANSDUCER designed to receive an input from an electrical system and provide an output

Electrocardiograph

to an acoustical system, or vice versa. Well-known examples are the microphone and loudspeaker.

Electrocardiograph. An instrument which amplifies and records the very small fluctuating electrical potentials set up between various parts of the body as a result of the beating heart. It consists of a high-gain electronic amplifier, to the output of which is coupled a direct-writing chart recorder. As the pulses are of a low frequency the amplifier is usually designed for a linear response from about 1 to 50 Hz, although some have a wider range from 0 to 100 Hz. The record produced is known as an *electrocardiogram*.

Electroceramic. Material made from powders of inorganic compounds chosen on the basis of their electrical and magnetic properties.

Electrochemical Equivalent. The weight of a substance liberated by the passage of one coulomb of electricity.

Electrochemical Machining. The machining of metals by electrolytic action. The item to be machined is made the anode in a fast-flowing electrolyte (commonly caustic soda). By the use of a shaped cathode and a heavy current, metal is removed at a high speed.

Electrochemical Series. A classification of the elements in order of their potential when immersed in a solution of normal ionic concentration.

Electrode. A conductor to or from which a current flows: for example, the ANODE and CATHODE of an electron tube or electrolytic bath. An electrode may be the probe of an electromedical apparatus, the terminal of an arc in a welder, or the metal rod driven into the ground to provide an electrical system with a direct earth connection.

Electrode Bar, Electrode Wheel. Of a resistance seam-welding machine, alternatively known as CONTACT BAR, CONTACT WHEEL (q.v.).

Electrode Boiler. A steel tank containing water in which are placed three or more electrodes, the electrodes being connected to an a.c. supply. Current is passed between the electrodes through the water and so the water is heated.

Electrode Potential. Difference in potential between a metal electrode and an electrolyte in which it is immersed. Two opposing forces are set up at the interface of a metal and liquid. On the one hand, metal atoms tend to ionize and pass into the solution, thus

Electrodynamic Instrument

making the metal surface negatively charged, and on the other the positively charged ions produced, and those already in the solution, form a positively charged layer in the electrolyte. As a result of this double-layer of attractive forces, a potential difference is set up between the metal and the solution, known as the electrode potential or *single potential* of the metal.

Electro-deposition. The deposition of a metal, alloy or compound by electrolysis.

Electrodialysis. An electrolytic method of desalting water. A direct current passes through an ionized solution by means of ion migration. Charged semi-permeable membranes separate positive sodium from negative chloride ions.

Electro-Diesel. A railway locomotive incorporating a Diesel engine and electric traction motors, capable of working either direct from an external electric power supply or with the Diesel generator supplying the traction motors.

Electrodynamic Instrument. An indicating instrument, similar to a MOVING-COIL INSTRUMENT in which the permanent magnet is replaced by a fixed coil system (see Fig. E–1). Both fixed and moving

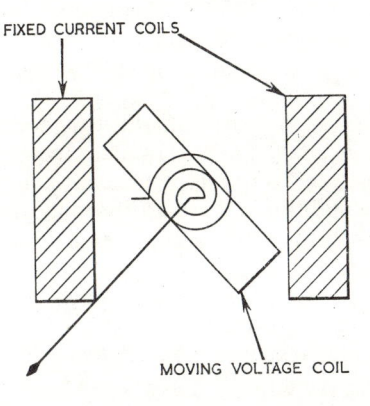

FIG. E–1. ELECTRODYNAMIC WATTMETER.

The electrodynamometer instrument can be used on both a.c. and d.c. circuits with great accuracy. Both fixed and moving coils are connected to the supply to be measured.

coils are connected to the supply to be measured, rendering the instrument of use for low-frequencies. In some fixed laboratory instruments the range may be up to about **20 kHz**. Pivoted instruments are of particular application as wattmeters, but as voltmeters and ammeters they have no advantage over the moving-iron type for power frequencies, and they are more expensive. Electrodynamic instruments may have an air core or an iron core.

Electroencephalograph

The latter is more recent, employing nickel-iron of low loss, and additional magnetic flux is provided by the windings through the reduced reluctance, resulting in a higher torque for a given consumption. The scale shape is linear for a wattmeter but square-law for an ammeter and a voltmeter. True r.m.s. value is indicated. The instrument is very useful as an a.c./d.c. transfer instrument in the laboratory.

Electroencephalograph. An instrument for recording, in the form of an *electroencephalogram*, the electrical activity of the brain. This activity can be detected in the form of tiny fluctuating potentials occurring between various points on the scalp. The instrument consists of a number of high-gain amplifiers each feeding a pen on a recording chart. This arrangement permits simultaneous recording from various parts of the brain. The frequency response of the amplifier is normally from 0 to about 75 Hz.

Electroextraction. The direct recovery of a metal from a solution of its salts by electrolytic means. The process is also known as *electro-winning*.

Electroforming. The production or reproduction of an article by the electrolytic deposition of a metal, alloy or compound. The process is sometimes known as *galvanoplasty*.

Electrographitic Brush. A BRUSH combining the low friction coefficient and good high-speed running and current-carrying capacity of the NATURAL GRAPHITE BRUSH with the toughness and mechanical properties of the ordinary CARBON BRUSH. During manufacture the material is subjected to a very high temperature under controlled atmospheric conditions, so transforming the original carbon and the binder into practically pure artificial graphite. The material is produced with various degrees of graphitization. The various grades cover not only the usual applications of the two types mentioned above but extend to the largest d.c. machines with the highest peripheral speeds and to the more difficult conditions generally.

Electrohydrodynamic Generator. A generation device in which a stream of gas is ionized, the positive ions being carried away by the stream while the electrons are collected by an electrode ring causing a current to flow through a wire between the ring and a collecting grid placed downstream.

Electroluminescence. The emission of light from a phosphor under the direct action of an electric field. In practical devices the phosphor, in the form of a fine crystalline powder dispersed in a

Electrolytic Cell

resinous dielectric, is sandwiched in a thin layer between two conducting surfaces to form a capacitor. One surface (or electrode) is transparent, to permit the light from the phosphor to emerge. In one form of construction the supporting member is a glass sheet, one surface of which is suitably treated to form the transparent conducting electrode. The other essential layers are built up on this substrate, as shown in Fig. E–2.

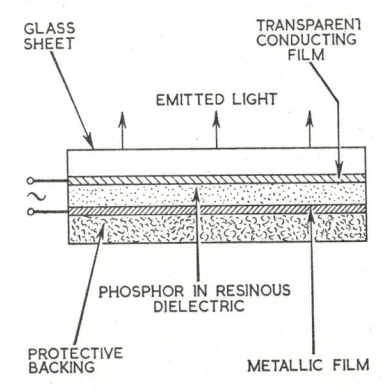

FIG. E–2.
ELECTROLUMINESCENT PANEL.
For the purpose of clarity, the vertical thickness in this diagram is exaggerated.

Electrolysis. The decomposition of an ELECTROLYTE by the passage through it of a direct current. It is accompanied by a transfer of ions between the two electrodes. The primary chemical reaction at the *anode*, or positive electrode, may include the dissolution of the metal of the anode or the evolution of oxygen or chlorine or some other gas. At the *cathode*, or negative electrode, the primary reaction may be the evolution of hydrogen or, if the electrolyte is the solution of a metallic salt, the deposition of the metal component.

Electrolyte. A substance which conducts by virtue of its ION content. It is usually either a liquid or a substance which is first dissolved in a suitable solvent, usually water.

Electrolytic Capacitor. A fixed capacitor in which the dielectric is a thin film of oxide deposited upon aluminium foil which forms the positive plate, the effective negative plate being a non-corrosive electrolyte. The great advantage of this type of capacitor is that a large capacitance can be obtained in a component of small dimensions.

Electrolytic Cell. A vessel containing a system of electrodes and an electrolyte for the purposes of electrolysis.

Electrolytic Diaphragm. A partition used in electrochemistry that allows the passage of ions but prevents conducting solutions from mixing freely.

Electrolytic Dissociation. The separation of certain substances into oppositely charged ions, or IONIZATION.

Electrolytic Meter. An INTEGRATING METER, the operation of which depends on electrolysis.

Electrolytic Rectifier. A type of rectifier now generally displaced by the METAL RECTIFIER. Its operation is based on the property of unilateral conduction possessed by certain combinations of metallic or non-metallic electrodes. One form is the ALUMINIUM RECTIFIER.

Electrolytic Tank. Strictly, a bath containing an electrolyte and electrodes, e.g. for electrochemical processes; more usually, a device for the plotting of electric conduction fields.

Electromagnet. A magnet excited by a current in a coil surrounding a ferro-magnetic core, in contrast to a PERMANENT MAGNET. The majority of practical magnets are of the current-excited type, because the current gives a simple control of the amount and sense of the magnetic flux produced.

Electromagnetic Force. The mechanical force between currents, developed through the medium of a magnetic field. Mechanical forces are developed in such a way that any resulting movement increases the flux-linkage with the electric circuit, or lowers the magnetomotive force necessary for a given flux. In the former case an increase of linkages requires more energy from the circuit, making mechanical work available; in the latter, stored magnetic energy is released in mechanical form.

Electromagnetic Induction. The production of an electromotive force in a circuit by means of a change in the magnetic flux through the circuit. An induced current can be produced in a closed circuit in a variety of ways: (*a*) by moving a magnet near the circuit, (*b*) by moving the circuit near a magnet, (*c*) by altering the size of the circuit, (*d*) by varying the strength of the magnetic field. All these situations are covered by FARADAY'S LAW OF ELECTROMAGNETIC INDUCTION.

Electromagnetic Machine. An energy-converting device in which the principles of electromagnetic force and induction are utilized. Such devices include mechanisms of the relay type, electromagnetic pumps, loudspeakers, etc., as well as rotating machines.

Electromedicine

Electromagnetic Pump. A device for pumping liquid metals. The good electrical conductivity of these makes it possible to establish electromagnetic forces directly within the liquid itself, so avoiding the need for mechanical impellers. The field may be alternating or steady, and the current either supplied directly from an external source (see CONDUCTION PUMP) or induced by the field (see INDUCTION PUMP).

Electromagnetic Radiation. The wave propagation of energy in electromagnetic form. The Faraday concepts of the electric and magnetic field as stress distributions in an ambient medium were translated by Maxwell into mathematical language, and with the displacement-current hypothesis, led to his prediction of electromagnetic wave propagation in 1865. Hertz produced experimental confirmation in 1887. It is now known that radio waves and visible light have the same essential physical nature, and both take a place in a wide ELECTROMAGNETIC SPECTRUM.

Electromagnetic Spectrum. The range of wavelengths of ELECTROMAGNETIC RADIATION. These wavelengths cover a very great range (see Fig. E–3). Those at present having engineering significance are, in order of decreasing wavelength: *radio waves* (produced by the rapid movement of free electrons in conductors, i.e. by alternating or rapidly changing electric currents), *infra-red waves* (produced by the small oscillations of atoms and molecules of a hot body), *visible light*, *ultra-violet* and *X-rays* (produced by the movement of electrons in atoms), and *gamma-rays* (generated by events taking place in the nuclei of atoms of radioactive substances).

Electromagnetic Unit. A unit based on the C.G.S. SYSTEM, with the force between unit magnetic poles 1 cm apart in a vacuum equated to 1 dyne, the magnetic space constant being taken as unity.

Electromechanical Braking. A system in which brake shoes are applied to the moving drum by means of springs or weights, and released electrically.

Electromedicine. The use of electricity in medicine. Electromedical apparatus can be divided into two categories. The first is remedial and includes all instruments designed specifically for treatment purposes. The second is diagnostic and refers to apparatus for measurement and detection. X-ray apparatus is a well-known example; one type is used for the destruction of tumours and cancerous growths, while the other less powerful type is employed

Electrometallization

in photographing the internal organs and bone structure. Another example is diathermic treatment (see SHORT-WAVE DIATHERMY, SURGICAL DIATHERMY).

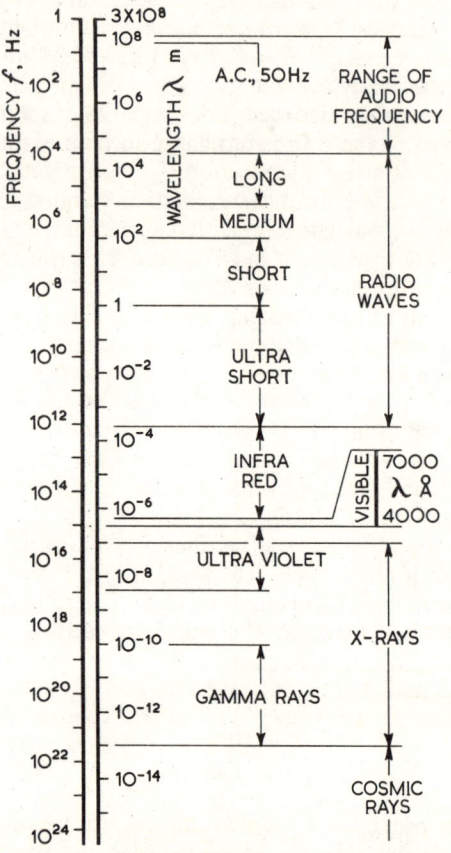

FIG. E–3. RADIANT-ENERGY WAVELENGTHS OF ENGINEERING SIGNIFICANCE.

Electrometallization. The deposition, by electrolytic means, of a metal on a non-conducting base for decorative or protective purposes.

Electrometallurgy. The smelting, melting and refining of metals by the application of electrical energy. Electrometallurgy

Electrometer Valve

formed one of the earliest uses of the electric current and made possible the production of many metals which had hitherto resisted thermal production processes. These new developments were a result of the high temperatures attainable by the electric arc or the adoption of electrolysis. See also DIRECT-ARC FURNACE, INDIRECT-ARC FURNACE, INDUCTION FURNACE.

Electrometer. Instrument for the measurement of electric charge. It comprises two electrodes capable of relative movement and connected to the source of charge to be measured. Normally one electrode is fixed and the other is suspended or otherwise mounted so as to be attracted or repelled under constraint. For maximum sensitivity only very limited movement is allowed and this, commonly a rotation, may be read by a microscope, or with a lens, mirror, lamp and scale.

A basic form of electrometer, shown in Fig. E–4, has two pairs of fixed electrodes in the form of quadrantal boxes. One pair is connected to a known potential V_0 (usually earth) and the other to

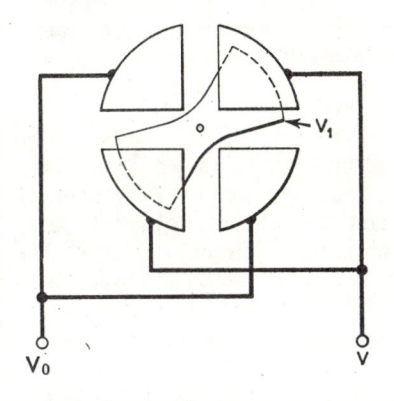

FIG. E–4. QUADRANT ELECTROMETER.

V is unknown potential; V_0 is known low potential; V_1 is known high potential.

an unknown potential V. A movable vane or needle, suspended to lie within the flat box-like space between the fixed quadrants, is connected to a known and relatively high potential V_1. Then the torque on the vane is proportional to

$$(V-V_0)[V_1 - \tfrac{1}{2}(V-V_0)]$$

which is proportional to $(V-V_0)$ if $V_1 \gg V$ and V_0.

Electrometer Valve. An amplifying valve with very high insulation resistance and impedance and very small grid current.

Electrometer Voltmeter. A form of valve voltmeter using an ELECTROMETER VALVE.

Electromotive Force. The force that tends to cause a movement of electricity in a conductor. The direction is taken as that of the movement of positive electricity. It is equal to the line integral of the VOLTAGE GRADIENT along a specified path in an electric field. It is commonly measured in volts, and is abbreviated *e.m.f.*

Electromyograph. Instrument for amplifying and recording the electrical signals associated with muscular action. These very small voltages are detected either by means of a special needle inserted into the muscle or by two contacts spaced about one inch apart on the surface of the skin over the muscle.

Electron. A fundamental constituent of matter which possesses a wave-particle dualism. In practice, it may be regarded as (*a*) an electrically charged particle, (*b*) a mass point with no spatial extension, or (*c*) an element of charge possessing inertia. Nuclear-bombardment experiments involving high-energy electrons indicate a major dimension of the order of 10^{-14} m. The charge is negative and of magnitude $e = 1 \cdot 6021 \times 10^{-19}$ C. The rest mass is $m_0 = 9 \cdot 109 \times 10^{-31}$ kg, and the charge-to-mass ratio is

$$e/m_0 = 1 \cdot 7588 \times 10^{11} \text{ C/kg}.$$

Electron-Beam Furnace. A melting furnace in which an accelerated stream of electrons in a vacuum impinges on the material to be melted. It can be used to melt and refine metals with very high melting points, such as niobium and tantalum.

Electron Diffraction. The diffraction of electrons obtained when a beam is passed through very thin metal foil, the metal crystals forming a diffraction space lattice. The diffraction of electrons is a natural outcome of their wave properties; the diffraction peaks arise from the reflection of electron waves by planes of atoms in the crystal. Techniques similar to those used with X-rays are employed, but since electrons react more strongly with matter than X-rays (owing to their charge), and have therefore less penetrating power, electron-diffraction methods are confined to surface investigations such as contaminant films on metals.

Electron Lens. An electric or magnetic field distribution, or combination of the two, which has an effect on an electron beam similar to that of an optical lens on a beam of light. An *electrostatic lens* consists of an electric field produced by two or more electrodes of suitable shape and dimensions and maintained at suitable potentials. A *magnetostatic lens* has a narrow current-carrying coil so shielded magnetically that the magnetostatic equipotentials are confined to a short axial distance within the coil.

Electron Microscope. Microscope employing an electron beam instead of the more common light beam. High-velocity electrons may pass through atoms without suffering deflection or other interference, and magnetic or electric fields may be used to render a beam of electrons convergent or divergent (see ELECTRON LENS). The particular advantage of an electron microscope is the enormous magnification of which it is capable. The wave nature of light limits the resolution attainable in visible-light microscopy to approximately $0.2\ \mu$, but the electron microscope can attain a resolution of $10^{-3}\ \mu$.

Electron Mobility. The drift velocity of free electrons in a conductor per unit voltage gradient.

Electron-Volt. The energy acquired by an electron when it is accelerated by a potential difference of one volt. $1\ \text{eV} = 1.6 \times 10^{-19}\ \text{J} = 4.4 \times 10^{-26}\ \text{kWh}$.

Electronegative. See ANODIC.

Electronegative Gas. A gas whose molecules possess affinity for free electrons with which they combine to form negatively charged ions. This property makes it applicable for use as an arc-extinguishing agent in high-voltage circuit-breakers. An outstanding example is sulphur hexafluoride.

Electronic Oscillator. A device for producing a.c. power at some required frequency. It comprises as essential constituents an electronic valve and a frequency-determining circuit.

Electronics. The study of the conduction of electricity in a vacuum, in gases and in semiconductors, and the design and application of devices whose actions are consequent upon this feature. In fact, the boundary line between electronics and "light" electrical engineering is hazy, to say the least.

Electroparting. The separation of metals by electrolytic means.

Electrophoresis. Alternative term for CATAPHORESIS (q.v.).

Electrophotoluminescence. An effect allied to ELECTROLUMINESCENCE. It includes varying degrees of enhancement or quenching by electric fields of both fluorescence and phosphorence in zinc sulphide and other phosphors.

Electroplating. The deposition, by electrolytic means, of a metal on a metallic surface for decorative or protective purposes.

Electroprecipitation

Electroprecipitation. The deposition of a colloidal substance by CATAPHORESIS.

Electroscope. Instrument for indicating an electric charge or potential difference by electrostatic means. A simple form consists of two strips of gold or aluminium foil supported in a container by an insulated wire. When the wire is charged, the strips separate.

Electrosmosis. The passage of a liquid through a diaphragm under the stimulus of a difference of potential. It is also known as *electro-endosmosis*.

Electrostatic Generator. A generator depending upon electrostatic action (see VAN DE GRAAFF GENERATOR, WIMSHURST MACHINE). It may also be known as an *influence machine* or *static machine*.

Electrostatic Instrument. Indicating instrument that employs the electrostatic attraction between oppositely charged conductors or the repulsion between similarly charged conductors to measure relatively high voltages. The attraction type is commonest, generally comprising an interleaving system of fixed inductors or quadrants and a moving system of vanes forming a variable capacitor; this is suitable for medium voltages. Two opposed discs or plates, one fixed and one free to move axially forms the common very-high-voltage instrument. Both direct and alternating r.m.s. values are indicated, but in general the type is less robust than others and so it has limited use. Its great advantage for d.c. use is that no current is taken from the circuit. On alternating current, the small current taken is a capacitance current having no power component.

Electrostatic Lens. See ELECTRON LENS.

Electrostatic Precipitator. An apparatus for the removal of suspended matter out of a gas stream. The gas to be cleaned is passed through an electric field established between so-called active or discharge electrodes and receiving electrodes. The active electrodes (usually wire) are maintained at a high negative direct voltage (10–100 kV) and take a current of about 10–1,000 mA. The high electric field strength in the vicinity of the wire causes electrical breakdown of the gas. Negative ions and electrons are thus created near the wire, travelling towards the receiving electrodes and charging the dust particles negatively. The charged particles are electrostatically attracted by the receiving electrodes and deposited on them. Rapping mechanisms of various types are fitted to dislodge the collected dust which then drops by gravity into a hopper.

Electrostatic Unit. A unit based on the C.G.S. SYSTEM, with the force between unit point electric charges 1 cm apart in a vacuum equated to 1 dyne, the electric space constant being taken as unity.

Electrostatics. The study of the behaviour of electric charges and potentials. The basic phenomenon of electrostatics is the force exerted between two charged bodies.

Electrostriction. Electrical phenomenon analogous to MAGNETOSTRICTION. Polycrystalline barium titanate and lead zirconate are electrostrictive materials, and vibrate when subjected to an alternating electric field.

Electrotherapy. The treatment of diseases and other physical disorders by electric currents or by electrically produced radiation. It is also known as *physiotherapy*.

Electrothermics. The application of electrical energy to the production of heat in chemical or metallurgical processes.

Embedded Temperature Detector. A thermocouple or resistance-temperature type of temperature-sensitive device, built into an electrical machine for monitoring hot-spot temperatures.

Enamel. A thick oil paint containing resin and having good insulating qualities. It is extensively used for covering fine wires.

Encapsulation. The encasing of electrical components in solid blocks of resin. Polyester or epoxy resins are used, the latter giving better results. The procedure is sometimes known as *potting*.

Enclosed-Ventilated Equipment. Equipment in which the live parts are protected by covers having openings for ventilation.

Enclosure. Structure containing an electrical machine. B.S. 2613 specifies 19 types of enclosure of which the chief are OPEN MACHINE, SCREEN-PROTECTED MACHINE, PIPE-VENTILATED MACHINE and TOTALLY ENCLOSED MACHINE.

End Bracket. An open structure at the end of the frame of a machine to support a bearing.

End Cell. Alternative name for REGULATOR CELL (q.v.).

End Plate. A thick plate between two of which the laminations of a laminated structure are clamped.

End Shield. A cover wholly or partially enclosing the end of a machine and possibly equipped to support a bearing. It is distinct from an *end guard*, which does not carry a bearing.

End Spring

End Spring. A spring of hard lead placed in a LEAD-ACID CELL between the outer negative plates and the end of the container to prevent the plates from spreading.

End Winding. The part of a winding which lies beyond the slots, serving simply to connect the two active coil sides and contributing nothing to the output of a machine. It is also known as an *overhang*.

Energy. The capacity to do work. The fundamental unit of work or energy is the JOULE. The basis of the definition of the joule is in terms of mechanical work done, but the concept of energy is so wide that there are numerous other ways in which energy has been defined, depending on the branch of physical science concerned. The spread of the SYSTÈME INTERNATIONAL has simplified the picture with most units defined in terms of the joule. However, older systems of units may still be encountered, and these are shown in Fig. E–5, the central column being in electrical terms, with thermal units to the left and mechanical units to the right. The numerical values marked show the relation between the units above and below.

Energy Component. Alternative name for ACTIVE COMPONENT (q.v.).

Energy Level. A term describing the state of excitation of an electron in an atom or molecule.

Epoxy Resin. A synthetic thermosetting resin. It is particularly suitable as a casting resin not requiring high pressure, for encapsulation, as an adhesive, and as a solventless varnish. Polymerization occurs at moderate temperatures, enabling it to be cast readily in moulds. Mixed with suitable fillers (e.g. glass fibre or asbestos), or applied to sheet material, it will give castings or laminates of varying degrees of electrical and mechanical strength, sometimes in very large pieces not readily made by ordinary moulding methods. It is especially useful for high-voltage power switchgear insulation.

Epstein Square. An assembly of strips of core-plate material, arranged in the form of a hollow square interleaved at each corner. Each limb is provided with primary (exciting) and secondary windings for the measurement of CORE LOSS. Cf. LLOYD-FISHER SQUARE.

Equalizer. Connection between two points of a winding, so that the potential at the two points is the same. The electromotive force generated in one path of a lap-connected d.c. armature winding may differ from that in another because of magnetic

Equipotential Connection

asymmetry. Even small differences may give rise to large internal circulating currents, with consequent sparking at the brushes and excessive heating of the winding. These effects may be mitigated by the use of a sufficient number of equalizer rings of low resistance,

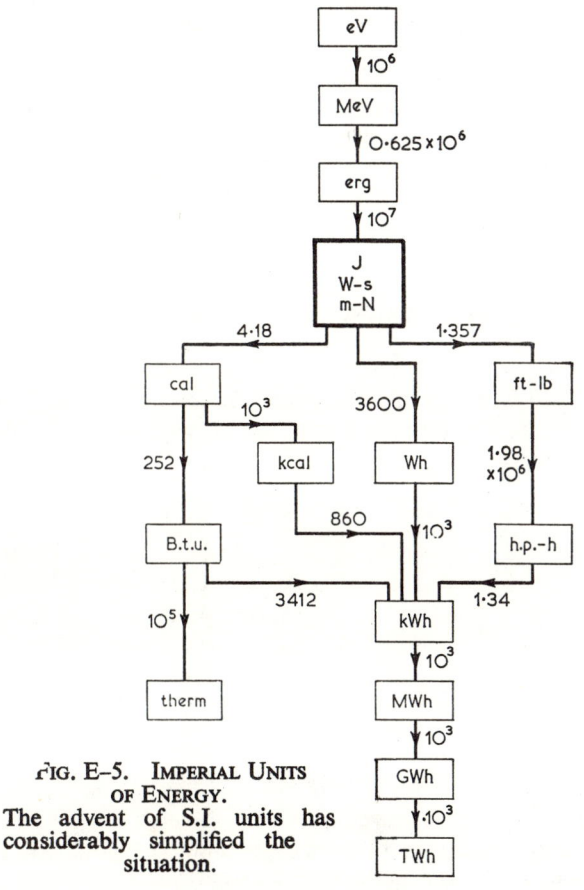

FIG. E–5. IMPERIAL UNITS OF ENERGY.
The advent of S.I. units has considerably simplified the situation.

which connect together all points on the winding that, ideally, should have no difference of potential between them. These may also be known as *equipotential connections*.

Equipotential Connection. Alternative name for an EQUALIZER (q.v.).

Equipotential Surface

Equipotential Surface. A surface within which there is no potential difference between any points.

Equivalent Circuit. An electrical circuit set up to represent an energy system, electrical or otherwise. It gives results in terms of measurable (or calculable) voltages, currents, charges, linkages, etc., that can be employed to represent analogous quantities in another physical system or network.

Equivalent Conductance. Electrolytic conductance of one gramme-equivalent of a substance when a potential difference of 1 V exists between electrodes 1 cm apart. It is equal to the conductivity multiplied by the number of millilitres per gramme-equivalent. *Maximum equivalent conductance* applies to a solution when this is at infinite dilution in its own solvent.

Equivalent Generator. Device employed in circuit analysis. An electrical energy source (such as a microphone, a battery, a synchronous generator or a valve amplifier) may be represented by an equivalent generator comprising a source of e.m.f. E_0 and an internal impedance Z_0 in series with it. On occasion (e.g. with pentode valves) it is more real to consider the source as producing a constant current I_0 with an internal admittance Y_0 connected across it (see Fig. E–6).

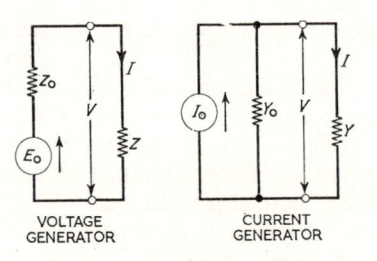

FIG. E–6. EQUIVALENT GENERATORS.

Equivalent Sine Wave. A SINE WAVE having the same r.m.s. value and fundamental frequency as the given wave.

Erg. The C.G.S. unit of energy, equal to the work done by a force of one dyne acting through the distance of one centimetre. The erg is the equivalent of 0.1 μJ or 10^{-7}J.

Ether. All-pervading, hypothetical, non-material medium, the existence of which was proposed to explain the transmission of

Exponential Function

electromagnetic waves. In the light of modern concepts, its assumption is now unnecessary.

Ettingshausen Effect. The appearance of a temperature gradient in a single material due to the interaction of an electric current and a magnetic field.

Eureka. Trade name for cupro–nickel alloy used in the manufacture of resistance wires. It has a small temperature coefficient and will withstand high temperatures.

Excitation. The magnetomotive force producing the magnetic flux in the field system of an electrical machine or in an electromagnet. The term may be applied to the force itself or to the effect produced.

Exciter. Any equipment used to provide the field current of electrical generators. Most exciters are rotating machines, although on occasion rectifiers have been used to convert a suitable variable-voltage a.c. supply to direct current. Since it is unusual to include a variable resistance in the main-field circuit, any change in excitation current is obtained by adjustment of the applied voltage. Consequently it must be possible to vary the exciter voltage over a wide range.

Excitron. A single-anode mercury-arc rectifier. It is made in both glass and steel, but is usually pumpless.

Exploring Coil. A coil employed for measuring magnetic field flux.

Explosion Pot. A small vessel of insulating material surrounding the fixed contact of an oil circuit-breaker, its outlet being sealed by the moving contact. An arc in this pot produces high gas pressures, and when the moving contact clears the outlet the arc is extinguished by a blast of oil and gas.

Explosion Vent. A frangible diaphragm fitted to a transformer, generally in the lid, to save the tank from bursting in the event of an internal explosion that might result from an electrical fault. It is often shrouded with a pipe to throw escaping oil clear when it operates, and varies in diameter with the size of the transformer, from about 15 to 30 cm (6 to 12 in.).

Exponential Function. A function

$$e^x = \exp(x) = 1 + x + x^2/2! + x^3/3! \ldots$$

If $x = 1$, then

$$\exp(1) = 1 + 1 + \tfrac{1}{2} + \tfrac{1}{6} + \tfrac{1}{24} + \ldots = 2 \cdot 7182818 \ldots = e^1 = e.$$

Expulsion Fuse

For negative values the series becomes
$$\exp(-x) = 1 - x + x^2/2! - x^3/3! + \ldots$$
Because $e^0 = \exp(0) = 1$, then $\exp(-x)$ converges and its sum to infinity is less than unity for any positive x.

The exponential function occurs in many analytical branches of electrical engineering. It expresses, for example, the change of temperature of an item of plant (such as a transformer) as determined by its thermal capacity and emissivity; the rate of energy decay of a system (such as a field coil or cable) when the supply is withdrawn; or the attenuation of a signal when travelling along a uniform transmission line.

Expulsion Fuse. A fuse in which the extinction of the arc is facilitated by an increase in length of the break in the link. This is achieved by expulsion of part of the fusible material.

Expulsion Gap. A SURGE DIVERTER in the form of a type of EXPULSION FUSE which may be connected in series with a gap across insulator strings of an overhead transmission line. On a voltage surge, expulsion of the hot gases and ions effectively interrupts any follow current and prevents disruption of the supply. The device is sometimes known as a *protector tube*.

Extrinsic Conduction. A process of *impurity conduction*, occurring in semiconductors, where it predominates over INTRINSIC CONDUCTION. Extrinsic conduction results from the presence in the base material of a small number of atoms of other elements. The foreign element may be a *donor*, in that it gives one possible free electron per atom to increase conduction and to yield n-type properties. Alternatively, an *acceptor* element may give rise to one positive hole per atom, endowing the base material with p-type properties (see Fig. E-7).

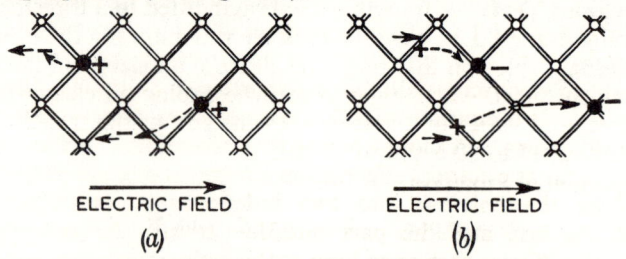

FIG. E-7. EXTRINSIC (IMPURITY) CONDUCTION. Donor impurities (*a*) cause excess-electron (n-type) conduction. Acceptor impurities (*b*) cause hole (p-type) conduction.

Extrusion. Production of rods, tubes, etc., by forcing hot material through a suitable die by means of a ram. Extrusion may be used for sheathing electrical cables, when the temperature used is limited by the presence of the insulating papers.

F

F-Class Insulation. One of the seven classes of insulating materials for electrical machinery and apparatus defined in B.S. 2757 on the basis of thermal stability in service. Class F insulation is assigned a temperature of 155°C. It consists of materials such as mica, glass fibre or asbestos suitably impregnated or coated.

Faceplate Breaker Starter, Controller. A FACEPLATE STARTER, or controller, which includes an interlocked contactor to remove the task of breaking or interrupting the circuit from the starter or controller contacts.

Faceplate Starter, Controller. A starter or controller which has the switch contact parts arranged upon a plane surface (cf. DRUM STARTER, CONTROLLER).

Fahy Simplex Permeameter. A YOKE PERMEAMETER which has no compensation. A bar specimen is clamped between the ends of a single rectangular yoke that carries the magnetizing winding. The search-coils for measuring B and H ballistically extend along the whole length of the specimen, the H-coil being mounted between two massive soft-iron posts clamped to the ends of the specimen.

Farad. The practical and M.K.S. unit of capacitance (abbreviation F). For a potential difference of 1 V, a capacitor of 1F will store a charge of 1 C and an energy of $\frac{1}{2}$ J.

Faraday. A unit employed in electrochemistry. It is the quantity of electricity required to deposit one gramme-equivalent of any ion. It is equal to 96,490 coulombs.

Faraday Cage. A closed box with a conducting surface which serves, so far as electrical effects from charges at rest are concerned, to divide the universe into two independent parts—the part within the box and the part outside. It may be constructed adequately of wire netting so long as the instruments or persons to be screened and the charges setting up the field are not closer to the netting than about ten times the mesh-opening. Whether of wire mesh or of a continuous unbroken conducting surface, the Faraday cage will not screen static or slowly changing magnetic fields. It will, however, reduce the intensity of electromagnetic waves if their wavelength is appreciably longer than the width of the mesh-opening.

Faraday–Neumann Law. A law of electromagnetic induction. Faraday showed that an electric current is induced in a closed circuit when a magnet is moved near it. The induced current can be developed also by moving the circuit near the magnet, by changing the circuit area, and by changing the magnetic field intensity. A more comprehensive statement by Neumann is: An electromotive force is set up in a circuit when the magnetic flux linking the circuit is changed in any manner. The magnitude of the e.m.f. is proportional to the time rate of change of flux-linkages with the circuit, and has such a direction that any current it produces tends to oppose the change of flux-linkage. The law may be expressed:

$$e = -\,d\psi/dt$$

where e is the total e.m.f. and ψ is the flux, or the effective flux-turn product. It is often possible to use the relation $\psi = N\phi$, where ϕ is the average flux per turn, and N is the number of turns of the circuit. This is the more familiar engineering approach.

Faraday's Law of Electromagnetic Induction. A law derived basically by Faraday, and in a more comprehensive form by Neumann (see above).

Faraday's Laws of Electrolysis. During the flow of an electric current through an electrolyte, the mass of ionic material set free at either electrode, deposited thereon, or dissolved therefrom, is proportional to the quantity of electricity that passes and the ionic weight of the ion and inversely proportional to the ionic valence of the material being electrolysed. It is independent of other factors such as the voltage, electrode size, current density and temperature.

Faradism. Electromedical treatment employing a pulsating current, derived from an induction coil, for stimulating muscles and nerves.

Fast Reactor. A NUCLEAR REACTOR employing a high proportion of pure fissile material. The neutrons resulting from fission of a fissile material are emitted with very high energies, e.g. 2 MeV, and at the high speeds corresponding to these energies they are not likely to cause fission of other nuclei unless very many

Ferrimagnetism

fissile nuclei are present. A chain reaction using these fast neutrons requires a fast reactor with a working core of nearly pure fissile material such as uranium-235 unmixed with uranium-238. Such a reactor is characterized by a very small heat-producing core giving, say, **15 MW/ft^3 (1 ft^3 = 0·028m^3)**.

Fault Current. The current which flows from one conductor to another or to earth, because of a fault in the insulation.

Faure Plate. A plate of a LEAD-ACID CELL which is made mechanically by the application of lead-oxide paste onto a grid. It is also known as a *pasted plate*.

Fechner Law. If brightness appears to be increased in equal steps, it is actually being increased logarithmically. The minimum fractional difference in brightness that can just be detected is known as *Fechner's fraction*.

Feedback. The injection of a fraction of the output signal or response of a system into the input which is externally excited by an input signal or stimulus. The feedback is positive (*regenerative*) or negative (*degenerative*), depending on whether a cophasal or antiphasal component is present in the fed-back signal (see POSITIVE FEEDBACK, NEGATIVE FEEDBACK).

Feedback Oscillator. A tuned feedback amplifier in which the feedback voltage has an amplitude and phase such that sufficient positive feedback is produced for oscillations to be sustained.

Feeder. A line supplying a point of a distributing network, and not having any intermediate connections.

Feeder Box. Alternative name for a JUNCTION BOX (q.v.).

Feeder Pillar. A pillar containing switches, links or fuses for connecting a FEEDER with distributing mains.

Feet Switch. Alternative name for a TROPICAL SWITCH (q.v.).

Ferranti Effect. A transmission-line phenomenon. In a line with length substantially less than one half-wavelength but long enough to have appreciable shunt capacitance, the voltage across the receiving end on no-load (or open circuit) can be greater than that applied at the sending end. The effect is the result of electric and magnetic energy inter-conversion imposed by the open-circuit conditions at the receiving end.

Ferrimagnetism. Phenomenon exhibited by materials in which two different kinds of ion, one have a greater magnetic moment than the other, are in opposition. It is frequently displayed by a FERRITE.

Ferrite

Ferrite. A crystalline oxide having the general formula $Fe_2O_3.MO$ where M is a divalent metal. Some (e.g. M = zinc) exhibit ANTIFERROMAGNETISM; others (e.g. M = copper, nickel, iron) exhibit FERRIMAGNETISM. All are less dense than iron and behave as "soft-iron" materials. They are of use in transformer cores. Because they have very high resistivity, their eddy-current losses at high frequencies are small compared with those of metals.

Ferrodynamic Instrument. An ELECTRODYNAMIC INSTRUMENT in which the action is augmented by the presence of ferromagnetic material.

Ferroelectricity. The dielectric analogue of FERROMAGNETISM. Ferroelectrics are made up of small regions called *domains*, with a spontaneous polarization that disappears at the Curie point, above which temperature the material behaves like a normal dielectric.

Ferromagnetism. Phenomenon exhibited by materials having a permeability which is considerably greater than unity, and which varies with the flux density. Examples of such materials are iron, steel, cobalt and nickel.

Ferro-resonance. A condition of resonance in a series inductance—capacitance circuit, the inductor having a ferromagnetic core. The combined current/voltage characteristic will show a point of instability at which a change in voltage causes a discontinuous jump in the characteristic. Ferro-resonance may occur with an unloaded transformer connected to a cable system, or it may be exploited for relay operation.

Feussner Potentiometer. A convenient form of practical POTENTIOMETER incorporating five or more dials. The overall resistance remains constant at all settings, as three double-contact dial switches take out resistance from one circuit as they add an equal resistance to the other. Two single-contact dial switches are connected to the unknown test circuit, through a galvanometer in series, as is seen in Fig. F–1. The reading is made directly from the dials, which are in decades.

Fibreglass. The trade name for a lightweight material made of fine glass filaments bonded together with a polyester resin product (see GLASS FIBRE).

Field. A concept devised to provide a basis of explanation for some physical phenomena which might otherwise be considered as representing action at a distance. A *scalar field* represents the distribution in space of a scalar quantity, such as temperature (which has a magnitude but no directional properties). A *vector field* describes the distrubtion in space of the direction and magnitude of a vector quantity, such as velocity, current flow, electric force, magnetic flux density, or heat flow. Such fields

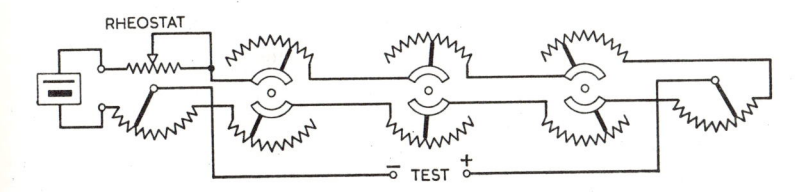

FIG. F–1. PRINCIPLE OF FEUSSNER POTENTIOMETER.

can be mapped in terms of flow-lines which give the direction at various points of the flux function concerned, and which by their concentration give some idea of the density of flow. In other words, a field in a given region of space comprises the aggregate of physical quantities that are functions of position within the region. Many fields conform with the Laplace equation

$$\frac{\partial^2 \phi}{\partial x^2} + \frac{\partial^2 \phi}{\partial y^2} + \frac{\partial^2 \phi}{\partial z^2} = 0$$

where ϕ is the flux function, and *xyz* are the co-ordinates of space.

Field Coil. Current-carrying coil employed to energize the FIELD MAGNET of a machine.

Field Control. A method of motor speed control in which the excitation is varied by shunting the series winding, cutting out some series turns, or inserting resistance into the shunt field circuit.

Field-Discharge Switch. A switch for controlling the field circuit of a motor or generator. It is connected in series with the field windings and has additional contacts so that a discharge resistor is connected across the windings when the supply to the field circuit is interrupted. It is also known as a *field-breaking switch*.

Field Emission. Alternative name for COLD-CATHODE EMISSION (q.v.).

Field-Failure Relay

Field-Failure Relay. A motor protective device that opens a shunt motor starter should the motor field collapse. Its application is usually restricted to drives where loss of the field would result in a dangerous condition such as overspeeding or the loss of dynamic braking. The operating coil of the relay is connected in series with the field, while its normally open trip contacts are in the undervoltage release circuit.

Field Magnet. The electromagnet or permanent magnet that provides the magnetic field in an electrical machine.

Field Regulator, Rheostat. A variable resistance for varying the current in the field winding of an electrical machine.

Field Spider. A rotating spoked structure that carries, at its periphery, the pole pieces of an electrical machine.

Field Spool. The structure on which a FIELD COIL is carried. The term is sometimes used to refer to the field coil itself.

Field Suppression. Reduction of the field of a generator in the shortest possible time in the event of an internal fault. The field may be reduced to zero by disconnecting the main field from its exciter and discharging it through an appropriate resistor, by open-circuiting the field winding, or by exciter reversal.

Fieldistor. A TRANSISTOR that uses surface conductivity control near a *p–n* junction and is characterized by extremely high input impedance.

Field's Test. Test for a series motor, similar to a BACK-TO-BACK TEST, but having the generator output dissipated in a resistance instead of returned to the motor. The generator field is in series with the motor to ensure that the generator always excites itself (see Fig. F–2).

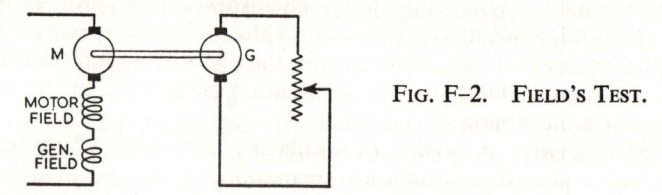

FIG. F–2. FIELD'S TEST.

Filament. Of a lamp, the thread-like conductor which is made incandescent by the passage of an electric current. It is usually made of carbon or a metal such as tungsten.

Filament Lamp. A lamp containing a fine FILAMENT in an evacuated or gas-filled glass envelope.

Fisher Loop Test. A modified LOOP TEST that employs two sound conductors which may be of unknown lengths and different cross-section from the faulty conductor. In Fig. F-3, if a_1 and

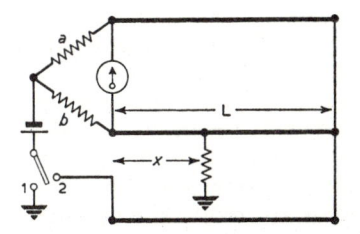

FIG. F-3. FISHER LOOP TEST FOR A FAULT IN A CONDUCTOR (LENGTH L).

b_1, a_2 and b_2 are the values of the resistances for balance in the appropriate positions of the switch, and x is the distance of the fault along the cable,

$$x = L\frac{b_1(a_2+b_2)}{b_2(a_1+b_1)}$$

Fixed Loss. The sum of the core loss, windage, bearing friction and brush friction of an electrical machine, occurring with no load on the machine. It does not include the exciting-circuit or exciting-current loss.

Fixed-Trip Circuit-Breaker, Switch. A circuit-breaker or switch, the automatic release mechanism of which cannot function independently of the closing mechanism.

Flame-Cored Carbon. A type of CORED CARBON. The channel fillings are a mixture which colours the arc.

Flameproof. Term applied to enclosures to signify ability to withstand, without injury, any explosion of a specified inflammable gas that may occur within the enclosure under normal conditions, and ability to prevent ignition of the gas in the surrounding atmosphere.

Flash Barrier. A screen of insulating and non-inflammable material employed to inhibit arc formation, or at least to reduce the effects of an arc.

Flash Bulb. Transparent envelope usually containing a metal foil in contact with a low-voltage filament in an atmosphere of oxygen. When the filament is heated the foil burns away very rapidly, producing a high-intensity flash of considerable use in photography.

Flash Butt Welding

Flash Butt Welding. A variety of BUTT WELDING in which the temperature is raised by flashing away a certain amount of metal until welding temperature is reached, when a high upset pressure is applied.

Flashing Fault. A fault which is not evident at low voltages, but which causes repeated flashover when a higher voltage is applied.

Flashover. The accidental occurence of an arc between two conducting parts of a machine or apparatus.

Flashover Test. A test applied to equipment designed to operate at high voltages. The voltage is raised until flashover of the test object occurs. The measured flashover voltage must be not less than a specified minimum value. Both dry and wet flashover voltage tests may be specified, and they usually follow on consecutively from the appropriate WITHSTAND TEST.

Flat-Compounded. Term applied to a compound-wound generator in which a series winding on the field system, connected to act cumulatively to the shunt winding, is used to neutralize the voltage drop which would otherwise occur as the load builds up. It is also known as *level-compounded*.

Flat Pressure Cable. An OIL-FILLED CABLE having three cores laid side-by-side and sheathed with lead. The lead is reinforced with metal tapes so that the flat side of the sheath can act as a loaded beam. The cable is useful for submarine cables, and for land cables up to 66 kV and occasionally 132 kV. See also MØLLERHØJ CABLE.

Flat-Rate Tariff. A TARIFF which involves a single charge proportional to the number of units of energy supplied.

Fleming's Rules. A method of indicating the relative directions of current, mechanical force and magnetic field in the case of a current lying in a magnetic field. The thumb and forefinger and second finger are arranged roughly at right-angles in three directions corresponding to co-ordinate axes. Then the Forefinger gives the direction of the Field, the sECond finger that of E.M.F. or Current, and the thuMb gives the direction of Motion under action of the force.

With the *left* hand, the directions are then indicated for *motor* action; with the *right*, for *generator* action.

Flit Plug. A detachable cable sealing box for attaching to apparatus and coupling cables.

Floating Battery. A battery connected in parallel with a direct-voltage supply whose normal voltage is a little higher than the open-circuit voltage of the battery. A small charging current then flows to maintain the battery in a fully charged condition.

Floodlight. High-intensity light obtained by positioning a lamp at the focal point of a reflector or lens system. A BUNCH FILAMENT provides the compact light-source necessary.

A $B1$ bulb is practically spherical so that the lamp cap is only a short distance from the filament, ensuring not only more accurate positioning of the filament but also a more compact unit. The $B2$ type uses a bulb with a longer neck and therefore requires a larger fitting. A general-service FILAMENT LAMP up to 1,500 W may be employed where a short range and wide beam are required.

Floor Warming. A form of SPACE HEATING in which the heat is applied to a concrete floor mass. The two principal types are (*a*) *withdrawable*, in which a heating cable rendered waterproof by suitable sheathing is drawn into a tube itself embedded in the concrete, or a preformed hole in the floor; heat is conveyed to the concrete mass partly by conduction, and partly by convection through the air heated in the tube, and (*b*) *directly embedded*, where the heating cable is itself completely embedded in the concrete, heat dissipation being by conduction; because of the direct contact with the mass of the floor an even heat distribution is maintained, and the temperature of the heating cable is less than with the withdrawable system.

Fluorescent Lamp. Lamp in which a discharge through low-pressure mercury vapour is used to generate ultra-violet radiations, which in turn are changed into visible light by the fluorescent effect of a coating on the inside of the glass tube. By suitable choice of the chemical coating, the quality of the light output may be varied.

Flux-Cutting Law. If a conductor, of length l, is moved at right-angles to a magnetic field of uniform density B at velocity u, the electromotive force induced is

$$e = Blu$$

Flux-Diversion Relay. A polarized relay in which a permanent flux exists. The application of operating current produces a flux which counteracts the permanent flux, diverting it to a magnetic circuit which includes the armature and an air-gap.

Flux Linkage

The armature is therefore attracted and is held by the permanent-magnet flux even if the current is subsequently disconnected.

Flux Linkage. The product of the number of lines of magnetic flux and the number of turns of a coil or circuit through which they pass. The unit of *linkage* is one unit of magnetic flux passing through one turn of the coil or circuit.

Fluxmeter. An instrument that measures magnetic flux by means of a search coil that may be moved into or out of a magnetic field, the induced quantity of electricity in flux-turns being fed to a moving-coil instrument.

The fluxmeter is a galvanometer in which the controlling torque is made as low as possible by employing a suspension or ligament above and below the coil of soft metal, not phosphor-bronze as is normal for galvanometers. In addition, the damping is made high by the use of a relatively thick aluminium former on which the coil is wound, swinging in the permanent-magnet field (see Fig. F–4). A long period is thereby attained.

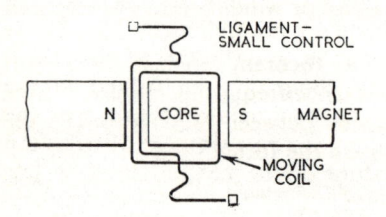

FIG. F–4. A SIMPLE FLUXMETER.

Follow Current. The current that succeeds a high-voltage surge through a discharge path after the discharge has been initiated.

Foot-Candle. The luminous flux corresponding to 1 lumen per square foot. It is the flux received by a 1 ft² surface normal to a source of 1 candela at a distance of 1 ft. The S.I. unit is the lux, and 1 foot-candle = 10·764 lx.

Foot-Lambert. The measure of the brightness of a surface in terms of the average luminous flux emitted per square foot.

Forced-Circulation Boiler. Boiler having pumped water flow to overcome problems of steam and water separation that may be encountered in high-pressure generating plant.

Forced-Draught Ventilation. A form of machine ventilation in which the ventilating air is supplied under pressure by some means external to the machine itself (cf. INDUCED-DRAUGHT VENTILATION, SELF-VENTILATION).

Forced Oscillation. Condition established when a system inherently oscillatory is energized by means of a periodic driving function.

Form Factor. The ratio (r.m.s. value/half-cycle mean value) for a sustained periodic function such as current or voltage. If a non-sinusoidal alternating current has fundamental and harmonic peaks of $i_1, i_3, i_5 \ldots$ then its form factor is

$$K_f = \frac{\pi}{2\sqrt{2}} \cdot \frac{\sqrt{(i_1^2 + i_3^2 + i_5^2 + \ldots)}}{i_1 + \frac{1}{3}i_3 + \frac{1}{5}i_5 + \ldots}$$

For a pure sine wave, $K_f = \pi/2\sqrt{2} = 1 \cdot 11$.

Formative Lag. In a charged spark gap, the time between the occurrence of an electron in a position suitable to initiate breakdown and the actual passage of the spark (see TOWNSEND DISCHARGE).

Formed Plate. A plate of a LEAD-ACID CELL, prepared by electrolytic conversion. It is also known as a *Planté plate*.

Former. A tool for forming a coil or winding into the required shape.

Foster Reactance Theorem. A theorem applied in circuit synthesis to give a required reactance/frequency relation. In a general network of pure reactances between two terminals, the impedance at the driving point take the forms

$$Z = j\omega H \frac{(\omega^2 - \omega_2^2)(\omega^2 - \omega_4^2) \ldots}{(\omega^2 - \omega_1^2)(\omega^2 - \omega_3^2) \ldots}$$

or

$$Z = \frac{H}{j\omega} \cdot \frac{(\omega^2 - \omega_1^2)(\omega^2 - \omega_3^2) \ldots}{(\omega^2 - \omega_2^2)(\omega^2 - \omega_4^2) \ldots}$$

where H is a positive constant, and $0 \leqslant \omega_1 \leqslant \omega_2 \ldots < \infty$. The expressions represent the variation of impedance with the frequency ω at which the network is supplied.

Four-Terminal Network. A network with two input and two output terminals. It may be a physical network designed, for example, as an attenuator, filter or transmission line, or an analytic representation of these. See QUADRIPOLE.

Fourier Series. A series employed in HARMONIC ANALYSIS. If a finite, univalued function $g(x)$ recurs, or may be considered to recur, over successive intervals of 2π in the variable x, so that $g(x) = g(x + 2m\pi)$ for all integral values of m, then it is possible

Fractional-Pitch Winding

to represent it by a synthesis of sinusoidal waves of fundamental period 2π and of various integral harmonic orders. Thus, a function $y = g(x)$ can be written as either:

$$y = a_0 + a_1 \cos x + a_2 \cos 2x + \ldots + a_n \cos nx + \ldots$$
$$+ b_1 \sin x + b_2 \sin 2x + \ldots + b_n \sin nx + \ldots$$

or

$$y = a_0 + c_1 \sin(x + \alpha_1) + c_2 \sin(2x + \alpha_2) + \ldots$$
$$+ c_n \sin(nx + \alpha_n) + \ldots$$

Each of the expressions above is a *Fourier series*, where

$$c_1 = \sqrt{(a_1^2 + b_1^2)}, \qquad c_n = \sqrt{(a_n^2 + b_n^2)}$$

are the amplitudes of the fundamental and nth harmonic, and the phase angles are $\alpha_1 = \arctan(a_1/b_1)$, $\alpha_n = \arctan(a_n/b_n)$, and a_0 is a constant.

Fractional-Pitch Winding. Winding that is accommodated in a number of slots not divisible by the product of the number of phases and poles. It has a fractional value for the number of slots per pole, and is also known as a *fractional-slot winding*. For the purpose of reducing harmonic e.m.f.s, an armature winding may have a span greater or (more usually) less than a pole-pitch. In a three-phase machine the coil-span is arranged generally to reduce the fifth- and seventh-harmonic e.m.f.s, the third, ninth, etc., being suppressed by the three-phase connection.

Frame Fault Protection. An earth-fault BUSBAR PROTECTION scheme applied to metalclad switchgear, the arrangement shown in Fig. F–5 being for a two-section switchboard. To afford protection, the switchgear framework is lightly insulated from the ground. It is then earthed through a current-transformer primary conductor. The arrangement shown discriminates between faults on the two busbar sections. An insulated barrier is placed between the bus-section switch framework and the right-hand section framework. The two frameworks are separately earthed, faults in the left-hand section initiating operation of relay R_1, and in the right-hand section R_2. For high-speed fault clearance of all faults, insulation of the switchgear framework is provided on both sides of the bus-section switch, and the bus-section framework is separately earthed. The bus-section circuit-breaker is then tripped for current in any of the three earth connections, and current in the bus-section earth is arranged to initiate tripping of both of the main sections.

Frequency Modulation

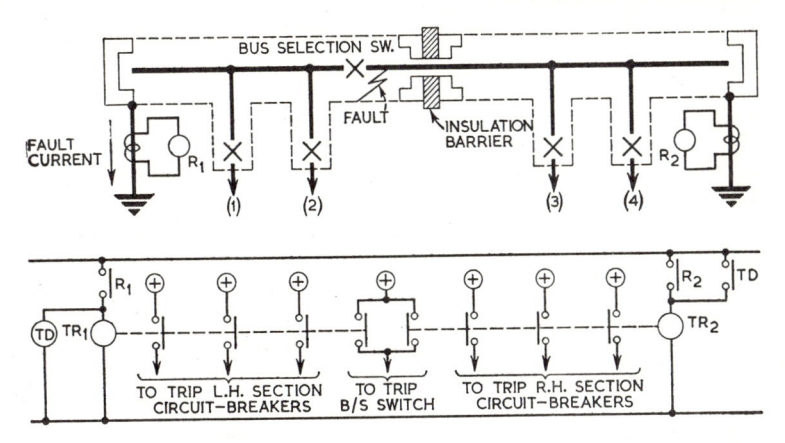

FIG. F-5. CONNECTIONS FOR TWO-ZONE FRAME FAULT PROTECTION.

Francis Turbine. A reaction-type WATER TURBINE used for medium heads of water up to approximately 300 m. The water flows through a spiral casing inwards through the wheel and acts on a series of blades running between the hub and an outer wheel ring. Use is thus made of both the kinetic and the potential energy in the water. The complete installation includes a casing with a ring of movable guide vanes round the circumference of the runner. Mechanical linkages, operated by servomechanisms, operate these vanes to regulate the flow of water to the turbine.

Frequency. The number of complete cycles of a periodic function (e.g. of current or voltage) in one unit of time. The S.I. unit of frequency is the HERTZ.

Frequency Changer. A machine for converting alternating current at one frequency to alternating current at another.

Frequency Converter. A PHASE ADVANCER in which the slip rings are fed with voltages at supply frequency which are converted by the action of the commutator to slip frequency. The injected e.m.f. is thus independent of slip.

Frequency Distortion. A change in waveform, owing to the input components being altered in their relative magnitude by a frequency-dependent response in the system. This is also known as *attenuation*.

Frequency Modulation. The imposition of a signal on to an alternating carrier wave in such a way that the amplitude of the carrier remains constant while its frequency is varied proportionally to the amplitude of the signal (cf. AMPLITUDE MODULATION).

Frequency Transformer. Transformer for converting alternating current at one frequency to alternating current at another. This term should always be used where a static transformer is employed for this purpose without a rotating machine, to distinguish it from a *frequency changer*.

Frosted Lamp. An electric lamp having its bulb partially or wholly etched or sandblasted to diffuse the light. An internally frosted bulb is the principal feature of a *pearl lamp*.

Fuel Cell. A device for converting the chemical energy of fuels into electricity without going through the intermediate stages of heat and mechanical energy involved in thermal generating plant. Two typical fuel cells are shown diagrammatically in Fig. F–6. The first of these is an H_2–O_2 cell (*hydrox cell*) of the type developed by Bacon, operating at 200°C and 400 kN/m^2, (600 lb/in^2) while the second uses air and almost any gaseous or vaporized fuel and operates at about 550–700°C. There is considerable interest in the development of *carbox cells*, which would burn liquid-hydrocarbon fuels. Fuel cells are irreversible.

Fulchronograph. An instrument for measuring the variation with time of the current in a flash of lightning. The latter is allowed to energize an electromagnet, between the poles of which move magnetic surge-current indicators.

Full-Pitch Coil. Coil of a winding in which the span is exactly a pole-pitch.

Fundamental Unit. A selected unit from which, in conjunction with other fundamental units, a physical quantity is derived. Length, mass and time are the traditional physical concepts. Electrical science needs four concepts and four arbitrarily defined units (not necessarily the same four), and at least one must be electrical in nature.

For the fourth electrical concept, that of *charge* has been widely adopted out of several claimants. The choice of a fourth physical standard, to rank with (kg), (m) and (s), is less obvious. Charge would be experimentally inconvenient; resistance (in terms of a mercury column) or current (in terms of electro-deposition) would be eminently suitable. However, the International Electrotechnical Commission has fixed on the magnetic space constant μ_0 thus settling the question. For dimensional equations, the electrical concept selected is current.

Fuse Link

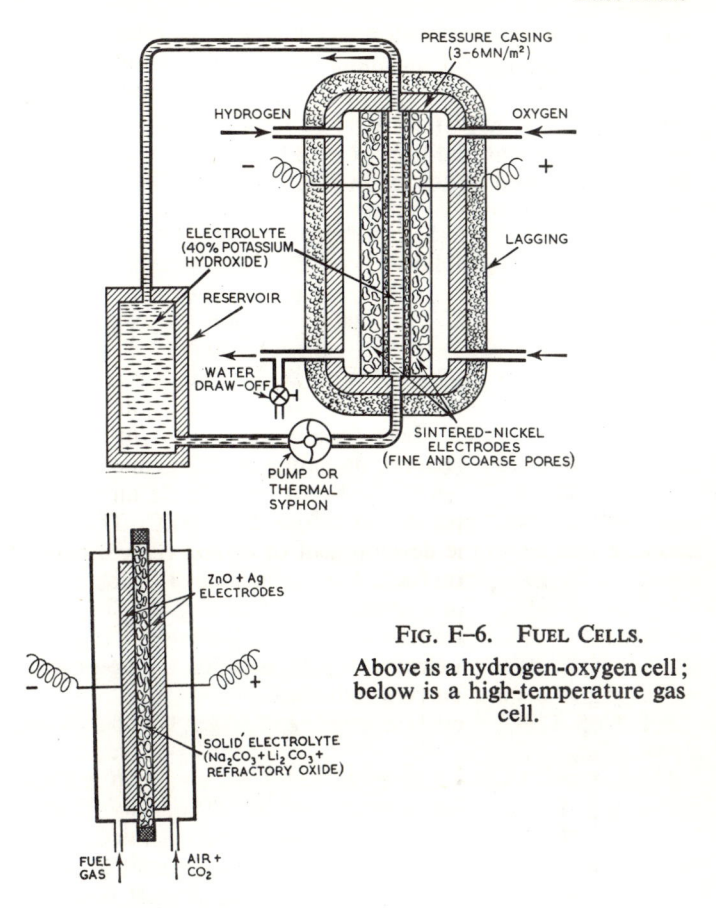

FIG. F–6. FUEL CELLS.

Above is a hydrogen-oxygen cell; below is a high-temperature gas cell.

Furnace (Electric). See RESISTANCE FURNACE, INDUCTION FURNACE, ARC FURNACE.

Fuse. A device that protects a circuit against damage from excessive current flowing in it by the melting of a fuse element. When the current melts the fuse element, the circuit is opened. The fuse comprises all the parts that form the complete device.

Fuse Element. Part of a FUSE that consists of a conductor designed to melt and thus open a circuit.

Fuse Link. Part of a FUSE that comprises a fuse element and a cartridge or any other container, and is capable of being attached

Fuse-Switch

to fuse-contacts or is itself fitted with fuse-contacts, i.e. contacts which are suitable for engaging with a fixed contact.

Fuse-Switch. Strictly, a switch that carries one or more fuses on its moving part (cf. SWITCH-FUSE). However, with high-voltage equipment, the term is used loosely to cover any combination of a fuse and switch, as in Fig. F–7.

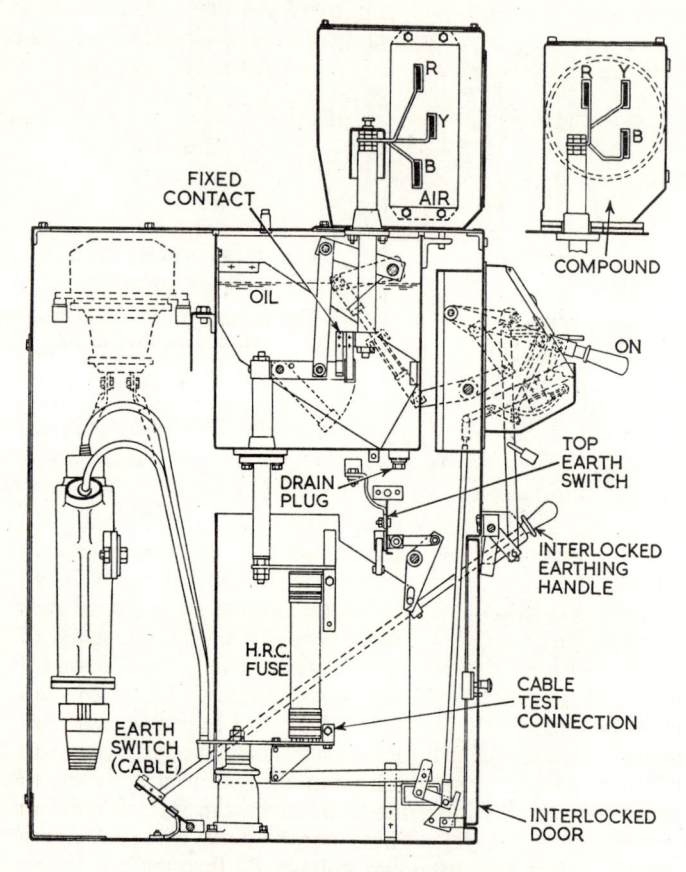

FIG. F–7. AN 11 KV FUSE-SWITCH.
This schematic diagram shows the principal components. The fuse-links are air-insulated.
(*G.E.C.*)

134

Fusing Factor. The ratio greater than unity of the MINIMUM FUSING CURRENT to the current rating.

G

Gall Potentiometer. An a.c. potentiometer that provides two variable components of electromotive force that are in quadrature with one another. These components are summed and balanced against the unknown e.m.f. In Fig. G–1, the supply voltage E is

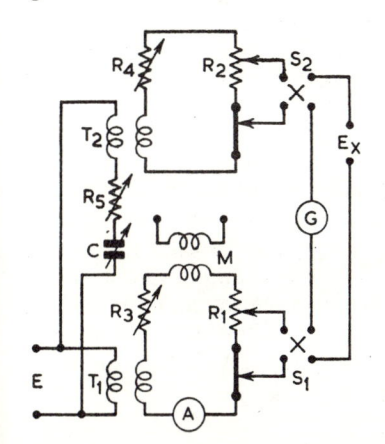

FIG. G–1. GALL CO-ORDINATE POTENTIOMETER.

The current in R_1 is in quadrature with that in R_2.

applied to the primaries of the isolating transformers T_1 and T_2. The secondary of T_1 supplies a current, approximately in phase with the supply voltage to R_1, which consists of a tapped resistor and slide wire in series. The current in this circuit may be adjusted by the rheostat R_3. The current also passes through the primary winding of the fixed mutual inductor M. Any desired value of in-phase voltage within the range available may be obtained from the tappings on R_1 shown. The primary of transformer T_2 is provided with a series rheostat R_5 and a variable capacitor C which may be so adjusted that the current in R_2 is in quadrature with the current in R_1. These components in series are then balanced against the unknown voltage E_x through the vibration galvanometer G.

Galloping. Said of overhead conductors when they oscillate owing to the aerodynamic effect of ice accretions or of their profile. They may also be described as *dancing*.

Galvanism

Galvanism. The treatment of diseases by the use of direct current. It is also known as *medical electrolysis*.

Galvano-magnetic Effect. An effect that depends primarily on the fact that an electrically charged particle, moving at right-angles to a magnetic field, experiences a force in a direction perpendicular to both the field and the motion. The magnitude of this force is proportional to the velocity of the particle. Examples are the ETTINGSHAUSEN EFFECT and the HALL EFFECT.

Galvanometer. An instrument to detect or indicate the presence or absence of a small current. Most galvanometers are of the moving-coil type for direct current, high sensitivity being attained by the use of one or two light phosphor-bronze control suspensions. The two-suspension or bifilar type includes, as well as control by torsion, a form of gravity control, as the coil is raised slightly as it is deflected either way from zero. With this suspension the movement is usually indicated by means of a mirror on the moving system which reflects a beam of light projected on to it from an external lamp. The ballistic moving-coil galvanometer has a small controlling torque and minimum damping: it reads impulses in flux-turns or microcoulombs and the reading is taken at maximum deflection. To obtain accuracy the discharge must take place into the coil quickly, before the movement commences to deflect. See Fig. G–2.

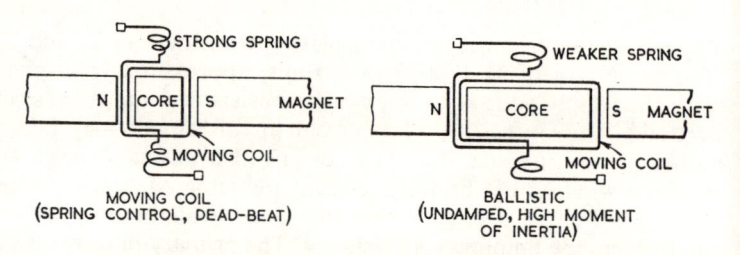

FIG. G–2. MOVING-COIL GALVANOMETERS.

Galvanoplasty. An alternative name for ELECTROFORMING (q.v.).

Gamma Radiation. An electromagnetic radiation equivalent to short-wavelength X-rays. Sources of gamma radiation may be used in IRRADIATION processes, and the radiation can be detected by a variety of special measuring devices.

Gas-Filled Cable

Gap. (1) The space (normally air-occupied) between the stationary and movable parts of an electromagnetic machine (see AIR-GAP).

(2) The distance between the electrodes of a high-voltage measuring or limiting device such as a sphere or rod gap (see SPARK GAP).

Gas-Blast Circuit-Breaker. A high-power circuit-breaker in which a blast of gas is directed across the contacts at the instant of separation to extinguish the arc (see Fig. G–3). A recent development uses SULPHUR HEXAFLUORIDE as a dielectric and arc-quenching medium (see also AIR-BLAST CIRCUIT-BREAKER).

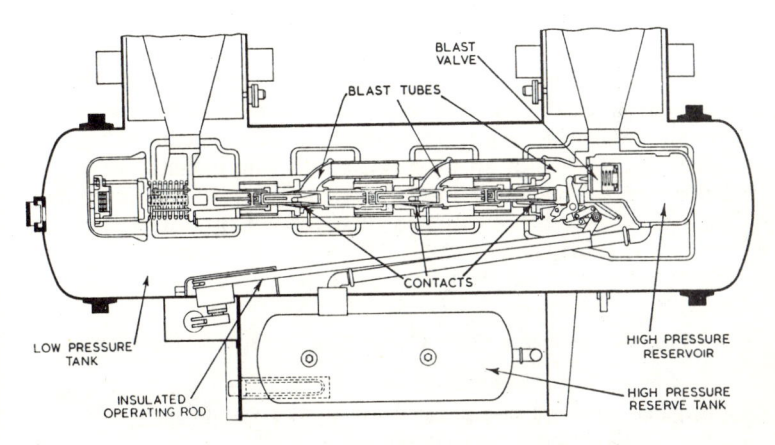

FIG. G–3. ONE PHASE OF A HIGH-VOLTAGE SULPHUR HEXAFLUORIDE CIRCUIT-BREAKER.
(*The General Electric Co. Ltd.*)

Gas-Bubble Protective Device. See BUCHHOLZ RELAY.

Gas-Filled Cable. An internal pressure cable in which pre-impregnated paper tapes are applied to the conductor in such a manner as to ensure that spaces of pre-determined dimensions are included throughout the dielectric. The compound used to impregnate the tapes has a petroleum-jelly base so that it does not flow and block the spaces at the continuous operating temperature. The spaces are charged with nitrogen gas at about 1 MN/m^2 (150 lb/in^2) after the cable has been installed. This ensures that the

Gas-Pressure Cable

creation of low-pressure voids with heat cycles is impossible. Gas-filled cables are used mainly at 33 kV.

Gas-Pressure Cable. A cable employing gas under pressure to prevent ionization in the dielectric. A cable employing *external pressure* is a COMPRESSION CABLE; one employing *internal pressure* is a GAS-FILLED CABLE.

Gas Turbine. A form of thermal generating plant able to provide rapid start and shut-down, which has found application in peak-load provision and for standby duties. The simplest form of gas turbine consists of a compressor, a combustion chamber and a turbine as shown in Fig. G-4. Driven by the turbine, the

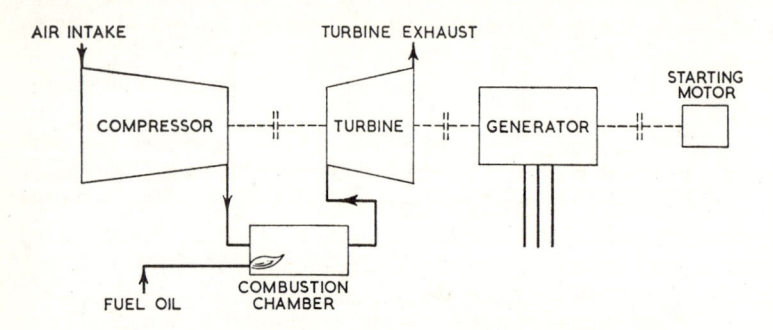

FIG. G-4. SIMPLE OPEN-CYCLE GAS-TURBINE GENERATOR.

compressor delivers compressed air to the combustion chamber where there is continuous combustion of the injected fuel oil at constant pressure. The heated air from the combustion chamber, including the products of combustion, is expanded in the turbine and is then exhausted to the atmosphere. The power developed in the turbine is partly used up in driving the compressor and the remainder is available for the generation of electricity. Like all turbines, the gas turbine is essentially a constant-speed machine. The simple open-cycle plant is not flexible and operates with highest efficiency at full load and constant speed. The addition of an efficient heat exchanger has a pronounced beneficial effect on efficiency, and is usually regarded as essential for land and marine plants.

Gassing. The evolution of gas in an accumulator, which occurs near the end of a charge.

Gate-End Switchgear. An electrically operated contactor switch interlocked electrically with a main isolator in a separate compartment. The contactor can be controlled locally or remotely. The remote control system is used also to interlock electrically the cable plug so that if this is withdrawn the power supply is cut off and the contactor trips. The name derives from the original application of the unit in the gate roadway of a mine at the entrance to the coal face.

Gauss. The electromagnetic C.G.S. unit of magnetic flux density equal to 1 maxwell per square centimetre (abbreviation: Gs). It is commonly called a C.G.S. line per cm^2. The mechanical force on 1 cm length of conductor carrying 1 abampere ($= 10$ A) lying at right-angles to a magnetic field of density 1 G is 1 dyne. The corresponding M.K.S. unit is 1 Wb/m^2 = 10,000 G.

Gauss' Law. The total number of lines of electric force emerging from a closed surface in an electric field is Q/E, where Q is the total obvious charge enclosed by the surface, and E is the permittivity of the dielectric in which it lies (see MAXWELL LAWS).

Gaussmeter. Instrument for measuring field strength at a point. A probe element consisting of a moving coil carrying constant current is inserted into the field at the point at which measurement is required. The coil rotates against the restoring torque of hair springs, and a pointer indicates the deflection which is proportional to the flux density at the point. Alternatively, the moving coil is replaced by a small permanent magnet, also controlled by hair springs. Rotation to the point of maximum deflection is necessary.

Gearless Motor. A traction motor which has its armature mounted concentrically on the driving axle of an electric locomotive.

Geiger Counter. A radiation detector. The Geiger, Geiger–Müller or G–M counter is similar in construction to the IONIZATION CHAMBER, but it is gasfilled with a high multiplication factor and used at higher voltages so that a pulse of radiation is indicated or recorded as a pulse of current. With this detector no pulse-amplitude discriminator is necessary and as the output may be several volts no amplifier is needed.

Geissler Tube. A form of discharge tube working at a moderately low pressure to show the luminous effects of discharges through various gases.

Generating Set

Generating Set. Equipment comprised of one or more generators driven by a prime mover.

Generating Station. A combination of equipment and its associated housing, for producing electrical energy from some other form of energy. It is often known as a *power station*. See HYDROELECTRIC GENERATING PLANT, THERMAL POWER STATION, WIND-POWER GENERATOR.

Generator. A machine for converting mechanical energy into electrical energy. Its design is based on the fundamental laws of electromagnetic induction.

Consider the simple arrangement shown in Fig. G–5, where a single turn of wire is rotated about its own axis in the air-gap of a two-pole magnet system. Since the two sides of the coil are moving

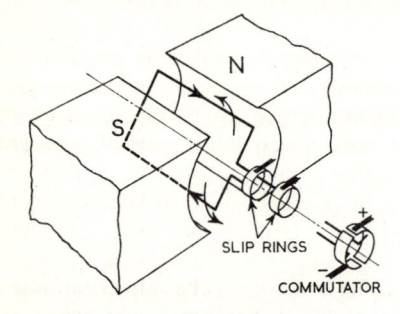

FIG. G–5. BASIC A.C. AND D.C. GENERATORS.

The slip-rings will give an alternating supply. The commutator will provide a unidirectional output.

in opposite directions in relation to the flux the voltage induced each will be of opposite sign, i.e. they will be additive in respect to the slip rings connected to the ends of the turn. Further, when the coil has rotated through 180° the slip-ring voltage will have the same value but the polarity will be reversed. Thus, the arrangement is an a.c. generator or *alternator*. If now the two ends of the coil are connected to two segments insulated from each other and spanning nearly 180°, it will be seen that during rotation one brush is always connected to the positive end of the coil and the other to the negative. Thus, a simple commutator has been added to the generator, and a d.c. supply can be taken from the brushes. The voltage generated between the brushes could be increased by multiplying the number of turns between the two commutator segments, but this arrangement has serious practical limitations and usually a number of coils and commutator segments are connected in series to generate the required voltage.

Generator-Field Control. Method of controlling an electric motor, such as that powering an electric lift, by the use of a motor-generator. A direct voltage applied to the armature of the lift motor is varied by altering the strength and polarity of the generator field. This is also known as *variable-voltage control*. See also WARD LEONARD CONTROL.

Generator Protection. See DIFFERENTIAL PROTECTION, OVER-CURRENT PROTECTION, BACK-UP PROTECTION.

Geothermal Generating Plant. Generating equipment that employs the internal heat of the earth as the source of energy. It is in use at a few places where the earth's crust is sufficiently thin for the hot water or steam to be tapped, e.g. Italy, New Zealand, Mexico, U.S.A. and Japan.

German Silver. A resistance alloy containing copper, zinc and nickel.

Germanium. A tetravalent metallic element. It is a semiconductor having low conductivity at room temperature and increasing conductivity with rising temperature. When alloyed with very small but accurately controlled proportions of trivalent or pentavalent metallic impurities, thus producing p-type or n-type germanium respectively, it is used in the manufacture of CRYSTAL DIODES and TRANSISTORS. Germanium has a melting point around 950°C.

Germanium Rectifier. A rectifier employing the semiconducting properties of germanium. One of the difficulties in the practical application of germanium (and silicon) rectifiers is that fact that the actual junction is very small, and the heat which is produced by the forward and reverse loss must be transferred to the outside without undue rise of temperature. This involves rather elaborate methods of cooling. Because of the very small separation between portions of the slice of opposite polarity, edge leakage effects are also important and complete hermetic sealing of each junction is essential. It will be seen from the typical characteristic curve (Fig. G-6) that the reverse current is greatly increased at higher temperatures. This limits the voltage per cell that can be employed in practical applications.

Getter. A chemically active metal such as barium or magnesium used to increase the vacuum in electronic valves, lamps, etc. After the component has been sealed off from the vacuum pump, the getter material inside it is volatilized by heating it electrically. During volatilization, the getter combines with the residual gases

Giga-

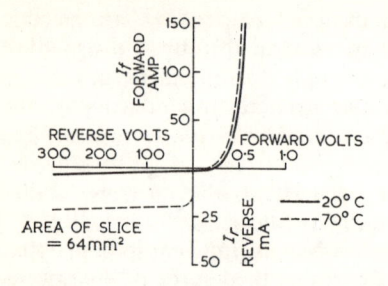

FIG. G–6. D.C. CHARACTERISTIC OF GERMANIUM JUNCTION RECTIFIER.

in the vessel so that they are fixed as chemical compounds deposited on the walls.

Giga-. A decimal prefix indicating multiplication by 10^9 (abbreviation: G).

Gilbert. The electromagnetic C.G.S. unit of magnetomotive force. One gilbert is equal to $10/4\pi$ ampere-turn.

Giorgi System. System of units adopted internationally in a *rationalized* form as the M.K.S. SYSTEM. Giorgi observed in 1901 that the confusion of electrical unit systems (i.e. the co-existence of electromagnetic and electrostatic C.G.S. systems, together with the "practical" system) could be resolved by adopting the metre and kilogramme in place of the centimetre and gramme as units of length and mass respectively. Such a system would incorporate the practical units and extend them.

Glass. A hard, amorphous brittle substance made by fusing one or more of the oxides of silicon, boron or phosphorus with basic oxides. It is used in electrical engineering for insulating envelopes and line insulators. Its melting point lies between 800° and 950°C; the specific resistance of soda-lime glass is 5×10^{11} and of Pyrex glass, 10^{14}.

Glass-Bulb Rectifier. A form of MERCURY-ARC RECTIFIER. The bulb may have three or six arms into which the anodes are sealed. Glass-blowing techniques and the problem of cooling set a limit to the size of bulb that can be manufactured. A rough limit is 500 A at 600 V d.c. output, but several bulb units are commonly operated in parallel as a bank, enabling higher outputs to be obtained.

Glass Fibre. A lightweight material made of fine glass filaments bonded together. In the electrical industry it is used as a base for flexible insulation, sleevings, winding wires and cables.

Graphite

Pre-impregnated glass-fibre tapes of high tensile strength are being produced to replace the steel-wire banding of armatures and similar binding applications.

Glow Discharge. A SILENT DISCHARGE through a gas causing luminosity.

Glow-Discharge Lamp. A tube containing neon, argon or krypton inert gas and provided with two electrodes. The light output may be as low as 0·3 1m/W, with a spectral character corresponding to the filler gas enclosed. The glow occurs around the negative electrode, where free electrons strike gas atoms and cause them to radiate.

Goliath Edison Screw Cap. See EDISON SCREW CAP.

Graded-Gap Machine. Electrical machine in which the distance between the armature surface and the pole face varies from point to point along the pole arc in such a way as to produce at no-load an unsymmetrical distribution of flux. The maximum value is on that side of the centre line of the pole opposite to that to which it tends to be moved by the action of the armature ampere-turns on load; under full load conditions, therefore, these two effects substantially counteract each other.

Grading. See INSULATION GRADING.

Grading Shield. A concentric circular conductor designed to improve the voltage distribution along a string of insulators.

Grain-Oriented Steel. Silicon steel in which the individual crystals are aligned by suitable treatment to give strongly marked directional magnetic properties. Sheets of this steel have higher permeability and lower iron loss in the direction of the grain than untreated sheets.

Gramme. The unit of mass in the C.G.S. unit system, defined as one-thousandth of a kilogramme (abbreviation: g).

Gramme Winding. Alternative name for RING WINDING (q.v.).

Graphic Instrument. An instrument for producing a graphic record of the quantity measured. It usually consists of a pointer in the form of pen which moves over a paper chart. It is also known as a *chart-recording instrument*, a *grapher*, a *recorder* or a *recording instrument*.

Graphite. An allotropic form of carbon, produced artificially by heating coal or coke in an electric furnace. As a mineral, it is found widely distributed in a variety of rocks. Graphite brushes for collecting the current from the commutator of an electric machine have higher conductivity and better lubricating properties

Graphite Brush

than ordinary carbon brushes. Very pure graphite is used for anodes in thermionic tubes.

Graphite Brush. See NATURAL GRAPHITE BRUSH, ELECTRO-GRAPHITIC BRUSH, METAL-GRAPHITE BRUSH.

Grassot Fluxmeter. A FLUXMETER comprising a sector-shaped horizontal-scale pointer instrument.

Grenz Ray. X-ray produced at tube voltages of 5–20 kV peak.

Grid. (1) A cast or stamped unit forming part of a resistor.

(2) The framework in a lead-acid accumulator supporting the active material of a pasted plate.

(3) An electrode in a discharge tube for controlling the current flow between two other electrodes.

(4) Colloquial description of British network of high-voltage transmission lines (see GRID SYSTEM).

Grid System. An electrical power transmission network connecting together, within a given area, all the sources of power and all the points of bulk supply to consumers. It is an essential feature of a grid system that the control of generation should be completely co-ordinated under one authority.

Growler. Device for detecting short-circuits in a winding coil. It consists of a stack of stampings wound to suit the supply voltage, and acts as a transformer with no secondary winding. It is used to induce a flux in the component under test. A short in a coil between the poles of the growler causes a distinct pull on a light steel feeler held over the slot.

Guard Wire. In overhead-line construction, an earthed conductor arranged to protect physically the line conductors. It may be run beneath the line conductors to prevent them from falling to the ground, or above them to prevent other conductors from falling on them.

Gyro-machine System. An electrical method of driving vehicles by means of the energy stored in a flywheel. A *gyrobus* has a large, heavy flywheel mounted horizontally under the floor. Charging columns at the roadside at less than six-mile intervals enable a three-phase supply to be connected for two or three minutes to a squirrel-cage induction motor that accelerates the flywheel to over 3,000 rev/min. The motor then absorbs the stored energy to operate as a self-excited generator, driving traction motors on the vehicle.

H

H-Class Insulation. One of seven classes of insulating materials for electrical machinery and apparatus defined in B.S. 2757 on the basis of thermal stability in service. Class H insulation is assigned a temperature of 180°C. It consists of materials such as silicone elastomer and combinations of materials such as mica, glass fibre and asbestos suitably impregnated or coated.

H-Type Cable. See HOCHSTÄDTER CABLE.

Half Cell, Half Element. Of an electrolytic cell, one electrode and the portion of the electrolyte in contact with it. It is also known as a *single-electrode system*.

Half-Life. A measure of the disintegration rate of a radioactive substance. It is the time taken for the activity rate to fall to one-half of the initial rate.

Half-Power Point. Point on a response characteristic that represents half the power intensity of that at the maximum point. It is also known as a *3 dB point*.

Half-Value Layer. A measure of the quality of a heterogeneous X-ray beam, defined as the thickness of a specified material that reduces the intensity of the beam to 50% of its original value.

Half-Wave. A term used particularly in connection with rectifiers to describe a system in which one half-cycle of an alternating voltage produces conduction current, while the other half-cycle produces a comparatively negligible or no current.

Hall Angle. Angle through which equipotential surfaces in a conductor are tilted owing to the influence of mutually normal electric and magnetic fields (see HALL EFFECT).

Hall Effect. The combined action of mutually normal electric and magnetic fields on the flow of conduction electrons in a conductor. For electron flow from left to right along a thin strip and for the magnetic field acting vertically upwards, electrons will be deflected towards the front edge and there will be set up between opposite points on the edges a potential difference known as the HALL VOLTAGE. In effect, the equipotentials are tilted through an angle known as the HALL ANGLE.

Hall Voltage. Potential difference between opposite edges of a conductor caused by the influence of mutually normal electric and magnetic fields (see HALL EFFECT).

Hanger

Hanger. (1) A fitting used for separating and insulating the overhead contact wire of a traction system from a transverse wire or structure.

(2) An insulating plate standing on edge in an accumulator and supporting the accumulator plates by means of lugs.

Harmonic. One of the sinusoidal components of a complex periodic waveform, having a frequency that is an integral multiple of the repetition (or fundamental) frequency of the waveform.

Harmonic Analysis. The reduction of a given waveform to its mean value, fundamental, and harmonic series.

Harmonic Distortion. A change in waveform, due to the inclusion of additional frequency components.

Harmonic Response. The ratio output/input of a dynamic system (such as an automatic control system servomechanism or electrical machine) for an input signal of sinusoidal waveform and specified frequency.

Hay Bridge. A real-product A.C. BRIDGE of the form shown in Fig. H–1. Usually the bridge is balanced by adjustment of the resistor in series with the capacitor, and of one of the non-reactive arms. The balance depends upon the frequency. Cf. MAXWELL BRIDGE.

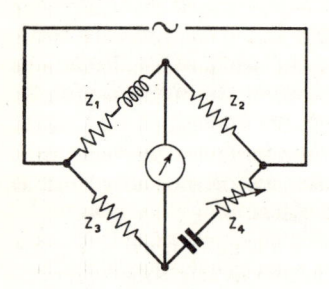

FIG. H–1. BASIC CIRCUIT OF A HAY BRIDGE.

Heat Pump. Device for transforming low-grade heat (as from the air, rivers, sea or soil) into higher. Heat is extracted by a refrigerant which is then compressed and transfers its heat to air or water that can be used for space or process heating.

Heat Run. Test of an electrical machine under operating conditions with full-load current flowing through the windings to determine the temperature rise under these conditions. It is frequently performed with two machines back-to-back as in the HOPKINSON TEST.

Helmholtz–Thévenin Theorem

Heating Inductor. Primary work circuit employed to induce high-frequency currents for the INDUCTION HEATING of conducting material.

Heating Resistor. Wire or other resistance material used as the source of heat in an electrical *heating element*.

Heavyside–Campbell Bridge. An A.C. BRIDGE that introduces mutual inductance, the primary winding of the mutual inductor being connected in the supply lead. Two balances are obtained, without and with a test coil RL in circuit (Fig. H–2). Then with $P = Q$, the values of the unknown are $R = S_1 - S_2$ and $L = 2(M_1 - M_2)$.

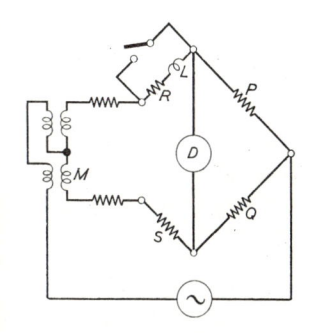

FIG. H–2. BASIC CIRCUIT OF A HEAVYSIDE–CAMPBELL BRIDGE.

Helmholtz–Norton Theorem. The dual of the HELMHOLTZ–THÉVENIN THEOREM. The voltage across any branch of admittance Y in a network that contains several current sources is given by $V = I_0/(Y + Y_0)$, where I_0 is the current that flows in the branch when short-circuited, and Y_0 is the admittance of the network viewed from the branch terminals. The theorem in effect converts the network (apart from the branch Y) into an equivalent current generator of current I_0 and shunt admittance Y_0. The applicability of the theorem is the same as that specified for the Helmholtz–Thévenin theorem (see below). The stages in the application are illustrated in Fig. H–3. When Y_0 is found, all current sources are replaced by their internal shunt admittances.

Helmholtz–Thévenin Theorem. A particularly useful method of network solution. It states that the current in any impedance Z, forming a branch of a network containing one or more sources of e.m.f., is $I = E_0/(Z + Z_0)$, where E_0 is the voltage that appears across the branch when open-circuited, and Z_0 is the impedance

Henry

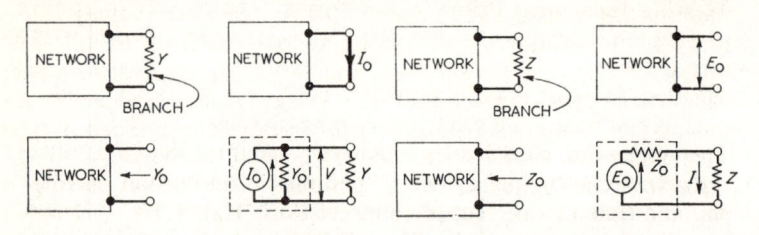

FIG. H–3. APPLICATION OF HELMHOLTZ–NORTON THEOREM.

FIG. H–4. APPLICATION OF HELMHOLTZ–THÉVENIN THEOREM.

of the network looking in at the terminals of the branch, as shown in Fig. H–4. If Z is specified for a given frequency, then the sources must all be of that frequency, but the theorem is applicable to transient cases when Z is also expressed in transient form.

For the theorem to be applicable, the network must have linear (i.e. constant) bilateral parameters. In effect it reduces the network (apart from the branch Z) into an equivalent voltage generator with an internal impedance Z_0 and an e.m.f. E_0. In specifying Z_0, the internal impedance of all sources must be included.

Henry. The practical and M.K.S. unit of inductance (abbreviation H). For a current of 1 A, an inductor of 1 H will have a total magnetic-field energy storage of $\frac{1}{2}$ J, and the flux-linkage will be 1 Wb-turn. For a rate of change of current of 1 A/s the e.m.f. induced in an inductor of 1 H will be 1 V.

Héroult Arc Furnace. An INDIRECT-ARC FURNACE. Direct current was originally used, but modern equipment invariably employs three electrodes connected to a three-phase supply (see Fig. H–5). Power is supplied by a step-down transformer with variable voltage to suit the requirements of the smelting or refining

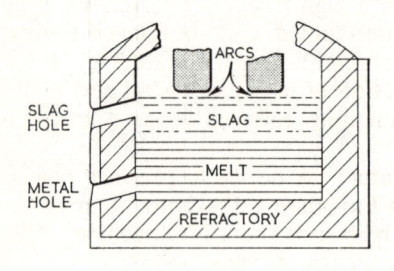

FIG. H–5. PRINCIPLE OF THE HÉROULT ARC FURNACE AS USED FOR STEEL MAKING.

operation. Currents range up to 10 kA, the lower values being used for the refining stages. Voltages may be from 65 to 90 V during refining, and between 150 and 180 V for the melting stage. The furnace consists essentially of a vertical steel cylinder with an appropriate lining. Separate control is normally applied to the three electrodes, so that they may be operated individually.

Hertz. The frequency of a periodic phenomenon having a periodic time of one second (abbreviation: Hz). $1 \text{ Hz} = 1 \text{ sec}^{-1}$. It is also known as a *cycle per second* (c/s).

Heteropolar Machine. An electromagnetic machine in which the conductors pass successively through magnetic fields of opposite sense (cf. HOMOPOLAR MACHINE).

Hexagon Voltage. The voltage between any two consecutive lines of a symmetrical six-phase system. Also known as *line voltage* or *mesh voltage*.

High Energy Rate Forming. The forming of metals by means of high-intensity shock waves which are usually generated in water and develop pressures approaching 7000 MN/m². The energy source may be the discharge of a high-voltage capacitor, the magnetic field of a coil placed parallel to the work-piece, or an explosive. The magnetic technique is considered the most promising, but the spark machine with a capacitor is cheaper.

High-Frequency Heating. The production of a temperature rise in a material by causing high-frequency currents to flow in the material. Very rapid heating can be achieved as compared with the method of producing a temperature rise by transference of heat energy into the material through the surface, as when heated by a flame or by contact with a hot body. INDUCTION HEATING is employed for a conducting material, and DIELECTRIC HEATING for an insulator.

High-Frequency Induction Furnace. An INDUCTION FURNACE which makes use of the fact that a high-frequency current at high voltage passing through a water-cooled coil of copper tubing induces eddy currents in the metal within the coil (see Fig. H–6). It is largely used for the production of high-grade steel and high-melting-point alloys. It is sometimes referred to as a *coreless induction furnace*.

High-Frequency Power Generator. A device to produce power up to about 100 kW for HIGH-FREQUENCY HEATING. It may be electronic, or may comprise a high-speed rotating machine. The latter is commonly called an INDUCTOR ALTERNATOR.

High-Rupturing-Capacity Fuse

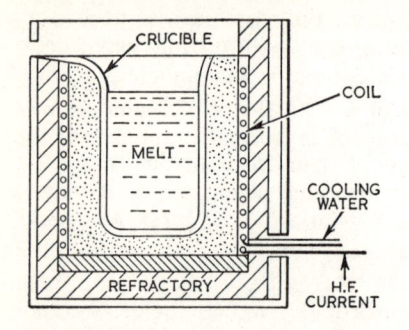

FIG. H-6. HIGH-FREQUENCY, OR CORELESS, INDUCTION FURNACE.

High-Rupturing-Capacity Fuse. A high-performance cartridge fuse, known also as a *high-breaking-capacity fuse* (h.r.c. or h.b.c). Its minimum rupturing capacity is normally in excess of 16,500 A. It is designed to limit current to a safe value (i.e. "cut-off") and to act rapidly.

High Voltage. A term which, when employed in official documents, normally implies a voltage exceeding 650 V.

Hilborn Loop Test. A modified LOOP TEST similar to the FISHER LOOP TEST. It is of value when two sound conductors of cross-section different from that of the faulty conductor have to be used.

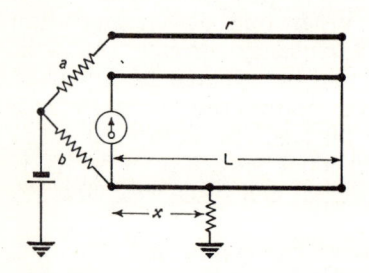

FIG. H-7. HILBORN LOOP TEST FOR A FAULTY CONDUCTOR.

In Fig. H-7 at balance, x, the distance of the fault along the cable is given by

$$x = L\frac{b}{a+b+r}$$

Resistance r may be ignored if small compared with $a+b$.

Ho and Koto Oscillograph. A mechanical oscillograph employing an electrostatic element. It is otherwise similar in operation to the DUDDELL OSCILLOGRAPH.

Hochstädter Cable. A screened multicore cable used for electricity supply. The insulation of each conductor is separately enclosed in a conducting film so that there is a radial electric field surrounding the conductor (Fig. H–8). The metallic sheath of the

FIG. H–8. HOCHSTÄDTER MULTICORE SCREENED CABLE.

cable is connected electrically with each of the films. In this way, tangential stresses are eliminated, so enabling the design to be used up to 50 kV.

Hole. Concept of a positive charge carrier in conduction by a *p*-TYPE SEMICONDUCTOR. It represents the deficiency of a valency electron in an atom of ACCEPTOR IMPURITY.

Hollow Conductor. Tubular conductor used (*a*) for large-current high-frequency conductors where the skin effect would render any interior regions inoperative, (*b*) in large turbo-generators to facilitate direct cooling by air, hydrogen or water, and (*c*) on high-voltage transmission lines where the diameter of the conductor required for elimination of corona loss is significantly greater than that required for current conduction or mechanical strength.

Home-Office Switch. A switch in which all the metal parts are covered by insulating material or earthed, to guard against the possibility of electric shock.

Homopolar Machine. An electromagnetic machine, rarely used, in which there is only one active pole, compared with the two (or multiple of two) in the common HETEROPOLAR MACHINE. In principle the machine consists of a conductor so disposed in a magnetic field that, in motion, it cuts through the magnetic field continuously in one direction.

Hopkinson Tariff. A maximum-demand tariff, i.e. TWO-PART TARIFF comprising a kilowatt or kVA charge and a unit charge.

Hopkinson Test

Hopkinson Test. A BACK-TO-BACK TEST for two similar machines. Fig. H–9 shows the connections for testing two d.c. machines. These machines are coupled mechanically, and joined electrically in parallel to a d.c. supply which compensates for the losses. One machine acts as a motor to drive the other as a generator with equal currents in the two machines.

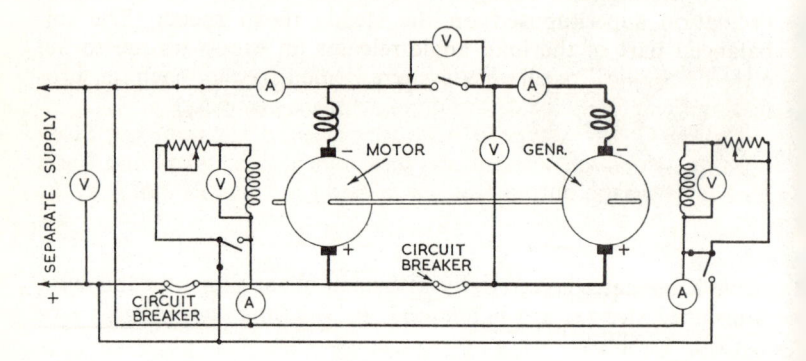

FIG. H–9. DIAGRAM OF CONNECTIONS FOR HOPKINSON TEST ON TWO D.C. MACHINES.

Horizontal Plugging. Method of switchgear busbar selection in which the connections are carried on a framework that moves horizontally (cf. VERTICAL PLUGGING).

Horn Gap. A ROD GAP having extended electrodes up which the arc may travel owing to its heat, thereby increasing its length so that it becomes self-extinguishing.

Horsepower. A practical unit of power, equal to 33,000 ft-lb/min, 42·42 Btu/min or about 746 **W (J/s)**.

Hot-Cathode Rectifier. Type of MERCURY-ARC RECTIFIER (q.v.).

Hot-Wire Instrument. Indicating instrument incorporating a fine wire that carries the current to be measured, the expansion produced being mechanically amplified by allowing the sag to operate a spring-controlled pointer. It has a square-law scale. True r.m.s. value is measured on alternating current; it also reads direct current. It is therefore of application up to high radio frequencies, but its limited robustness and its liability to burn-out restrict its use. It is also known as a *thermal-expansion instrument*.

Hunter. A form of SYNCHRO having three-phase windings on both stator and rotor. It is used in servo systems especially for operating hydraulic valves and indicating differences in angles.

Hunting. An oscillatory phenomenon with particular reference to synchronous machines. Under steady load, the driving and load torques of a synchronous machine balance for a given load angle. Any sudden change of condition will set up a transient oscillation superimposed on the steady mean speed. The unbalanced part of the load angle releases an excess torque to act on a mechanical oscillatory system formed by the machine, producing the possibility of hunting (or phase-swinging).

Hybrid Coil. A type of transformer used for reducing SIDETONE in telephone subscribers' sets, and also in telephone lines. In Fig. H–10 the output from amplifier 1 induces an e.m.f. in the

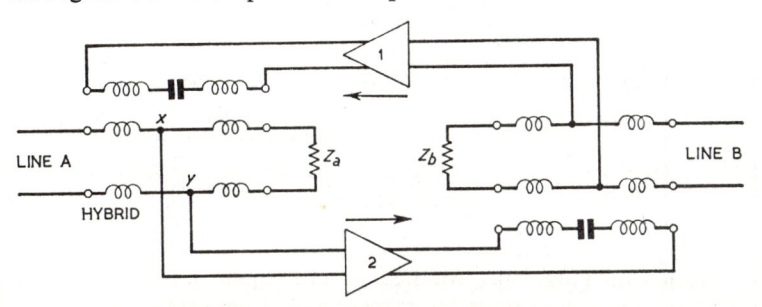

FIG. H–10. APPLICATION OF HYBRID COILS.

secondary, which impresses equal voltages on line A and its balancing impedance Z_a. There will therefore be no voltage across points xy, and no signal will be fed into amplifier 2. A voltage arriving on line A will, however, be transferred to the amplifier 1 for transmission into line B. The output of one amplifier is thus isolated from the input of the other, at the expense of half the output power because it is equally shared by the line and its balance impedance.

Hydro-Alternator. Alternator used in conjunction with a WATER TURBINE in hydroelectric generating plant. As a water turbine is a low-speed machine compared with a steam turbine, the hydro-alternator is much larger and heavier than one of equivalent capacity for use in a steam generating station. It is frequently convenient to build the units with vertical shafts.

Hydroelectric Generating Plant. Generating equipment which is driven by WATER TURBINES, the water being supplied from a reservoir (possibly pumped-storage) or tidal barrage.

Hydrogen Cooling. A method of cooling large electrical machines of the turbo type which can readily be enclosed. Hydrogen, compared with air, has 1/14 of the density, reducing windage loss and noise; 14 times the specific heat; 1½ times the heat-transfer capability, so more readily absorbing and giving up heat; and 7 times the thermal conductivity, so reducing temperature-gradient. It also has corona-reducing properties, and will not support combustion so long as the proportion of hydrogen to air in a mixture exceeds 3:1. As a result, hydrogen cooling at 100, 200 and 300 kN/m² (1, 2 and 3 atmospheres) can raise the rating of a machine by 15, 30 and 40% respectively.

Hydrometer. Instrument for determining the specific gravity of a liquid, e.g. the electrolyte of a battery.

Hysteresis. A phenomenon in which the effect due to an impressed stimulus is two-valued; a lower value when the magnitude of the stimulus is rising, and a higher one when the stimulus has a falling magnitude. The effect is observed in gears ("backlash"), springs, some insulating materials, and particularly in the magnetization of ferromagnetic materials.

Hysteresis Loop. See B/H CURVE.

Hysteresis Loss. The loss in energy when magnetic material is taken through a complete cycle of magnetization. Its magnitude is proportional to the area of the hysteresis loop.

I

I-Pot. Alternative term for INDUCTIVE POTENTIOMETER (q.v.).

I^2R Loss. The rate of energy loss as heat in a conductor of resistance R carrying a current I. The current I generally represents, for simple a.c. cases, the r.m.s. value, and the resistance R must then be strictly constant, or such value as will properly give the mean rate of energy loss.

Ideal Transformer. A transformer devoid of loss in windings and core, with perfect coupling between primary and secondary windings, and with a magnetic circuit of zero reluctance. It is a useful concept in the theoretical analysis of magnetically coupled electric circuits.

Idle Component. Alternative name for REACTIVE COMPONENT (q.v.).

Ignition Coil. Induction coil employed with internal-combustion engines to convert the l.t. current supplied by the battery into the h.t. current required by the sparking plugs.

Ignitron. A single-anode, pool-cathode MERCURY-ARC RECTIFIER. The ignitron differs from the other mercury-arc rectifiers (single and multi-anode) in the method used for ignition of the arc. As in the case of the thyratron, the arc, once started, cannot normally be extinguished except by interrupting the anode supply or by the anode-to-cathode voltage becoming less than the arc drop (as would happen every cycle with an a.c. supply). When ignitrons are used in a rectifier circuit, controlling the phase position of the igniter pulse relative to the positive half-cycle of the a.c. supply to the anode varies the conduction period as required. This is analogous to the grid control of a thyratron.

Ilgner System. A method of smoothing out the peak demands on a motor/generator system, used in conjunction with WARD LEONARD CONTROL. A heavy flywheel is mounted on the d.c. generator shaft and arranged to give up some of its stored energy during the peak-load period.

Image Impedance. Impedance which, when connected to the terminals of a network, equals the impedance presented to it. There are thus no reflection losses at the junction.

Immersible Switchgear. Switchgear capable of working indefinitely without detriment while submerged to a considerable depth in water.

Immersion Heater. A heater, usually for domestic water heating, in which the heat is supplied by a metallic resistor completely insulated from the water. It is fitted into the hot-water supply tank.

Impact Grinding. A form of machining in which a tool of the necessary profile is vibrated longitudinally and cuts the work by means of an abrasive. If the frequency is ultrasonic, using a magnetostrictive transducer driving a drill that vibrates longitudinally through a solid stub, the work can be drilled with an abrasive by a microscopic chipping process. A frequency of about 20 kHz is employed, the stub being tuned mechanically to vibrate at the designed frequency. The amplitude of vibration is of the order of 0·025 mm.

Impedance

Impedance. A network function defined as the ratio voltage/current (symbol Z; unit, ohm (Ω)). Where both current I and voltage V are sinusoidal and represented as complexors (vectors) in r.m.s. terms, the impedance operator is

$$Z = V/I + Z \angle \phi,$$

where ϕ is the angle by which the voltage leads the current. Then $Z \cos \phi = R$, the resistance, and $Z \sin \phi = X$, the reactance. The generality of the impedance as a voltage/current ratio is seen in the CHARACTERISTIC IMPEDANCE of a transmission line, the ITERATIVE IMPEDANCE of a quadripole, the INTRINSIC IMPEDANCE of a material in which an electromagnetic wave is travelling, or the WAVE IMPEDANCE of a material in which incident and reflected electromagnetic waves co-exist.

Impedance Bond. An inductive form of CONTINUITY BOND used at the extremities of a railway track circuit so that d.c. traction current will pass through it conductively unhindered, while the a.c. track-circuit currents are presented with a high reactive impedance.

Impedance Drop. The decrease in output voltage of a circuit due to the internal IMPEDANCE VOLTAGE within the circuit.

Impedance Triangle. A vector triangle formed by R, X and Z (resistance, reactance and impedance) as in Fig. I–1. Using complex algebra, the complex impedance operator

$$Z = R + jX$$

The modulus $Z = \sqrt{(R^2 + X^2)}$.

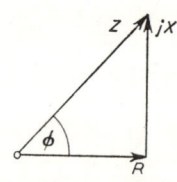

FIG. I–1. IMPEDANCE TRIANGLE.

Impedance Voltage. The voltage due to a current flowing through the impedance of a circuit. It is the vector sum of the RESISTANCE VOLTAGE and the REACTANCE VOLTAGE.

Impregnated Carbon. A carbon rod for use as an electrode in a carbon arc lamp. The carbon is mixed with other materials to give a required colour to the arc.

Impregnation. Of an insulating material, substitution of the air between its fibres by a suitable substance (such as a wax or varnish). The impregnating substance may not necessarily fill completely the space between insulated conductors.

Impulse Circuit-Breaker. Oil circuit-breaker, having a low oil content, in which an oil blast to extinguish the arc is produced mechanically by means of a spring or compressed-air operated piston. Simultaneously, the operating mechanism opens the contacts and the oil-blast is directed across the arc towards an insulating baffle where the arc is split and extinguished.

Impulse Clock System. A self-contained clock system which is characterized by the fact that the minute hands of the slave clocks move round in half-minute steps, under the control of a master-clock.

Impulse Generator. A device for producing high-voltage or heavy-current surges. It usually consists of a number of capacitors that are charged in parallel. For a high-voltage surge they are discharged in series; for a heavy-current surge they are discharged in parallel.

Impulse Ratio. The ratio between the breakdown voltage of some insulating material when an impulse voltage is applied to that when it is subjected to a sustained **50 Hz voltage.**

Impulse Test. The application of a single voltage pulse obtained by a capacitor discharge. Such a test is specified by the peak value of the voltage wave, the time in microseconds taken to reach 90% of peak value from 10% of that value and the time in microseconds for the voltage to decay to 50% of peak voltage. The classification of impulse tests is similar to that of tests using alternating voltages (i.e. WITHSTAND TEST, FLASHOVER TEST, PUNCTURE-WITHSTAND TEST), except that the duration of the test is specified by the number of impulses applied.

Impulse Turbine. (1) A STEAM TURBINE in which velocity compounding is employed with each stage of expansion, associated with two or more rows of moving blades separated by fixed guides to redirect the steam into the next row of moving blades without shock but with unaltered velocity (see Fig. I–2).

(2) A WATER TURBINE such as the PELTON WHEEL (q.v.).

Impulse Voltage, Current. A transient, unidirectional voltage or current lasting only a few microseconds. An impulse voltage is frequently used in the high-voltage testing of electrical equipment.

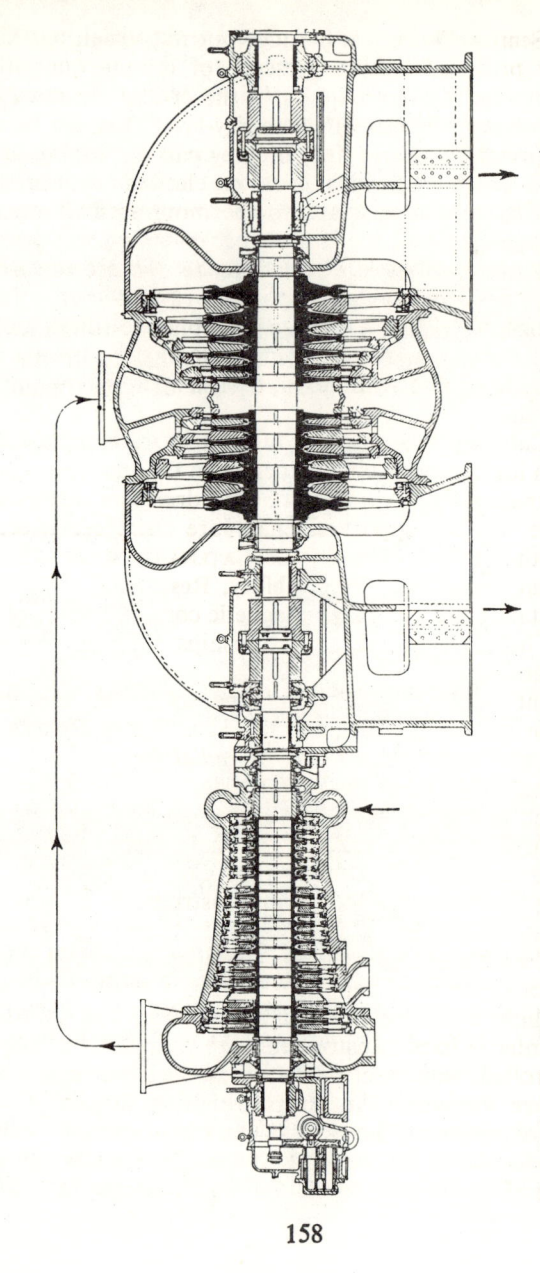

Fig. I-2. Sectional Arrangement through Two-Cylinder Impulse Steam Turbine of 60 MW Output. (*The General Electric Co. Ltd.*)

Indirect-Arc Furnace

Impurity Semiconductor. A SEMICONDUCTOR which has its low conductivity provided by the presence of minute quantities of foreign atoms or by deformations in the crystal structure. The impurities "donate" electrons of energy-level that can be raised into a conduction band (n-type), or they can attract an electron from a filled band to leave a hole, or electron deficiency, the movement of which corresponds to the movement of a positive charge (p-type).

In-Phase Component. Alternative name for ACTIVE COMPONENT (q.v.).

In-Situ Conduit. Passages formed in concrete foundations by means of inflatable pneumatic tubing. When the concrete is set, the tube is deflated and withdrawn, a "fish wire" being pulled in at the same time.

Inclined Catenary Suspension. In electric traction, a form of construction for overhead conductors when the support structures are on alternate sides of the track. The droppers attaching the contact wire to the support catenary are therefore necessarily inclined to the vertical. See SIMPLE CATENARY SUSPENSION.

Incremental Inductance, Permeability, Resistance. A measure of the inductance, etc., in a ferromagnetic core, inductor, or other component having non-linear relationships between the appropriate properties.

Independent Feeder. A FEEDER used solely for supply to a substation or a feeding point, and not as a TRUNK FEEDER. It is also known as a *dead-ended feeder* or *radial feeder*.

Independent-Trip Circuit-Breaker, Starter. A circuit-breaker or starter which is tripped with the aid of an auxiliary electricity supply, regardless of the state of the main circuit in which the breaker is connected (cf. DIRECT-TRIP CIRCUIT-BREAKER).

Indicating Instrument. Measuring instrument in which the value of the quantity being measured is indicated by the relative positions of a pointer and scale, or a similar means. It normally comprises an electrical system with a moving pointer and a fixed scale, but there are some instruments in which the scale moves and the pointer is fixed. Instruments are normally balanced and spring-controlled, with steel pivots in jewelled bearings.

Indirect-Arc Furnace. An electric melting furnace having a cylindrical or spherical body with axially disposed horizontal electrodes between which the arc is struck. The furnace and charge is rocked backwards and forwards, so tumbling the charge to

Indirect Stroke

present a fresh surface to the heat source and to wash molten metal over heated refractory. The furnaces are single-phase, and so special equipment is usually required to take a balanced load. Malleable iron, alloy steels, non-ferrous alloys, etc., are melted in this type of furnace. See MOISSAN ARC FURNACE, HÉROULT ARC FURNACE, RADIANT-ARC FURNACE.

Indirect Stroke. A stroke of lightning which, while not striking a transmission system, affects it by inducing a voltage in it.

Induced-Draught Ventilation. A form of machine ventilation in which the ventilating air is drawn through the machine by some means external to the machine itself (cf. FORCED-DRAUGHT VENTILATION, SELF-VENTILATION).

Induced E.M.F. Electromotive force produced in a circuit by a change in magnetic flux through the circuit (see ELECTROMAGNETIC INDUCTION).

Induced Moving-Magnet Instrument. An instrument consisting of a movable piece of magnetic material which is deflected by the resultant field produced by a fixed current-carrying coil and a permanent magnet at an angle to the coil.

Induced-Overvoltage Test. An overvoltage test in which a transformer is overexcited at an enhanced supply frequency, to a voltage about 67% greater than the nominal system voltage. As the overvoltage is internally produced, it is automatically graded along the winding in conformity with the grading of the insulation.

Inductance. The ratio of the magnetic flux-linkage of an electric circuit and a current. If the current concerned is that of the circuit itself, the inductance is a SELF-INDUCTANCE property; if the linkage results from the current in another circuit, the two circuits possess the property of MUTUAL INDUCTANCE.

Induction Coil. An early form of open-cored transformer for developing damped high-voltage transients from a d.c. primary supply by means of an interrupter. A descendant in common use is the IGNITION COIL employed in petrol engines. It is also known as a *spark coil* or *Ruhmkorff coil*, after the inventor.

Induction Furnace. An electric furnace for melting metals, the heat being produced in the charge itself. There are two types, the HIGH-FREQUENCY INDUCTION FURNACE and the LOW-FREQUENCY INDUCTION FURNACE. In both, the principle is the use of an alternating current in the primary circuit to induce a current in the metal to be melted, which acts as the secondary. The

Induction Instrument

advantages of the induction furnace are the ability to use special atmospheres or vacuum melting (avoiding contamination from electrodes or furnace gases), and the close control which can be exercised over the melting operation. The electromagnetic action of the alternating current gives efficient circulation of the melt.

Induction Generator. A GENERATOR similar in construction and operation to an INDUCTION MOTOR. An induction motor runs at a speed slightly below synchronism and draws power from the supply to drive its load. If the load is replaced by a prime mover and the speed is increased above synchronism, the machine will then operate as an induction generator and will supply power to the busbars whilst at the same time drawing its magnetizing current from them. In some circumstances the installation of such a machine offers several advantages over a synchronous generator, notably in the smaller sizes; as the output is increased the saving possible with the induction generator becomes of less importance and above about 5,000 kW the disadvantages of the scheme become more apparent.

Induction Heating. A form of HIGH-FREQUENCY HEATING applied to conducting materials. When electrically conducting material is placed in an alternating magnetic field, the voltage induced by transformer action causes alternating currents to flow in the conductor. These currents give rise to resistance losses that generate heat in the conductor. The power transferred into the work-piece depends on the coupling between it and the heating inductor. A close coupling allows about $1\frac{1}{2}$–6 mm ($\frac{1}{16}$ – $\frac{1}{4}$ in.) diametral clearance between the coil and the metal being heated. In these conditions, the power input to the work-piece is proportional to the square of the current in the heating inductor, and to the square root of the inducing frequency. See also INDUCTION FURNACE.

Induction Instrument. An indicating instrument in which two magnetic fields are produced in one or two adjacent iron circuits excited by current, the two magnetic fluxes having a small phase-angle between them to produce initial movement and so to indicate on a fixed scale. Only alternating current of power frequencies can be measured, but as a wattmeter (see Fig. I–3) the type is very useful, since the scale arc may be about 300° and so be readily readable from a distance. This long scale arc also is conducive to the use of the instrument as a voltmeter and an ammeter; otherwise it is more expensive than a moving-iron instrument.

Induction Motor

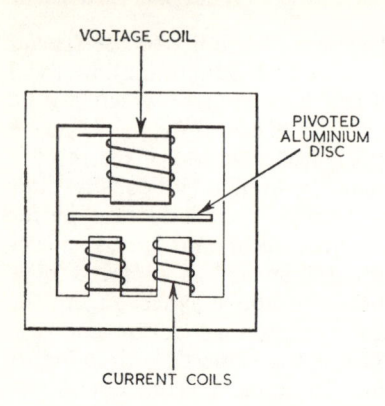

FIG. I-3. INDUCTION WATT-METER.

An induction instrument has the advantage of a long scale arc.

Induction Motor. An a.c. motor in which the current in the stator winding produces a rotating flux which induces a current in the winding of the rotor, thus producing the necessary torque. The stator winding to produce this rotating flux is fairly simple in the case of a polyphase motor, being based on a symmetrical winding. The two basic constructions are the SQUIRREL-CAGE MOTOR and the SLIP-RING MOTOR. These motors, particularly the cage type, are of simple construction, and since relatively little control gear is required, they are the types of electrical prime mover used most commonly. The majority of polyphase induction motors are designed for three-phase supplies, although occasionally two-phase electrical power supplies only are available.

A revolving flux can be produced in a single-phase motor by the combination of two alternating fluxes, displaced from each other in a circumferential direction, and out of time phase so that their peak values occur at different instants. The method of achieving this phase shift is obtained by different means in the CAPACITOR MOTOR, the SHADED-POLE MOTOR and the SPLIT-PHASE MOTOR.

Induction Pump. A type of ELECTROMAGNETIC PUMP in which the current in the liquid is induced by the field. Induction pumps are restricted to the handling of alkali metals. There are three main types operating on the induction-motor principle in which the liquid metal ("rotor") current is induced in proportion to the difference (or slip) velocity between the travelling wave of flux (created by the polyphase winding) and that of the liquid. The *spiral induction pump* shown in Fig. I-4 closely resembles the

Inductive Interference

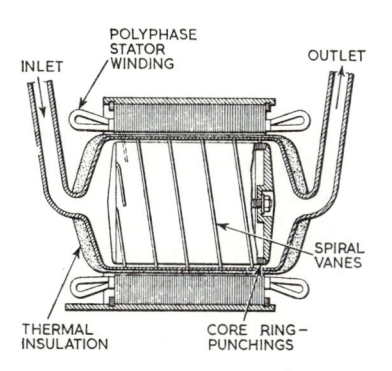

Fig. I–4. Spiral Induction Pump.

induction motor in appearance except that liquid metal, guided by spiral-vanes, fills the space between the stator and the now stationary "rotor" core-punchings. For higher flow applications the *linear* form of this pump is usually more suitable, in either the *flat* or the *annular* forms.

Induction Regulator. An electromagnetic device having a primary winding in parallel, and a secondary winding in series, with an a.c. supply. The voltage impressed on a load may be adjusted by changing the relative positions of the primary and secondary windings.

Induction Relay. A relay employing either a rotating disc or a rotating cup to operate the contacts. In the former case a metal disc is mounted so that it is free to rotate between the poles of two electromagnets. Torque is produced by the interaction of eddy currents, produced in the disc by one magnet assembly, on the main flux of the second electromagnet. The torques so produced by the two main fluxes with the respective eddy currents are additive.

Inductive Circuit. A circuit having self-inductance that is appreciable compared with its resistance.

Inductive Interference. Induction caused by a conductor, particularly of a power line, in a neighbouring conductor. It occurs because any conductor carrying a current and having a potential to earth is surrounded by magnetic and electric fields. The voltage *magnetically* induced in the neighbouring parallel conductor, which can be a telephone wire, depends upon the

Inductive Load

strength of the magnetic field linking the two conductors, the frequency, the resistivity of the earth, the length of the parallelism and the distance between the lines. *Electric* induction, however, is a function of the potential of the inducing conductor, the frequency, and the capacitances between the induced and inducing conductor and between the conductors and earth.

Inductive Load. See LAGGING LOAD.

Inductive Potentiometer. A precision-wound toroidal AUTO-TRANSFORMER with two sliding contacts. In its basic use the auxiliary contact is set for initial calibration, while the main contact is moved in accordance with some rotary mechanical input, to produce a corresponding voltage to an accuracy better than ±0·1%. Fig. I–5 shows the essential features of the "I-pot".

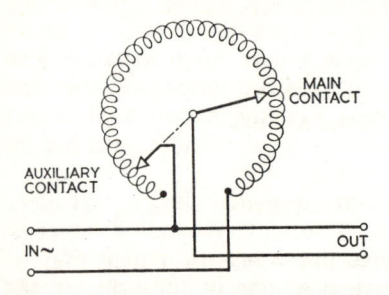

FIG. I–5. INDUCTIVE POTENTIOMETER.

Combinations of I-pots can be arranged to produce outputs that are fractions, multiples, and other functions, of the input.

It comprises in effect a low-loss variable auto-transformer of small leakage (and therefore low internal impedance). The input voltage can be divided accurately by the main sliding contact.

Inductor. A device whose primary characteristic is the property of INDUCTANCE. Its form is almost invariably a compact coil.

Inductor Alternator. A rotating high-frequency generator usually driven by a standard induction motor running at 3,000 rev/min. The two machines share a common frame, and are designed for outputs up to 500 kW and frequencies up to 50 Hz.

Industrial Capacitor. A capacitor distinguished by its large size, both physically and electrically. Two main forms of construction are employed, known as UNIT CAPACITORS and TANK CAPACITORS. The units are commonly standardized at 25 kVAr output single-phase, but there is a recent tendency to increase this to 50 kVAr, or even higher. Tank type capacitors are normally larger—up to at least 500 kVAr. In all these capacitors the dielectric is liquid-impregnated paper, the liquid being either a refined mineral oil or a synthetic chemical having suitable electrical properties.

Industrial Instrument. Measuring instrument of sufficient accuracy for normal commercial use. The description applies to instruments designed and constructed for panel-mounting or portable use in connection with general testing work where high accuracy is not required.

Inert Cell. A closed dry primary cell containing solid ingredients that form an electrolyte only when water is added.

Inertia Constant. A measure of flywheel effect. The inertia of the rotating parts of an electrical machine (with its prime mover or mechanical load) affects its stored kinetic energy when running. A large and heavy machine will therefore take a considerable time to start and stop. If sudden load changes are not to cause excessive change of speed, the rotating parts must have not less than a minimum inertia. The requisite flywheel effect is expressed in terms of an inertia constant H, defined as

$$H = \frac{\text{stored energy, joules}}{\text{rating, volt–amperes}}$$

The values of H normally required are between 2 and 9 sec.

Infinite Busbar. A busbar system of rigidly constant voltage and frequency. The behaviour of a single synchronous or asynchronous machine connected to a supply system fed by an extensive network of generators may, to an approximation, be considered as working on a infinite busbar system. Then no change of load will affect the terminal conditions of voltage or frequency. The conditions of an infinite busbar are never actually achieved in practice except under rare, rigidly constant conditions.

Influence Machine. Alternative name for ELECTROSTATIC GENERATOR (q.v.).

Infra-red Heating. Production of heat by means of infra-red rays. These rays may be produced electrically by utilizing a "dull-emitter" resistance element operating at black heat, but it is more usual to employ a special infra-red tungsten filament lamp, designed to produce "light" at a wavelength within the infra-red band; such lamps are usually of 250 W rating. The principal advantage of the infra-red lamp is that heat is transferred directly to the object to be heated without raising the temperature of the

Inherent Regulation

intervening atmosphere; consequently the heating of the object is rapid.

Inherent Regulation. The change in output terminal voltage of a generator or converter when the load is reduced from the full rated value to zero, all other conditions remaining constant. It is usually expressed as a percentage of the normal value.

Inhibited Oil. Transformer or switch oil, the deterioration of which during its working life is retarded by the use of anti-oxidants, particularly oxidation *inhibitors*. These are usually of the phenolic or amino types, which convert chain-forming molecules in the oil into inactive molecules, being gradually consumed in the process.

Instantaneous Value. The magnitude, at any given instant in time, of a time-varying quantity such as voltage, current, charge, flux, etc. A succession of instantaneous values defines a waveform.

Instrumentation. The whole science and art of measurement. The term embraces both the production of the instruments themselves and their application to a measuring problem, with or without automatic control.

Instrument Transformer. A TRANSFORMER used to reproduce the voltage and current values of a circuit, usually a power or test circuit, in true proportion and phase relationship but of dimensions suitable for application to measuring or protective instruments (see CURRENT TRANSFORMER, VOLTAGE TRANSFORMER).

Insulance. Alternative term for INSULATION RESISTANCE (q.v.).

Insulated System. A distribution system that has no point connected to earth.

Insulated-Return System. A supply system for electric traction in which both the outgoing and the return conductors are insulated from earth.

Insulating Material. Material offering high resistance to the passage of electric current. Other important electrical characteristics of an insulating material are its DIELECTRIC STRENGTH and PERMITTIVITY. Insulating materials are of all forms: gaseous, liquid and solid; organic and inorganic; natural, refined, manufactured or synthesized. They vary from air to expensive and rather rare minerals, such as MICA. Organic insulating materials range from wood, used in a dry state, to products of complicated conversion processes (e.g. rayon) and others entirely synthetic, such as PHENOLIC RESIN. An insulating material may also be known as a *dielectric* or *insulant*.

Integrating Wattmeter

Insulating Oil. Oil used as an insulating medium in transformers and switchgear. It is a pure hydrocarbon mineral oil, clean and free from matter likely to impair its properties, and without additives. It may, if agreed between seller and buyer, contain an oxidation inhibitor.

Insulation Classification. The classification of insulating materials in terms of their operating temperature limitations.

Insulation Grading. The adjustment of an electric field in dielectric materials in order to secure a better utilization of the properties of the materials, especially as regards their electric strength.

Insulation Resistance, Conductance. The resistance or conductance offered by an insulating material separating two conductors having a potential difference, or a live conductor and earth.

Insulator. A component made of insulating material, used to separate a conductor from earth or from another conductor. It often, at the same time, supports the conductor. Overhead-line insulators are made from either porcelain or toughened glass. See SHACKLE INSULATOR, PIN INSULATOR, POST INSULATOR, DISC INSULATOR.

Integrating Frequency Meter. Alternative name for MASTER FREQUENCY METER (q.v.).

Integrating Meter. A measuring instrument which adds up the value of the quantity measured with reference to time. A single-phase meter for metering power supplied has a voltage system consisting of E-shaped iron laminations energized by a multi-turn voltage coil, mounted above a second set of U-shaped laminations energized by a current coil carrying the load current. Between the two systems is an air-gap in which a flat circular disc mounted on a vertical spindle rotates in special bearings.

There is a difference in both space and phase between voltage and current flux, creating a moving field which does work upon the rotor system. Fig. I–6 shows the simplicity of the working parts of a single-phase two-wire meter. It is the basic unit for energy measuring devices for more complicated supply systems, such as three-phase three-wire and three-phase four-wire, where two- and three-meter units, respectively, are employed, working upon common rotor systems.

Integrating Wattmeter. Alternative name for WATT-HOUR METER (q.v.).

Integrator Cube, Sphere

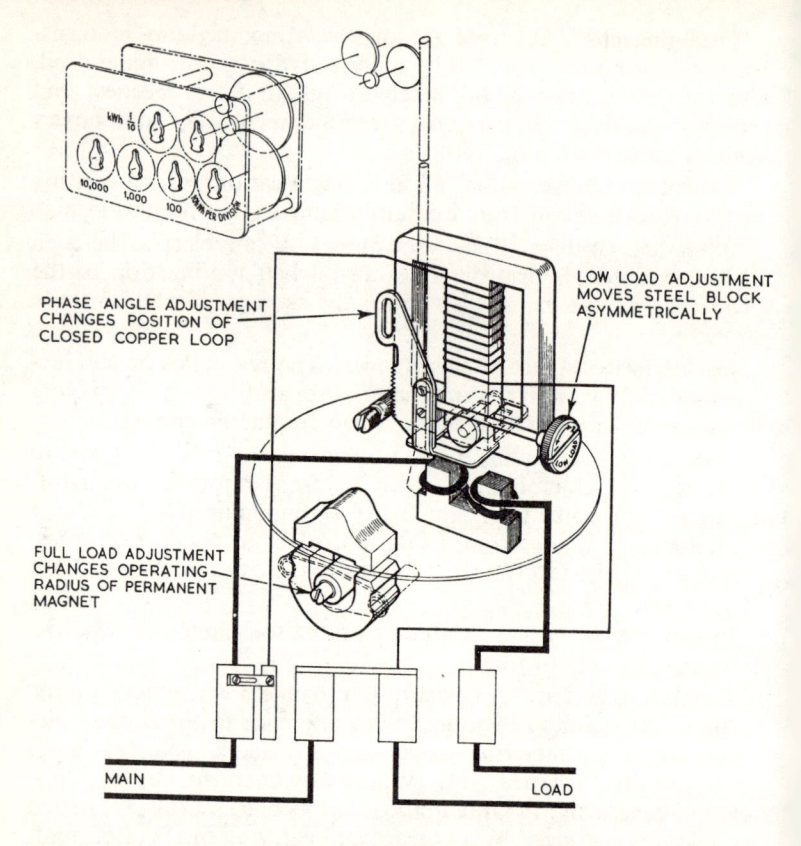

FIG. I-6. SINGLE-PHASE, TWO-WIRE, A.C. INTEGRATING METER.

Integrator Cube, Sphere. A hollow cube, or sphere, the smooth interior surface of which has a uniform flat, matt, white coating. If a lamp is suspended within the cube or sphere, the illumination at any point of the surface (including a small observation window) corresponds to the rate of total light-flux emission.

Intensity of Magnetization. Magnetic moment per cubic centimetre. Its symbol is M.

Interconnected Star Connection. Alternative name for a ZIGZAG CONNECTION (q.v.).

Interconnector. A cable or line which connects two sources of energy or two distribution networks. A distribution system has its reliability increased by multi-interconnections of feeders, but high capital cost restricts this to primary networks in densely loaded areas.

Interference Suppressor. Filter or attenuator fitted to apparatus that is capable of radiating interference signals, or to the terminals of the electric supply which might be a source of interference.

Intermodulation Distortion. A change in waveform, through the introduction of combination frequencies arising from two or more input frequency components, because of a non-linear response in the system.

International Ampere. The constant current which, when passed through a standard specified solution of silver nitrate in water, deposits silver at the rate of 0·001118 g/sec.

International Ohm. The resistance, at the temperature of melting ice, of a column of mercury of 14·4521 g mass, uniform cross-sectional area and 106·300 cm long. It is also known as a *standard ohm* or *B.O.T. ohm*.

International Volt. The potential difference which, when applied to a conductor of resistance one INTERNATIONAL OHM, produces a current of one INTERNATIONAL AMPERE.

Interphase Reactor. Alternative name for ABSORPTION INDUCTOR (q.v.).

Interpole. Auxiliary pole incorporated in a commutator motor, mounted on the magnet frame midway between the main poles and excited by a winding connected in series with the armature ciruit. As the commutator moves relative to the brushes, reversal of current occurs in those coils connected to the commutator segments that are short-circuited by the brushes. This reversal is delayed by the self-inductance of the coils and the effect on the current-sharing, between the short-circuited segments, of the resistance current characteristic of the copper–graphite brush material. A delay in current reversal leads to bad commutation indicated by sparking at the trailing edge of the brushes. This delay can be corrected by inducing into the shorted coils an e.m.f. in opposition to the self-induced e.m.f. This is achieved by the use of interpoles whose polarity is opposite to the main pole in advance. They are also known as *compoles* or *commutating poles*.

Intertripping. The operation of all the circuit-breakers connected with a unit by means of the functioning of a protective

Intrinsic Conduction

relay. It may relate to breakers situated in the same station or separated by many miles, as in the case of those associated with feeder transformers.

Intrinsic Conduction. Conduction in a SEMICONDUCTOR by equal numbers of electrons and holes, i.e. by electron-hole pairs, as shown in Fig. I-7. It is outweighed, in practice, by EXTRINSIC CONDUCTION.

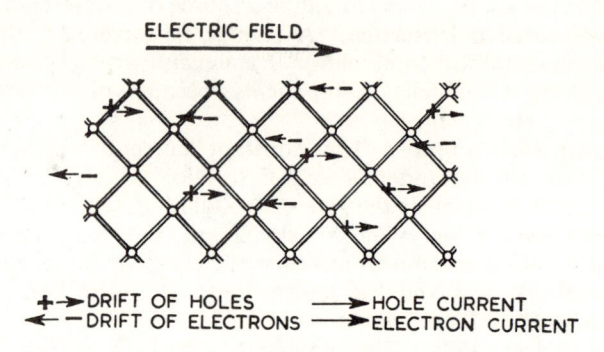

FIG. I-7. INTRINSIC CONDUCTION.
Plane diagram of a diamond-type crystal lattice.

Intrinsic Impedance. The ratio of the electric to the magnetic field intensities in a plane electromagnetic wave in any medium or in space. For a given medium of conductivity σ, permittivity ϵ and permeability μ, the intrinsic impedance for a sine-shaped travelling electromagnetic wave of angular frequency ω is

$$Z_0 = \sqrt{[j\omega\mu/(\sigma+j\omega\epsilon)]}.$$

Intrinsic Strength. The maximum possible DIELECTRIC STRENGTH of a material. It is the stress that would be withstood if breakdown from all causes (such as internal voids, contaminants and stress concentrations due to electrode configuration) were removed. In practice this value is never attained.

Intrinsically Safe. Term applied to apparatus or a circuit to signify that any sparking that may occur in it is incapable of causing ignition of any specified inflammable gases or vapours.

Inverse-Speed Motor. Alternative name for a SERIES-CHARACTERISTIC MOTOR (q.v.).

Ionization Chamber

Inverse Square Law. A law of common occurrence in electrical engineering. It applies to the force between point electric charges, the force between magnetic poles, intensity of illumination at a distance from a light source, etc. The force or intensity in these instances varies inversely as the square of the distance.

Inverse Time Lag. A device which delays the operation of equipment for a period which is inversely proportional to the magnitude of the operating force.

Inverse Voltage. Of a multi-anode three-phase RECTIFIER, the voltage, with respect to the common cathode, of the anode which, at a given instant, is not conducting. The rectifier must be able to withstand the peak inverse voltage without BACKFIRE.

Inverter. The reverse of a CONVERTER; it changes a direct-current input to an alternating-current output. High-power inverters are usually grid-controlled mercury-arc rectifiers connected so that the grid circuit and the d.c. output polarity are reversed from that necessary for normal rectification.

Ion. An atom or radical possessing a larger or smaller number of electrons than its normal content and consequently carrying a charge of negative or positive electricity. The conduction of electricity in material bodies is due to the motion of ions or electrons, or both.

Ionic Breakdown. A form of DIELECTRIC BREAKDOWN that occurs when the voltage stress is raised to such a level that a corona discharge occurs in small voids within the material. The bombardment of the sides of a void causes erosion of the walls at a rate depending on the energy imparted to the electrons.

Ionic Theory. In an electrolyte, the dissolved substance is dissociated into equal numbers of positive and negative ions, the liquid thus remaining electrically neutral.

Ionization. The production of IONS from electrically neutral atoms or groups of atoms by the addition or removal of electrons.

Ionization Chamber. A gas-filled enclosure used in particle physics. It comprises an enclosed vessel through which radiation is directed, and in which there are two electrodes. Commonly one is a central wire and the other a surrounding cylinder. This construction is effectively a capacitor, and the presence of radiation reduces the effective insulation between the two electrodes so that more current will pass when a voltage is applied. The current, which is of the order of 1 pA, requires considerable d.c. amplification for its measurement.

Ionization Current

Ionization Current. The current resulting from the movement of ions in an electric field.

Ionization Gauge. An instrument for the measurement of high vacua. The electrons from a heated filament in a vacuum bulb are accelerated towards a collecting electrode, and the positive ions formed during their passage are collected on a third electrode, the current recorded varying with the pressure inside the bulb.

Ionosphere. The upper part of the atmosphere where ionization, mainly due to radiation from the sun, is sufficient to have a marked effect on radio transmission.

Iron. A cheap and plentiful ferromagnetic material (symbol: Fe; atomic weight 55·85; melting point 1,535°C; boiling point 3,000°C). In pure form it is applicable to steady-flux applications such as machine frames, poles, yokes and rotors, electromagnet poles and yokes, relay cores, etc., where hysteresis and eddy-current losses are immaterial. Its low resistivity is a disadvantage in a.c. applications.

Iron Loss. The loss in an iron material due to varying magnetization (see CORE LOSS).

Irradiation. The exposure of materials to ionizing radiation in order to improve or alter their properties. Rubbers can be vulcanized, the thermal properties of plastics improved, foods sterilized, and insects controlled by suitable irradiation.

Irwin Oscillograph. A mechanical oscillograph employing a thermal element. It is otherwise similar in operation to the DUDDELL OSCILLOGRAPH.

Isolator, Isolating Switch. A switch for isolating a circuit or disconnecting it from the supply, which may be operated only when no current is passing through it.

Isotope. Atom, the nucleus of which has the same number of protons as another, but a different number of neutrons. For example, uranium atoms have always 92 protons, but isotopes of uranium may be found with total atomic mass (i.e. number of protons plus neutrons) between 230 and 240, the most common being 238.

Iterative Impedance. Of a quadripole, that impedance which, when connected across one pair of terminals, produces a like impedance across the other pair.

J

j. Operator employed to denote rotation of a vector through 90°. $j = \sqrt{-1}$.

Jar. A unit of capacitance once used in the Royal Navy. It is equal to 1/900 μF. The name is derived from the LEYDEN JAR capacitor.

Jet-Wave Rectifier. A basically mechanical device for commutating an alternating current. The moving member is a mercury jet, projected downwards at constant velocity from a nozzle.

Johnson Noise. The most fundamental source of "noise" in electric circuits, particularly those containing amplifiers. It is a consequence of the thermal agitation of free electrons in metal, and is also called *thermal noise*.

Joint Box. See JUNCTION BOX.

Jointing Chamber. A chamber at which cables may be jointed.

Joubert Disc. A device applied, in a point-by-point measurement of the amplitude of a voltage, at a series of selected instants in a repeated cycle. It comprises a circular plate of insulating material, carrying a narrow metallic insert at one point on the rim. The disc is rotated synchronously. The insert then forms an intermittent contact from which a sample voltage can be picked off at the same instant in successive cycles of a given wave.

Joule. The fundamental unit of energy (abbreviation: J). It is given by the work done by a force of one newton acting through the distance of one metre; or the work done by the transfer of an electric charge of one coulomb through a potential difference of one volt. The joule has been adopted internationally as the unit of mechanical, electrical and thermal energy.

Joule Effect. Magnetostrictive effect in which a specimen changes in length and cross-section when magnetized longitudinally. The Joule effect in rods can be employed to produce longitudinal resonance which may be used in oscillators, filters and transducers.

Joule's Law. The rate of heating in a resistor of resistance R carrying a current i is given by $i^2 R$, at every instant.

Jumper. In an overhead line, a non-tensioned conductor connected across two anchor clamps to maintain electrical continuity. In electric traction, a *jumber cable* is an insulated cable connected across the gaps in a conductor rail for the same purpose.

Junction Box

Junction Box. A closed box, usually underground, to which are brought the ends of feeders and distributing mains for connection and protection.

Junction Diode. Semiconductor diode in which the "anode" and "cathode" are in large-area intimate contact.

Junction Law. A primary law of electric networks which expresses the continuity of current (see KIRCHHOFF'S LAWS).

Junction Transistor. Transistor consisting either of a zone of n-type semiconductor sandwiched between two p-type zones, or a zone of p-type semiconductor sandwiched between two n-type zones.

K

Kalium Cell. A dry PRIMARY CELL with a mercuric oxide/potassium hydroxide/zinc system. It will hold its voltage for long periods and at high temperatures.

Kaplan Turbine. A type of PROPELLER TURBINE in which the blades have an adjustable pitch enabling them to be accommodated to a wide range of operating conditions. Kaplan turbines are used in large sizes for heads of water as low as 6 m.

Kapp Vibrator. A set of three d.c.-type commutator armatures with permanent-magnet fields, each connected into one rotor phase circuit of a slip-ring polyphase induction motor to raise the power factor.

Keeper. A piece of ferromagnetic material placed across the extremities of a permanent magnet to complete the magnetic circuit.

Kelvin Balance. Electrodynamic measuring instrument in which the electromagnetic forces are balanced against the force of gravity using a travelling weight.

Kelvin Double Bridge. An adaption of the WHEATSTONE BRIDGE which is useful for the measurement of low resistances and four-terminal shunts. From Fig. K–1, the currents in S and R are equal at balance, so that

$$Si = Qi_1 - Wi_2 \; ; \quad Ri = Pi_1 - Vi_2.$$

If

$$W/Q = V/P,$$

then

$$S/R = Q/P = W/V.$$

The bridge can be balanced with $Q = W$ both fixed, and P and V varied together.

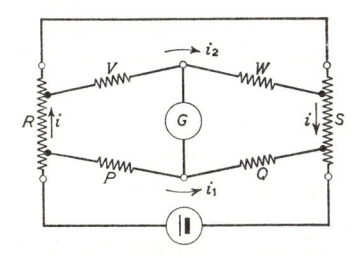

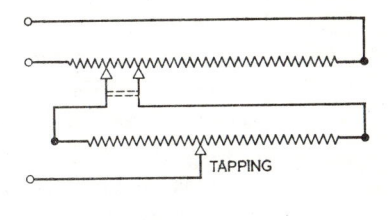

FIG. K-1. KELVIN DOUBLE BRIDGE.

FIG. K-2. TWO-DIAL KELVIN–VARLEY SLIDE.

Kelvin Effect. See THOMSON EFFECT.

Kelvin–Varley Slide. A form of electrical vernier, comprising a set of coils and a stud switch taking off a small fixed fraction of the overall voltage, which then feeds another set of coils with a single tapping, in the two-dial form (see Fig. K–2). The top dial has 101 coils each of 100 Ω and the lower has 100 coils of 20 Ω, totalling 2,000 Ω. The lower coils bridge two coils in the first dial so that the total resistance is effectively equal to 10,000 Ω. In this way sub-division to 1/10,000 of the main dial resistance (and potential difference) is simply obtained. Further subdivision is possible by the use of more dials similar to the first, in cascade.

Kelvin's Law. The most economic conductor area for a transmission circuit is that which makes the annual cost of the line losses equal to the annual charges on the capital cost of the line.

Kerr Cell. A pair of metal plates, forming a capacitor, sealed into a glass tank filled with nitrobenzine. If a beam of plane-polarized light is passed through the cell between the capacitor plates, with the cell flanked by a pair of crossed Nicol prisms, the field of view is dark. When a potential difference is applied to the capacitor, establishing an electric field across the path of the light, the plane of polarization is rotated and the field of view brightens.

Keyholder. A lampholder incorporating a switch. It is also known as a *switch-lampholder*.

Kilo-. A decimal prefix indicating multiplication by 10^3 (abbreviation: k).

Kilogramme. The reference standard of mass in the metric (M.K.S. and C.G.S.) unit system (abbreviation: kg). It is one of the seven primary units in the SYSTÈME INTERNATIONAL.

Kilowatt-Hour. A practical measure of energy. It is the energy expended in one hour when the power is 1,000 watts. It is equal to $3 \cdot 6 \times 10^6$ joules in S.I. units.

Kilowatt-Hour Meter. See INTEGRATING METER.

Kiosk Switchgear. The outdoor counterpart of CUBICLE SWITCHGEAR. It is used mainly for suburban distribution systems. It usually comprises a sheet-steel weatherproof kiosk which houses a step-down transformer with oil circuit-breaker for switching and protection, and a number of fuse-protected outgoing circuits for lower-voltage distribution. The kiosks are replaced or transferred as the load requirements change.

Kirchhoff's Laws. Two primary laws of electric networks.

The *Nodal* or *Junction Law* expresses the continuity of current, i.e. its circuital characteristic: the sum of the currents flowing into a circuit junction (or node) is zero. This is true at every instant, and can also be applied to r.m.s. values of current in complex (vector) form. Its statement is

$$\Sigma i = 0$$

with due regard to sign; i.e. if currents flowing towards the node are taken as positive, then those flowing away from it are negative.

The *Mesh Law* expresses the principle of energy conservation: the total source e.m.f. in a closed circuit (or mesh in a network) is equal to the sum of the voltage drops round the mesh, or

$$\Sigma e = \Sigma i Z.$$

In a single circuit with current i everywhere the same, ei represents the electrical output of the sources, and $i^2 Z$ the total rate of energy absorption by the rest of the circuit, and these must be equal.

Klydonograph. Apparatus on which a LICHTENBERG FIGURE is recorded. It is commonly built with multiple electrodes and with provision for both polarities, and will record for voltages of 3–15 kV.

Klystron. A velocity-modulated thermionic tube, utilizing the finite transit time of the electron in generating very-high-frequency oscillatory power. It comprises an electron gun (emitter cathode and focussing electrode), two tuned cavity resonators and a collector electrode.

Kraemer System

Korndorfer Starting. A method of starting motors which avoids the interruption of current that occurs with the simple AUTO-TRANSFORMER STARTER. This is achieved as shown in Fig. K–3. On the first step, (*a*), the motor accelerates at a reduced voltage determined by the transformer tapping. On the second step, (*b*), the star point of the transformer is opened so that the

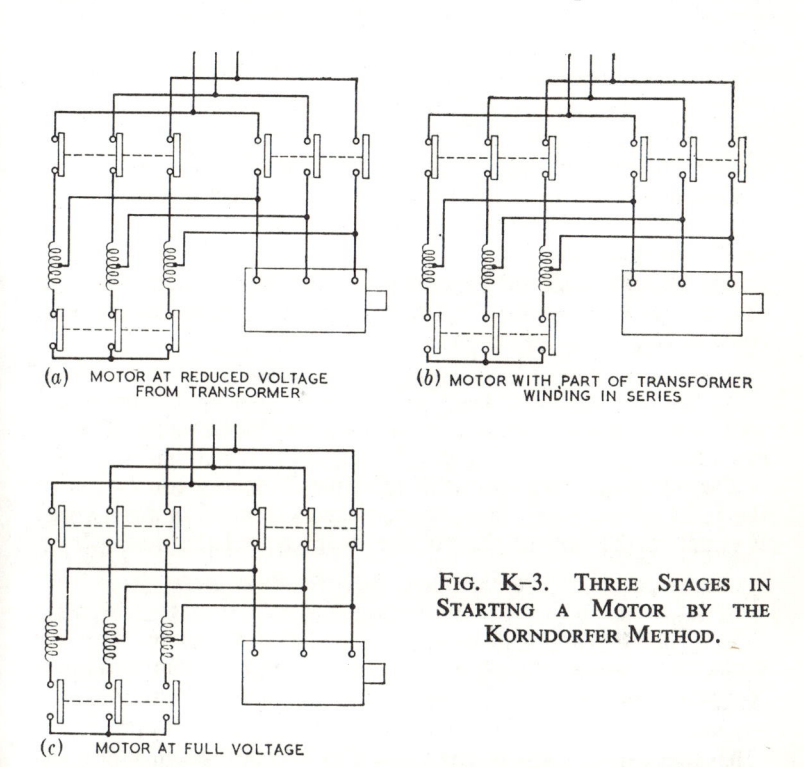

(*a*) MOTOR AT REDUCED VOLTAGE FROM TRANSFORMER

(*b*) MOTOR WITH PART OF TRANSFORMER WINDING IN SERIES

(*c*) MOTOR AT FULL VOLTAGE

FIG. K–3. THREE STAGES IN STARTING A MOTOR BY THE KORNDORFER METHOD.

motor continues to run with part of the transformer winding in circuit. Next, this part is short-circuited by the "run" contactor or switch, and finally the "start" contactor or switch is opened, (*c*).

Kraemer System. A system for regulating the speed of a large induction motor by using as an auxiliary machine a synchronous converter, to change the slip power from slip frequency to zero (i.e. direct current).

L

Lag. The fraction of a period, usually expressed as a corresponding fraction of the angle 2π, by which a given sinusoidal function (e.g. an alternating current) is displaced behind a reference function (e.g. an alternating voltage) of the same frequency.

Lagging Load. A reactive load in which the current at the terminals lags in phase behind the voltage at the same point. It is sometimes termed an *inductive load*.

Lagging Phase. In the two-wattmeter method of three-phase power measurement, the conductor in which the current at unity power factor lags behind the voltage associated with it in the wattmeter. The term is also applied to a three-phase circuit to indicate the phase whose voltage is lagging behind that of one of the other phases by 120°.

Lambert. Unit of brightness (abbreviation: L). It is the brightness of a perfectly diffusing surface (see below) emitting one lumen per square centimetre.

Lambert's Law. In illumination, a perfectly diffusing surface is one for which the luminous intensity per unit area in any direction varies as the cosine of the angle between the direction and the normal to the surface, so that it appears to be equally bright whatever the direction in which it is viewed.

Lamination. Of a machine or transformer, each of the thin, metallic sheets which form part of the core. The sheets are also referred to as *core plates*, *punchings*, or *stampings*.

Lamp. A means of converting electrical energy into light. There are three practical ways of doing this: (*a*) by means of the heating effect of an electric current which raises a fine filament of wire to a state of incandescence so producing light (see FILAMENT LAMP), (*b*) by an electric discharge through a gas which creates characteristic colours of light (see DISCHARGE LAMP), and (*c*) by a two-stage process whereby ultraviolet radiations are first generated and these in turn are changed into visible light through the process of fluorescence (see FLUORESCENT LAMP).

Landing Switch. A single-pole changeover switch with no "off" position, for use in a circuit that requires control from two or more positions. It is also known as a *two-way switch*.

Lap Welding. A form of RESISTANCE WELDING in which the parts being joined are overlapped. Varieties of lap welding include

Spot Welding, Projection Welding, Seam Welding and Roller-Spot Welding.

Lap Winding. A spread form of winding in an electrical machine in which successively connected coils, of approximately full pitch, overlap each other (see Fig. L-1). Cf. Wave Winding.

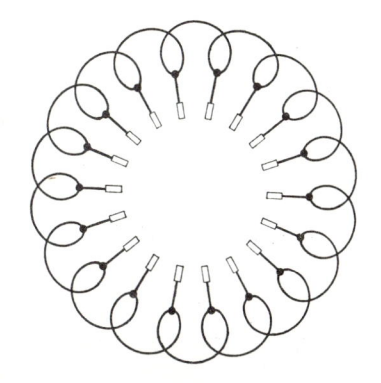

Fig. L-1. Lap-Wind Armature.

Laplace Equation. The equation in terms, for example, of the three rectangular co-ordinates x, y, z, which describes a function ϕ in the form

$$\frac{\partial^2 \phi}{\partial x^2} + \frac{\partial^2 \phi}{\partial y^2} + \frac{\partial^2 \phi}{\partial z^2} = 0$$

It applies to electric, magnetic and current-conduction fields, as well as to several other comparable physical systems.

Laplace Transform. Of a function $f(x)$ defined for all real positive x, the function $L(f) = F$ defined by

$$L(f)(s) = F(s) = \int_0^\infty f(x) e^{-sx}\, dx.$$

It is employed for transient analysis.

Larsen Potentiometer. An arrangement for the measurement of alternating potential differences. In Fig. L-2, where E_x is the unknown voltage, the primary winding of a variable mutual inductor and a resistor provided with tappings are connected in series to a source of supply E. E_x and E are usually derived from the same source since they must, of course, be of the same frequency. The current I may be regulated by a rheostat R_1. The induced e.m.f. in the secondary of the mutual inductor in series

Laser

with a tapped fraction of the potential difference on the resistor R is now balanced against the unknown E_x. The detector is conveniently a vibration galvanometer G, the sensitivity of which may be varied using R_2. If M and R are the readings of mutual inductance and resistance at balance, $E_x = I\sqrt{[R^2+(\omega M)^2]}$, since the components IR and $I\omega M$ are in quadrature. The phase angle ϕ between E_x and I is $\phi = \arctan(\omega M/R)$.

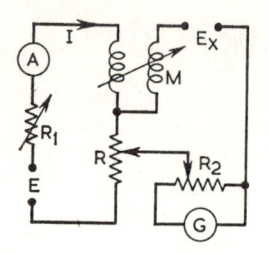

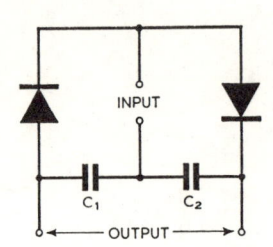

Fig. L-2. Larsen Potentiometer.

Fig. L-3. Latour Doubler.

Laser. Development of the MASER operating at light frequencies. Bursts of light illuminate a large crystal of ruby, stimulating the emission from one of the crystal faces of a very narrow and intense beam of *coherent light*, or light in a uniform plane and direction.

Latour Doubler. A basic form of VOLTAGE DOUBLER. The a.c. input is applied to the centre of the circuit in Fig. L-3. The potential differences across C_1 and C_2 each equal the peak input voltage. Therefore the output voltage across the two capacitors is double the peak input voltage.

Lattice Diagram. A graphical construction for the rapid assessment of surge propagation and reflection on a transmission network.

Lattice Network. A form of passive QUADRIPOLE.

Lattice Tower. A structure of metal bars to support, at a suitable height from the ground, the phase conductors of an overhead line. The sizes of towers necessary for grid lines of various voltages are indicated in Fig. L-4. It is popularly known as a *pylon*.

Lattice Winding. DISTRIBUTED WINDING in which the overhang is arranged to give diamond-shaped coils.

Fig. L-4. Double Circuit Lattice Suspension Towers.
The towers shown are, from left to right, 132 kV (175 mm² conductors) for 305 m (1000 ft.) span, 275 kV (2 × 175 mm²) for 366 m (1200 ft.), 400 kV (2 × 400 mm²) for 366 m (1200 ft.), and 400 kV (4 × 400 mm²) low height construction for 305 m (1000 ft.) span. The right-hand tower is for use near airfields or for amenity reasons.

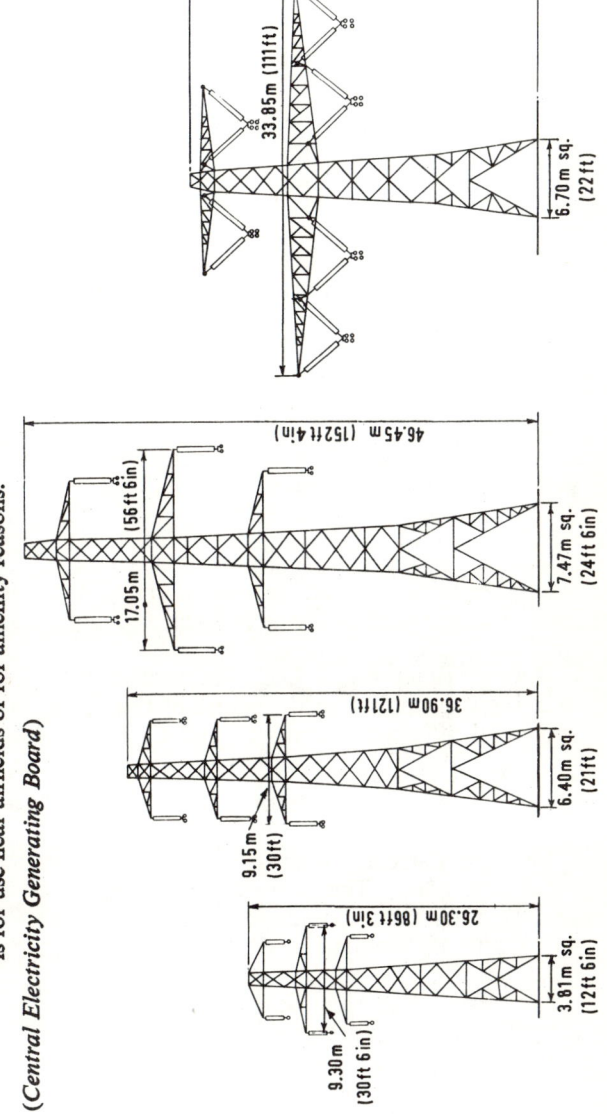

(*Central Electricity Generating Board*)

Lay. The axial length of one turn of the helix formed by a cable core or the strand of a conductor. The term is sometimes used to mean LAY RATIO.

Lay Ratio. The ratio of the LAY of a cable or conductor to the mean diameter of the helix.

Lead. The fraction of a period by which a given sinusoidal function (e.g. an alternating current) is displaced in advance of a reference function (e.g. an alternating voltage) of the same frequency. It is usually expressed as the fraction of the angle 2π.

Lead–Acid Cell. A storage cell or ACCUMULATOR in which the positive plates are lead peroxide (PbO_2), the negative plates pure lead in spongy form, and the electrolyte dilute sulphuric acid (H_2SO_4). During discharge the lead peroxide partially reduces, the spongy lead oxidizes, and both products combine with the sulphuric acid, to produce water and lead sulphate. During recharge the process is reversed:

$$PbO_2 + 2H_2SO_4 + Pb \leftrightarrow PbSO_4 + 2H_2O + PbSO_4$$

FORMED PLATES and PASTED PLATES are employed. The nominal voltage of a lead-acid cell is 2 V.

Lead-In Insulator. Alternative name for BUSHING (q.v.).

Leader Stroke. Of lightning, the discharge which initiates the track of the lightning channel.

Leading Load. A reactive load in which the current at the terminals leads in phase the voltage at the same point. It is sometimes termed a *capacitive load*.

Leading Phase. In the two-wattmeter method of three-phase power measurement, the conductor in which the current at unity power factor leads the voltage associated with it in the wattmeter. The term is also applied to a three-phase circuit to indicate the phase whose voltage is leading that of one of the other phases by 120°.

Leakage Current. A small FAULT CURRENT.

Leakage Flux. The part of the flux produced by a magnetic field that traverses non-useful paths. Leakage flux affects the field excitation required by d.c.-excited windings, and the leakage reactance of a.c. windings.

Leakance. The reciprocal of insulation resistance. It is also known as *leakage conductance*.

Leblanc Advancer. A PHASE ADVANCER comprising an armature with a commutator and three brushes per double pole-pitch spaced

120 electrical degrees. The brushes are connected to the slip rings of the main induction motor. The machine produces an electromotive force approximately proportional to the rotor current of the induction motor, and leading or lagging on it by 90° depending on the speed at which the armature is driven. No stator winding is needed, but a stator core may be provided to reduce the reluctance. Commutation is difficult, and as a consequence the advancer is limited in size.

Leblanc Connection. A method of interconnecting three single-phase transformers or windings for the conversion of three-phase voltages to two-phase, or two-phase voltages to three-phase.

Leclanché Cell. A PRIMARY CELL, orginally wet, having zinc and carbon electrodes in a solution of ammonium chloride, with a manganese depolarizer. A dry form of construction is now usual, in which the electrolyte is a thick paste or jelly of starch (see Fig. L-5). In its cylindrical form, it is widely used for torches.

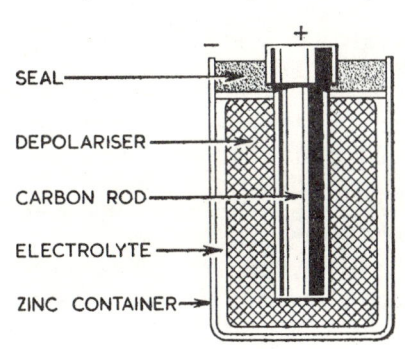

FIG. L-5. DRY LECLANCHÉ CELL.

Leg. Of a transformer, that part of the core surrounded by the windings. It is also known as a *limb*.

Lenz's Law. When an electromotive force is induced in a circuit by a changing magnetic flux, the direction of the induced e.m.f. is such that any current produced by it opposes the change of linkage.

Leyden Jar. A capacitor consisting of a jar, usually of glass, having its inner and outer surfaces coated with a conducting material.

Lichtenberg Figure

Lichtenberg Figure. A figure produced on a photographic plate separating two electrodes when an impulsive voltage is applied. The normal figure obtained (if the peak voltage is high enough) comprises a set of radial streamers, with a well-defined radius which is a measure of the peak voltage.

Lift Machine. The part of an electric lift system that includes motors, any reduction gear, brakes and winding drum or sheave. Modern electric lifts chiefly consist of "geared" or "gearless" traction machines. Winding drums are confined to lifts having a low rise.

Lifting Magnet. Electromagnet used in industry for handling iron and steel. For general lifting duties, the circular magnet is most common, while plate handling and the moving of long pipes is achieved with a rectangular magnet; a multi-polar type is employed for lifting coiled sheet steel.

Lightning. An electric discharge which takes place between two clouds charged to different potentials or between a charged cloud and the earth's surface.

Lightning Arrester. A SURGE DIVERTER employed to protect apparatus from surges arising from lightning.

Lightning Conductor. A continuous system of metallic conductors arranged to provide easy passage to earth from the highest point of a building for a lightning discharge. It comprises one AIR TERMINATION and one down conductor.

Limb. Of a transformer, that part of the core surrounded by the windings. It is also known as a *leg*.

Limit Switch. Switch fitted to moving equipment such as a lift or travelling crane to cut off the power supply if the equipment attempts to move beyond a particular point. It is used either as a safeguard against errors of judgment by an operator or as a necessary feature of automatic operation. If used to break the main supply it is referred to as a *series limit switch*; but when it operates in a control circuit, cutting off the supply by the action of a contactor or circuit-breaker, or initiating some protective altered action by energizing a trip coil or relay circuit, it is referred to as a *shunt limit switch*.

Lindemann Electrometer. A form of ELECTROMETER having a very small and light moving system providing a very short periodic time. The reading system incorporates a microscope. The quadrants of a basic electrometer are replaced by pairs of slotted flat plates, and instead of a vane there is a metallized quartz fibre carried on a similar fibre under tension to provide connection and constraint (see Fig. L–6).

Line Choke. An inductor or choking coil (see CHOKE) connected in series with electrical equipment to absorb (or reflect) the effects of high-frequency surges. It is sometimes known as a *screening reactor*.

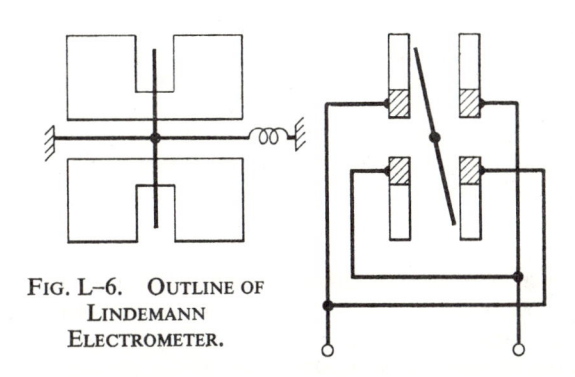

FIG. L–6. OUTLINE OF LINDEMANN ELECTROMETER.

Line-Drop Compensator. Device incorporating resistance and reactance, employed to maintain feeder voltage constant at a point remote from the regulated transformer. It passes a current proportional to the load current and produces at the control point a voltage equal to, and in phase with, the drop along the feeder.

Line of Force. The line of action of a force, such as an electric or magnetic force, or a line drawn to show at any point the direction in which a force acts.

Line Voltage. (1) The voltage between the two lines of a single-phase system.

(2) The voltage between any two lines of a symmetrical three-phase system.

(3) The voltage between any two consecutive lines of a symmetrical six-phase system.

Linear Accelerator. An indirect PARTICLE ACCELERATOR in which the particles travel in a straight line and arrive at gaps in the structure at the right phase of an r.f. excitation, or move in step with a travelling electromagnetic wave.

Acceleration of *protons* is accomplished by passing them through a resonant cavity containing a series of drift tubes, spaced and dimensioned in such a way that the electric field between gaps is nearly constant. Particles are accelerated between gaps and move between centres of successive gaps in one complete cycle of oscillation. An important machine for the acceleration of *electrons* is the travelling-wave accelerator using megawatt pulses of h.f. power at 3,000 MHz. The power is propagated along a "corrugated waveguide," a section of circular waveguide containing a series of irises. The travelling wave set up has an axial component of electric field which is used to accelerate the electrons.

Linear Motor. A motor designed to produce linear, rather than rotary, motion. It may be visualized as a conventional annular three-phase stator cut across at one point and laid out flat, the "rotor" (now moving in a straight line) being an equally conventional magnetic inductor. Induction machines are most common, but as many types of linear motor are possible as there are rotary machines.

Linkage. A measure of the product of the number of lines of magnetic flux and the number of turns of a coil or circuit through which they pass. The unit is one unit of magnetic flux passing through one turn of the coil or circuit.

Liquid-Quenched Fuse. A fuse employing a liquid to quench the arc. It is generally a CARTRIDGE FUSE, and may be referred to as simply a *liquid fuse*.

Liquid Starter, Controller. A rheostatic starter or controller in which the resistor is liquid.

Lissajous Figure. Formerly a closed graphical figure formed by the reflection of light rays from vibrating tuning forks; now generally that formed on the screen of a cathode-ray tube by the combined action of two cyclic voltages applied to pairs of deflecting plates arranged at right-angles.

Litzendraht. Stranded cable for high-frequency currents. In order to reduce skin effects, individual strands are insulated, and so woven that each occupies all possible geometrical positions in the cable section within a comparatively short overall length. It is also known as *litz wire*.

Live. Term applied to an object when either a difference of potential exists between it and earth, or it is connected to the "neutral" of a supply system in which that conductor is not permanently and solidly earthed.

Live-Line Indicator. An instrument for indicating whether an overhead transmission line is live, disconnected (but carrying a charge) or dead. The live-line tester usually comprises a relatively low-range electrostatic voltmeter or rectifier-type voltmeter with a high-insulation capacitor in series, the combination being mounted near the end of a long insulated pole which can be raised to the line from the ground, enabling a safe reading to be made by a user on the ground.

Lloyd-Fisher Square. An assembly of strips of ferromagnetic material which may be used for the measurement of CORE LOSS. The strips are arranged on edge in the form of a hollow square and interlinked with corner pieces (cf. EPSTEIN SQUARE).

Load. The power output of a generator, motor, transformer, etc., or the power carried by a circuit.

Load Angle. The electrical angle between the stator and rotor magnetomotive forces in an electrical machine. It is significant in all rotating machines, but its effects are most usually observable, and operationally important, in synchronous machines. It is sometimes known as the *power angle* or *torque angle*.

Load Curve. A graph in which power supplied or consumed is plotted as a function of time.

Load Discriminator. A device for diverting the series winding of a crane motor to obtain the optimum speed/torque ratio and the maximum acceleration.

Load Factor. The ratio of the number of energy units supplied during a given period to the number that would have been supplied had the maximum demand been maintained throughout the period.

Locked-Coil Conductor. A STRANDED CONDUCTOR having its outer wires so shaped that they are restrained from any radial movement.

Locus Diagram. A curve plotted on rectangular or polar co-ordinates, to show the variation of a system function (e.g. admittance, load, transfer function) when some parameter other than time is changed in steps. Each point on the curve represents a sinusoidal steady state for a given set of parameters, transient conditions being ignored.

Logarithmic Decrement. The natural logarithm of the DECREMENT of an oscillating system.

Lohys. A dynamo steel used for the cores of general-purpose industrial motors. It has a silicon content of $0\cdot2\%$.

Loop Test. Method of localizing a fault in a cable. The faulty conductor is looped to a sound conductor. Variations include the MURRAY LOOP TEST, WERREN OVERLAP TEST, OPEN AND CLOSED TEST, FISHER LOOP TEST and HILBORN LOOP TEST.

Loop Winding

Loop Winding. A method of winding a lap-connected armature (see also LAP WINDING).

Looping-In. A method of avoiding the use of a tee-joint in wiring installation. The conductor is taken to and from the point to be supplied.

Loss Angle. The angle by which the angle between the leading current and the voltage in a piece of insulating material differs from 90°. It is equal to zero for a perfect dielectric.

Loss Factor. Of a particular type of loss, such as the I^2R loss, the ratio of the actual number of units consumed in this loss over a given period, to the number of units that would have been consumed if the maximum loss had been maintained continuously over the whole period.

Loss Tangent. The tangent of the LOSS ANGLE in insulating material. It is approximately equal to the POWER FACTOR.

Low-Frequency Induction Furnace. An INDUCTION FURNACE that is particularly adapted to non-ferrous metals, especially brass and, more recently, aluminium alloys. It consists of a single- or double-yoke core with a primary winding in its centre to which current is supplied at **50–60 Hz.** The hearth is designed with a narrow V- or U-shaped channel in the lower part where melting occurs (see Fig. L–7).

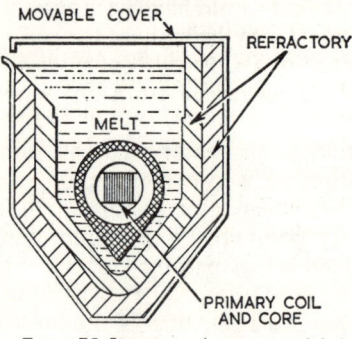

FIG. L–7. LOW-FREQUENCY INDUCTION FURNACE.

Low Voltage. A term which, when employed in official documents, normally implies a voltage not exceeding 250 V.

Lumen. The basic unit of luminous flux. An ideal source of light having a luminous intensity of one candela in all directions will emit one lumen in any unit solid angle with its origin at the source, and the total luminous flux will be 4π lumens corresponding to the total solid angle.

Lumen Factor. Means of determination of the luminous output of a light source from its polar curve. The candle power in each of a series of zones is multiplied by an appropriate factor and the results added.

Luminaire. A lighting fitting.

Lux. A measure of the luminous flux per square metre, being an illumination of 1 lm/m². It corresponds in metric units to the foot-candle (1 lm/ft²) but is only about one-tenth of the magnitude.

M

M.H.D. Generator. See MAGNETOHYDRODYNAMIC GENERATOR.

M.K.S. System. A system of physical units based on the metre, kilogramme and second (see also M.K.S.μ SYSTEM).

M.K.S.μ. System. A *rationalized* M.K.S. SYSTEM in which the unit of permeability is such that the permeability of free space is numerically equal to 10^{-7}. The practical electrical units are the ampere, volt, ohm, henry and farad; the mechanical are the joule, watt and newton. Rationalization means that the system is based on ideas of uniform fields rather than divergent fields from unit charges or poles. Consequently the factor 4π is removed from uniform field cases and transferred to cases of spherical geometry.

Magnavolt. Trade name for a type of single-stage ROTATING AMPLIFIER relying on the use of positive feedback.

Magnet. A piece of ferromagnetic material which has acquired, either permanently or temporarily, the power of attracting or repelling other pieces of similar material and of exerting a mechanical force on a neighbouring conductor carrying an electric current (see PERMANENT MAGNET and ELECTROMAGNET).

Magnetic Amplifier. Device consisting of one or more ferromagnetic cores with windings so arranged and connected that the alternating current flowing in one winding can be modified by causing saturation of the core by means of direct or low-frequency alternating current flowing in another winding (see Fig. M–1).

The basic amplifying element is known as a *transductor* or *saturable reactor*. The combination of one or more transductors together with other circuit elements for amplifying electrical signals form the magnetic amplifier.

Magnetic Blowout. The earliest type of ARC-CONTROL DEVICE, still in common use. By means of a coil carrying the current to be interrupted, it sets up a magnetic field across the arc. This

Magnetic Brake

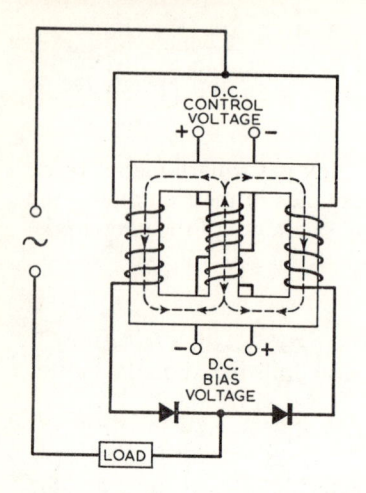

FIG. M–1. BASIC CIRCUIT FOR A FULL-WAVE MAGNETIC AMPLIFIER.

The iron-cored coil, that forms the basic amplifying element, is known as a transductor.

accentuates the natural tendency for the arc to lengthen, by means of the electromagnetic force resulting from the interaction between the field and the arc current, Fig. M–2.

Magnetic Brake. In its simplest form, an a.c. or d.c. solenoid arranged to operate, through a system of levers, a pair of brake shoes provided with friction linings. When the solenoid is de-energized, the brake wheel is held stationary by the brake shoes under the action of a counterweight or, more usually, a spring. But when the solenoid is energized, the brake shoes are lifted against the spring pressure and the wheel is free to move.

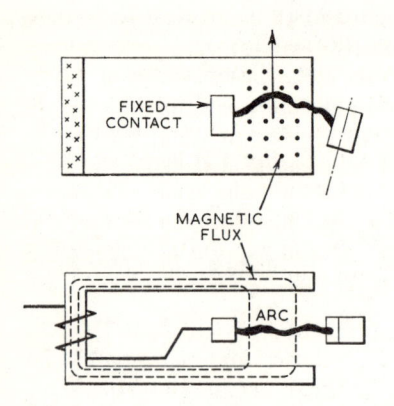

FIG. M–2. PRINCIPLE OF MAGNETIC BLOWOUT.

An arc between two contacts naturally tends to rise because of its temperature. This natural arc-lengthening process is artificially aided in this arc-control device.

Magnetic Flaw Detection

Magnetic Circuit. The closed path round which MAGNETIC FLUX passes. Magnetic flux cannot be confined as can an electric current: there is no magnetic insulator. The saturation phenomenon associated with ferromagnetic materials is another reason why analysis of magnetic circuits must be different from that of an electric circuit.

A circuit law relates the total magnetic-circuit flux ϕ to the magnetomotive force F producing it through a quantity called the reluctance S, in the form $\phi = F/S$. This expression is of rather limited application, because the reluctance is a function of the absolute permeability, which in useful ferromagnetics is variable. Further, the expression assumes that the flux density is constant over any given section. This is rarely the case. The practical approach to a problem requires the use of magnetization (or saturation) curves of the materials of the circuit; the initial assumption of a known flux; the assessment of the flux density at various parts of the circuit; and the summation of the m.m.f. necessary to produce the assumed flux round the complete circuit.

Magnetic Clutch. Clutch consisting usually of two parts: a field member containing an electromagnet and the necessary slip rings through which the current is supplied to the coil, and an armature member. It is usual for the field member to be mounted on the driving shaft and the armature on the driven shaft. The two halves of the clutch are held apart by means of a spring plate when the coil winding is not energized.

Magnetic Difference of Potential. A difference in the magnetic states existing at two points which produces a magnetic field between the two points. It is equal to the line-integral of the magnetic force between the two points, except in the presence of electric currents.

Magnetic Field. The state of the space in the vicinity of an electric current or a permanent magnet throughout which the forces produced by the current or magnet are discernible.

Magnetic Flaw Detection. A method of detecting flaws in metal castings; it is particularly effective for checking for surface cracks on ferromagnetic material. It depends for its application on the leakage of magnetic flux from the surface of the specimen in the vicinity of a crack-like discontinuity. Near the crack, lines of flux will pass through the air and any small magnetic particles near the edge of the crack will, therefore, collect in this region and indicate the presence of a crack.

Magnetic Flux

Magnetic Flux. Concept of the magnetic properties of a magnet as appearing to "flow" along definite paths termed *lines of magnetic force*, the total number of such lines of force issuing from a particular magnet pole being the magnetic flux. The M.K.S. unit is the weber. If the flux linking one turn in a circuit changes by one weber in one second an electromotive force of one volt will be induced in that turn. The C.G.S. unit is the maxwell; one weber $= 10^8$ maxwells.

Magnetic Hysteresis. Phenomenon observable in the magnetization of some materials in which their magnetic state at an instant is related to their previous state (see HYSTERESIS, B/H CURVE, HYSTERESIS LOSS).

Magnetic Leakage. The part of a magnetic flux that follows a path in which it is ineffective for the purpose desired.

Magnetic Link. Device that indicates the magnitude of a lightning surge through a conductor. It is usually in the form of a container, holding highly remanent steel wires or laminae, which is mounted adjacent to the conductor and indicates by a change in its magnetic state the passage of the lightning current. It is also known as a *surge-current indicator*.

Magnetic Moment. Of a magnet situated in a uniform field in a vacuum, the ratio of the torque on it when in the position of maximum torque to the magnetizing force of the field.

Magnetic Overload Relay. An OVERLOAD RELAY that uses the magnetic effect of the current for operation. When the current flowing in the coil reaches a predetermined value, a plunger housed in the dashpot is attracted into the coil. When near the end of its travel, it opens the normally closed trip contact. To obtain a time delay, the dashpot is filled with a suitable fluid, usually a mineral oil.

Magnetic Separator. Device for removing ferromagnetic materials from a mixture. It contains an electromagnet and may be used when material has to be handled in bulk.

Magnetic Sheet-Steel. The basic constructional material of all electric motors, generators and transformers. It consists of iron alloyed with 0·4–4·2% silicon, rolled into sheets or strips whose thickness usually lies between **0·35 and 0·63 mm.** Occasionally a small amount of aluminium is also added, but in general, other additional elements are reduced to very small proportions. In particular, carbon is a serious cause of increased magnetic losses; it is usually reduced to 0·02% in hot-rolled, and below 0·005% in oriented, sheets.

Magnetic Space Constant. An alternative (and technically preferable) term for the absolute permeability of free space. Its value in M.K.S. rationalized units is $\mu_0 = 4\pi/10^7$ H/m.

Magnetization Characteristic. The relation between the flux density and the magnetizing force for a prepared specimen of magnetic material; or the relation between the induced electro-motive force on no load to the field (magnetizing) ampere-turn excitation, for a given speed. The latter is often called the *open-circuit characteristic* (see OPEN- AND SHORT-CIRCUIT CHARACTERISTICS).

Magnetizing Coil. See FIELD COIL.

Magnetizing Force. The force at a point which produces or is associated with the flux density at the point. It is the magnetomotive force per unit length measured along the lines of force. The C.G.S. unit is the *oersted*.

Magneto. A magneto-electric GENERATOR, in which the magnetic flux is provided by permanent magnets. It is capable of producing high-voltage impulses and is sometimes used in the ignition circuit of internal-combustion engines.

Magneto-electric Generator. See MAGNETO.

Magnetohydrodynamic Generator. A device for converting thermal energy into electrical, by braking a stream of hot ionized gas. In the simplest type of *m.h.d. generator*, the gas stream passes through the poles of a magnet at right-angles to the magnetic field. With suitable pick-up arrangements, the electric field induced in the stream may be used to drive a current through an external load. The device is also known as a *plasmahydrodynamic generator*.

Another type of generator employs liquid metal as the working fluid. This avoids the containment problems associated with high-temperature plasmas.

Magnetometer. An instrument for the measurement of a weak magnetic field (originally the earth's field). Various patterns depend in general on the deflection or the oscillation of a small suspended magnet in a magnetic field. A more recent type employs wires of nickel–iron alloy, or a core of this material carrying a coil. In the a.c. magnetometer the action depends on the impedance of a thin nickel–iron wire to alternating current, as affected by the presence of a weak external field.

Magnetomotive Force. The force which causes a magnetic flux to exist in any magnetic circuit. Quantitatively, it is the line integral of the magnetizing force along a path.

Magneton

Magneton. The magnetic moment of a spinning ELECTRON. A rotating charge is equivalent to a circular current and gives rise to a magnetic field.

Magnetoplasmadynamic Generator. See MAGNETOHYDRODYNAMIC GENERATOR.

Magnetostatic Lens. See ELECTRON LENS.

Magnetostriction. Phenomenon whereby, when a magnetic material is magnetized, its dimensions are changed. Conversely, if the material is strained, changes occur in its magnetic properties. The property of magnetostriction is possessed in a marked degree by nickel, by some nickel–iron, cobalt–iron and aluminium–iron alloys, and also by ferrites of cobalt and nickel. Magnetostrictive effects include the JOULE EFFECT, the VILLARI EFFECT and the WIEDEMANN EFFECT.

Magnetostrictive Vibrator. A vibrator employing the phenomenon of MAGNETOSTRICTION. A low-frequency magnetostrictive vibrator is shown in Fig. M–3.

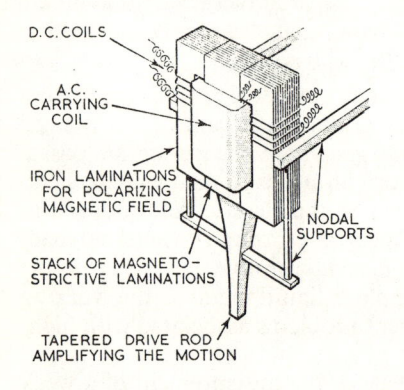

FIG. M–3. MAGNETOSTRICTIVE VIBRATOR.
(*W. Bryan Savage Ltd.*)

Magnetron. Valve for generating power at centimetric wavelengths. It utilizes the electron transit time between internal elements which take the form of resonant cavities.

Magnicon. Trade name for a form of ROTATING AMPLIFIER with cross-field excitation.

Magslip. A small precision-built electrical SYNCHRO applied to position-indication and comparable functions in automatic control systems.

Main. Conductor or conductors arranged for the transmission and distribution of electrical energy.

Making-Current. Of a switch, circuit-breaker, etc., the total maximum current peak which occurs immediately after the circuit is closed.

Manganin. A copper–manganese–nickel alloy possessing high resistivity and low temperature coefficient of resistance. It is frequently used in the manufacture of instrument resistors.

Manipulator. Equipment which simulates the actions of a human arm and hand, for use in situations where operations must be carried out remotely, for example on toxic or radioactive material.

Maser. Device based upon the fact that an elemental atomic magnet in a magnetic field may have one of two energy states, i.e. parallel or anti-parallel to the field. If electromagnetic radiation of suitable frequency is applied to the system, transitions of atomic magnets from one energy state to the other can take place, energy being absorbed if the transition is from the lower energy state to the higher, or radiated if the transition is from the higher to the lower. Employing this principle, practical amplifiers and oscillators have been produced. The name is derived from the term "Microwave Amplification by Stimulated Emission of Radiation."

Mass-Impregnated Cable. A paper-insulated cable in which the paper tapes are applied unimpregnated, the complete cable being subsequently dried and impregnated with compound as a whole.

Mass-Impregnated Gas-Pressure Cable. Cable in which the normal mass-impregnated dielectric is employed but the lead or aluminium sheath is applied with a small clearance. This space between the screened dielectric and the sheath is charged with nitrogen gas at high pressure. In this design the gas passes into and out of solution with the compound with variations of temperature.

Mass Number. The total number of PROTONS and NEUTRONS in the nucleus of an atom.

Mass Resistivity. The product of the VOLUME RESISTIVITY and the density of a material at a given temperature. It may be useful as expressing directly the resistance of a conductor of unit length and unit weight, or the weight of a conductor of unit length and unit resistance.

Mass Susceptibility. The quotient of SUSCEPTIBILITY and magnetic flux density.

Mass-Type Plate

Mass-Type Plate. Of an accumulator cell, a plate consisting of large blocks of active material held in a frame.

Master Clock. A clock which controls a system of other clocks by transmitting a current impulse at predetermined intervals.

Master Controller. A multi-way switch for controlling the operation of a set of contactors. It is sometimes called a *pilot controller*.

Master Frequency Meter. An instrument which sums the total number of cycles of an a.c. supply in a given time, thus enabling a comparison to be made with the required number. It is sometimes called an *integrating frequency meter*.

Matching. Adjustment of the effective impedance of a load with respect to that of the source, so as to ensure maximum power transfer.

Matching Transformer. A transformer coupling a source to a load, and designed to ensure maximum energy transfer in spite of the fact that the load impedance differs from the source impedance. For optimum matching the square of the turns ratio of the transformer should equal the ratio of the two impedances.

Maximum Demand. The highest value of the power, volt-amperes, or other quantity such as the current, taken within a demand assessment period (e.g. day, month or year). The appropriate quantity is assessed usually by integration over each successive demand integration period (e.g. half an hour). It is usual for the maximum demand to be applied only to power quantities.

Maximum-Demand Tariff. A Two-Part Tariff consisting of a kilowatt or kVA charge and a unit charge. It is sometimes known as a *Hopkinson tariff*.

Maximum Power Transfer Theorem. In order that a source of electromotive force E and internal impedance z shall deliver maximum power to a load of impedance Z, then z and Z must be conjugate; i.e. if $z = r+jx$ then $Z = R-jX$ with $R = r$ and $X = x$. With the reactances equal and opposite in sign, there is no resultant reactance in the circuit, and in this respect the current is greatest because the total impedance is $r + R$. If, further, $R = r$, then the load power is the maximum possible.

Maxwell. The electromagnetic C.G.S. unit of magnetic flux. The flux in maxwells is commonly referred to as the number of "lines". A magnetic flux linking a single-turn coil and changing at the rate of 1 maxwell/sec induces an e.m.f. in the coil of 10^{-8} V. In terms of the weber, 1 maxwell = 10^{-8} Wb.

Maxwell Bridge. A real-product A.C. BRIDGE of the form shown in Fig. M-4. The balance is independent of the frequency (cf. HAY BRIDGE).

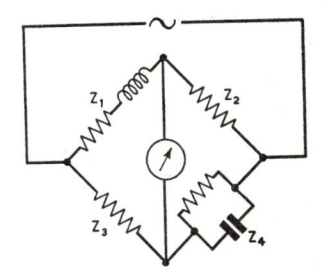

FIG. M-4. MAXWELL BRIDGE CIRCUIT.

Maxwell Laws. The basic laws of electromagnetism, formulated by Maxwell on the basis of the work of Gauss, Ampère and Faraday, and completed by his own concept of displacement current. The laws are, in integral form:

I. The magnetomotive force, or line integral of the magnetic field intensity H, round any closed path, is equal to the current enclosed by the path:

$$\int_o H \cdot dl = I = I_c + I_d$$

This is AMPERE'S LAW, with the provision that I includes both conduction current I_x and displacement current $I_d = \partial \psi / \partial t$, where ψ is the electric flux through the path.

II. The total electric flux emerging from a charge Q is equal to Q. Then, the surface integral over any closed surface of the outward electric flux density D is

$$\int_S D \cdot ds = Q$$

If there is no net charge enclosed within the surface, the integral is zero. This is GAUSS' LAW.

III. Round any closed path encircling a magnetic flux Φ changing with time, there is an electromotive force e which is the line integral of the electric field intensity E:

$$\int_o E \cdot dl = e = -\partial \Phi / \partial t.$$

This is FARADAY'S LAW OF ELECTROMAGNETIC INDUCTION.

Maxwell Theorem

IV. Magnetic flux is a solenoidal entity: that is, it always comprises a structure of closed loops and there is no "source" of magnetic field. The surface integral, over any closed surface in a magnetic field, of the flux density B must always be zero:

$$\int_S B \cdot ds = 0$$

To the above are added the *constitutive* equations, so called because they relate to the properties of the media in which electromagnetic fields may be produced. They relate the electric and magnetic vectors E and H to the permittivity ϵ, permeability μ, and conductivity σ of a medium:

$$B = \mu H; \quad D = \epsilon E; \quad J_c = \sigma E$$

The third of these relates conduction current density to the voltage gradient in a conducting medium.

Maxwell Theorem. A formalization of Kirchhoff's mesh law for the solution of networks. To each closed mesh in a network is assigned a circulating current. The node law is then automatically satisfied at each junction. Applying the mesh law to each closed mesh, a set of simultaneous equations can be written down for solution (see KIRCHHOFF'S LAWS).

Mean British Thermal Unit. A unit of heat, equal to 1/180th part of the quantity required to raise the temperature of 1 lb of water from 32° to 212°F. It equals 1·055 joules. Cf. BRITISH THERMAL UNIT.

Mechanical Rectifier. A full-wave rectifier that incorporates a synchronously rotating or oscillating commutator.

Medical Electrolysis. The treatment of diseases by the use of direct current. It is also known as *galvanism*.

Medium Voltage. A term which, when employed in official documents, normally implies a voltage between 250 and 650 V.

Mega-. A decimal prefix indicating multiplication by 10^6 (abbreviation: M).

Megger. Trade name for a range of portable insulation-resistance testers, consisting of a high-range ohmmeter and a hand-operated generator.

Melting Furnace. See INDUCTION FURNACE, ARC FURNACE, ELECTRON-BEAM FURNACE.

Mercury. White metallic element, liquid at normal temperatures. Symbol Hg; atomic weight 200·61; specific gravity at

Mercury-Vapour Lamp

20°C, 13·546; melting point −38·5°C; boiling point 356·7°C; specific resistivity 95·8 $\mu\Omega/cm^3$. It is used in MERCURY-ARC RECTIFIERS, MERCURY VAPOUR LAMPS, MERCURY SWITCHES, etc.

Mercury-Arc Rectifier. Rectifier in which rectification takes place in an arc discharge through mercury vapour at low pressure. GLASS-BULB RECTIFIERS, in which the arc occurs within a glass bulb, were the earliest and are still used for a wide range of applications. They are made in sizes from 10 to 500 A. *Steel-tank rectifiers*, in which the arc chamber is of steel, are essentially high-capacity rectifiers and are made for outputs of about 200 A and up to 6,000 A or more. A third type, the *hot-cathode rectifier*, combines the electron-emitting properties of a heated filament with the property of mercury vapour to reduce the resistance of the electron path. A fourth type is the IGNITRON.

Mercury Motor Meter. An INTEGRATING METER embodying a motor, part of the moving portion being immersed in mercury.

Mercury Switch. Enclosed switch that provides continuity between two conductors by immersing them in a common mercury pool. The contact resistance is lower than any obtainable by a pair of solid metal-to-metal surfaces. The liquid must be enclosed in a sealed bulb of hard glass or ceramic material to exclude air, which causes surface contamination. The bulb is filled with an inert or reducing gas to a pressure that assists extinction of the arc and increases rupturing capacity. As the break is enclosed the switch can be used in inflammable atmospheres. Types of mercury switch in general use are shown in Fig. M–5.

Mercury-Vapour Lamp. A DISCHARGE LAMP containing mercury vapour, useful light being emitted from or excited by the discharge through this vapour. Mercury-vapour lamps are generally divided into groups according to the current density and gas pressure of each group. *Medium-pressure lamps* (types MA) cover 250 W, 400 W, and 2,500 W sizes. They all have inner and outer tubular envelopes. The tubular inner member contains the mercury with a little argon gas to assist striking, together with the electrodes. *High-pressure lamps* (type MB) have higher gas pressures and current densities. The colour of the light is slightly whiter and the discharge more compact, creating a high brightness source. The inner tube is of quartz to withstand the greater temperature and pressure, while the outer envelope is similar to the general-purpose range of incandescent lamps, but is standardized in pearl finish owing to the brightness.

Merz-Price Protection

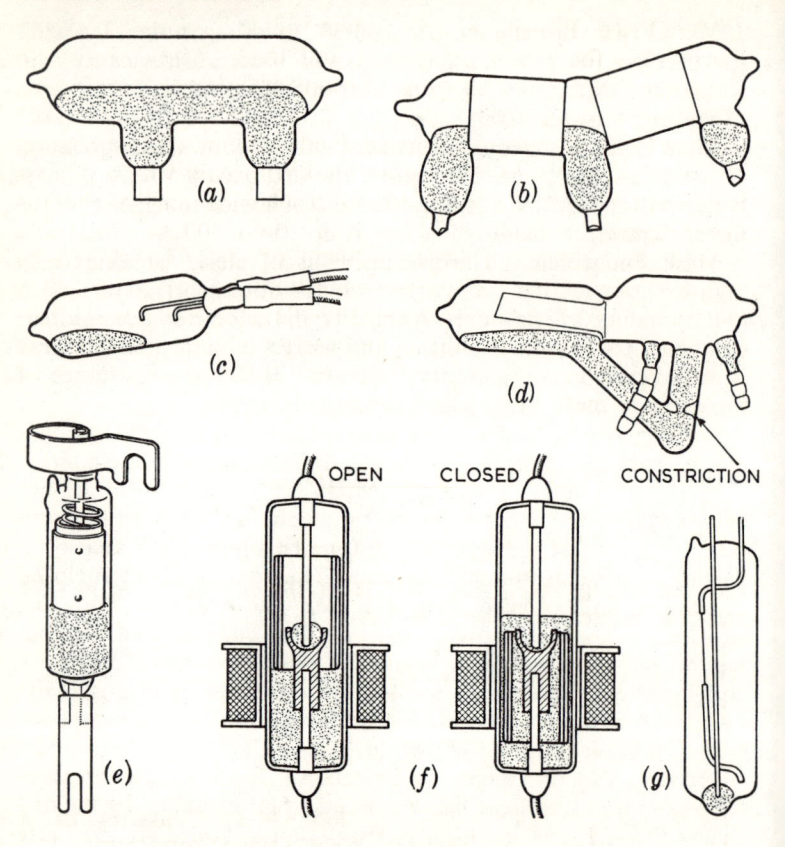

FIG. M–5. TYPES OF MERCURY SWITCHES.
Tilting switches may be simple on-off (*a*), changeover (*b*), bridge-type (*c*), or delayed-action (*d*). A displacement switch (*e*) operates by the movement of a plunger, which is drawn magnetically into a pool of mercury (*f*). A non-tilting switch may incorporate a movable electrode (*g*).
(*I.A.C. Ltd.*)

By combining a tungsten filament into a mercury-vapour lamp, the need for separate control gear is eliminated. Such filaments are usually of the double-rating type, i.e. a full resistance is in circuit when the lamp is striking up and a reduced value when the lamp is fully run up.

Mesh-Current Analysis

Merz-Price Protection. A system of BALANCED CURRENT PROTECTION for power equipment and transmission circuits in which the ingoing and outgoing currents are compared by means of current transformers.

Merz Unit. A maximum-demand indicator unit for attachment to an INTEGRATING METER registering kWh or kVAh. A pointer is moved forward by a spring-loaded mechanism operated by the meter register.

Mesh Connection. The arrangement of phase windings into a closed loop by the successive connection of the end of one to the beginning of the next. Normally the windings generate (or have applied to them) voltages, and carry currents, differing only in time phase in a symmetrical manner. If the phase voltages of an m-phase mesh interconnection are

$$V_1 = V \angle 0$$
$$V_2 = V \angle -(2\pi/m)$$
$$\cdots \quad \cdots$$
$$V_m = V \angle -(m-1)(2\pi/m)$$

then the voltage AB in Fig. M–6 is V_1, the voltage AC is $V_1 + V_2$, etc. The current leaving at line A is $I_m - I_1$, etc.

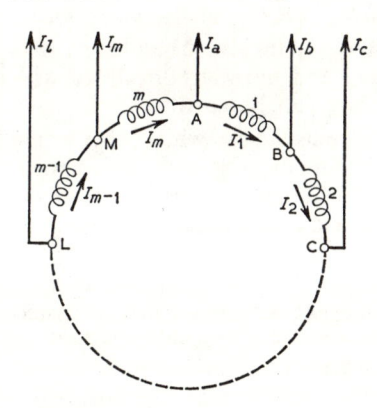

FIG. M–6. CURRENTS IN A MESH CONNECTION.

Mesh-Current Analysis. A systematic method of solving for the currents in a network. It is based on the MAXWELL THEOREM, in which each mesh or closed loop of a network is assigned a fictitious circuital current, and the assembly of all such currents conforms automatically to Kirchhoff's nodal law.

Mesh Voltage. The voltage between any two lines of a symmetrical three-phase system, or between two consecutive lines of a symmetrical six-phase system.

Mesons. Charged particles observed in cosmic rays, having rest masses greater than that of an electron but less than that of a proton. The three principal types are the μ-meson, the π-meson and the τ-meson, having rest masses of 215, 280 and 1,000 respectively referred to the mass of an electron as unity. Other types also exist. Neutral, positive and negative mesons are known; charged mesons carry a charge equal to that of an electron.

Metadyne Converter. A machine, basically similar to a METADYNE GENERATOR with the supplementary set of brushes connected to an external d.c. supply so that the output power does not require any appreciable mechanical power input to the transformer. Metadyne is a trade name.

Metadyne Generator. Trade name for a ROTATING AMPLIFIER of the cross-field excited type. It is a quick-response d.c. generator, which takes in power at constant voltage through one set of brushes and delivers it at constant current through a second set. It is similar in nature to an AMPLIDYNE.

Metal Arc Welding. Arc welding with a metal electrode. Additional metal is provided by the melting of the electrode.

Metal-Clad Switchgear. Switchgear in which all live parts are enclosed within a metal casing and normally insulated with oil or compound. The metal casing can be earthed.

Metal-Enclosed Switchgear. Switchgear in which the whole equipment is enclosed in a metal casing capable of being earthed.

Metal–Graphite Brush. A BRUSH for use on low and medium speed slip-rings and on low-voltage d.c. generators where commutation is not difficult and where a low contact resistance is desirable. Finely powdered copper or bronze is mixed with graphite and a suitable binder and the whole is then moulded into blocks under heavy pressure.

Metal Rectifier. A rectifying assembly of cells which consist of either (*a*) a semiconductor, such as selenium or cuprous oxide, in contact with a metal, or (*b*) two semiconductors of different types, usually formed from a single piece of pure germanium or silicon by alloying or diffusing two different materials into opposite faces. In either case, rectification takes place at the junction between the dissimilar materials. The essential feature of a metal

Meter. See INTEGRATING METER.

Metre. The M.K.S. (and S.I.) unit of length (abbreviation: m). The standard metre is defined as 1,650,763·73 wavelengths of the orange-red light emitted by krypton-86. This standard was adopted in 1960 in place of the bar of platinum-iridium preserved at the International Bureau of Weights and Measures. A metre is about 3 ft 3·37 in. The original metre (1791) was intended to be one ten-millionth part of the earth's polar quadrant passing through Paris.

Mho. The reciprocal ohm, a name sometimes used for the unit of conductance or admittance (abbreviation: ℧). Several alternatives have been suggested (such as *siemens*) but have not received much acceptance.

Mica. A complex alumino-silicate of potassium, magnesium and iron, or of partial combinations of sodium, lithium, titanium and vanadium. Micas, either in sheets or, more usually in small flakes or ground powders, are valuable heat-resistant electrical insulators. In flake form, mica needs a flexible cloth or paper backing.

Mica Cone. Of a commutator, a vee-shaped mica-compound ring which insulates a metal clamping ring from the bars. It is also known as a *mica V-ring*.

Micanite. A flexible insulating material formed of mica splittings bonded into a sheet by a flexible gum, bitumen or synthetic adhesive.

Microgap Switch. A switch for low-power low-voltage a.c. circuits in which the gap between the contacts when open is of the order of 0·125 mm (0·005 in.).

Micron. Unit of length, equal to one millionth of a METRE (abbreviation: μ or μm).

Microradiography. The examination of extremely thin sections of materials by means of X-rays, the resulting microradiograph being subsequently optically magnified before examination.

Microtron. An ORBITAL ACCELERATOR in which electrons are accelerated in a vacuum chamber between the poles of a fixed field magnet. The orbits consist of a series of discrete circular arcs which have a common tangent at a resonant cavity in which the electrons gain their successive increases of energy from a high-frequency electric field. Upper limits of energy achievable are

Mil

below 50 MeV with mean currents below 1 µA. It is also known as an *electron cyclotron*.

Mil. Unit of length, equal to one-thousandth of an inch.

Milker, Milking Booster. Preferably called a Milking Generator (q.v.).

Milking Generator. Low-voltage d.c. generator for charging individual cells of a battery.

Millman Theorem. A network theorem that, in suitable cases, is much more direct than the application, for example, of Kirchhoff's Laws. It is also called the *parallel-generator theorem*. It states that the common terminal voltage V of a number of sources connected in parallel is

$$V = I_{sc}Z,$$

where I_{sc} is the sum of the short-circuit currents of the generators, and Z is the parallel sum of all the impedances between the common points. The theorem is applicable when all generators have the same frequency.

Miniature Circuit-Breaker. Circuit-Breaker rated at about 60 A or less, that may be employed as protection in a domestic installation. It is commonly abbreviated *m.c.b.*

Miniature Edison Screw Cap. See Edison Screw Cap.

Minimum Fusing Current. The minimum current at which the fuse element in a fuse will melt.

Minority Carrier. Charge carrier forming only a small proportion of the total number of carriers in a "doped" semiconductor. In *p*-type material the minority carriers are electrons; in *n*-type material they are holes.

Minus Tapping. A tapping on a winding, having its position such that fewer turns are included in the active part of the winding than are required for the service voltage or current ratio.

Mirror Galvanometer. A Galvanometer having a mirror attached to the moving part, which reflects a beam of light on to a scale or reflects the image of a scale into a telescope.

Moderator. Material in the core of a Thermal Reactor for slowing down the fast neutrons so that they may be captured by the fuel. Common moderator materials are water, heavy water, graphite, beryllium and organic liquids.

Modulation. The process whereby a characteristic of a carrier (or propagated) wave of high angular frequency is varied in

Motor

accordance with the time variation of the intelligence to be transmitted (see AMPLITUDE MODULATION, FREQUENCY MODULATION, PHASE MODULATION).

Moissan Arc Furnace. An INDIRECT-ARC FURNACE in which an arc is struck between two carbon electrodes placed above the the material to be heated (see Fig. M-7). The effect is therefore

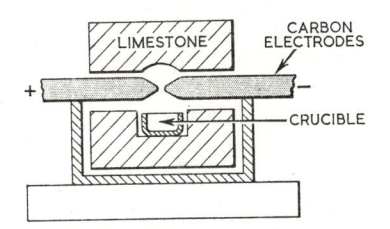

FIG. M-7. PRINCIPLE OF THE MOISSAN INDIRECT-ARC FURNACE.

entirely thermal without electrolysis, but the carbon may play a part in reducing oxides in the metal. The Moissan furnace is for use at about 60 V with currents of several hundred amperes, and the refractory may be limestone or other suitable material.

Møllerhøj Cable. A pressure cable in which the cores are laid side by side in flat formation (Fig. M-8). The lead sheath is

FIG. M-8. MØLLERHØJ CABLE.

supported by circumferential metal tapes together with corrugated metal strips bound onto the flat faces of the sheath by copper wires. An oil of low viscosity is used for the impregnating medium.

Motor. A machine for converting electrical energy into mechanical. A current-carrying conductor in a magnetic field is acted on by a force proportional to the current and the field strength. This principle is applied to the rotation of a wound rotor. D.C. motors may be divided into three classes: SERIES-WOUND MOTOR, SHUNT-WOUND MOTOR and COMPOUND-WOUND MOTOR. A.C. motors also are of three main types: INDUCTION MOTOR, SYNCHRONOUS MOTOR and COMMUTATOR MOTOR, the last including the VARIABLE-SPEED MOTOR, SCHRAGE MOTOR, REPULSION MOTOR, SERIES MOTOR, etc.

Motor Converter

Motor Converter. A CONVERTER consisting of a wound-rotor induction motor coupled to a SYNCHRONOUS CONVERTER. The energy transmission is partly mechanical and partly electrical, as the two armatures and their circuits are connected. The speed is usually half the synchronous speed of the induction motor.

Motor Generator. A CONVERTER consisting of an a.c. motor directly coupled mechanically to a d.c. generator. There is no electrical connection between the two machines.

Motor Meter. An INTEGRATING METER which embodies a motor of some type.

Moving-Coil Galvanometer. See GALVANOMETER.

Moving-Coil Instrument. The commonest type of indicating instrument, although by itself it measures only direct current. It comprises a fixed permanent magnet, in the air-gap of which is a coil carrying current, which, combined with the steady magnetic flux, produces motion, shown by a pointer (see Fig. M–9). The scale is linear and easy to read. Ohmmeters commonly have two moving coils and read the ratio of volts and amperes, thus indicating true ohms.

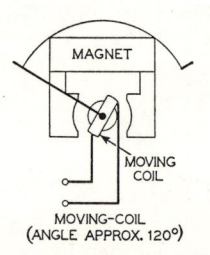

FIG. M–9. MOVING COIL INSTRUMENT.

Moving-Coil Regulator. A form of VOLTAGE REGULATOR. An AUTO-TRANSFORMER having straight limbs is provided with two similar windings connected in opposition, so that the flux through each follows the path indicated in Fig. M–10. Over these windings moves a short-circuited coil; if this coil were not present the voltage across each winding would be the same, and the point A would be midway in potential between the two supply terminals.

Moving-Iron Instrument

If the coil is moved to a position covering the upper winding, the magnetomotive force due to the currents in the short-circuited coil will oppose that of the upper winding, thus reducing the upper flux almost to zero. The potential of A will therefore be almost that of the upper supply terminal and the output voltage be nearly equal to the supply voltage. If the coil is moved to cover the lower winding the lower flux will be reduced almost to zero and the potential of A will therefore also be reduced almost to zero. By moving the coil a smooth output voltage variation is thus obtained without the use of sliding contacts.

Moving-Coil Relay. A relay consisting of a permanent magnet and a coil, the latter mounted to move in the permanent magnet flux against spring restraint, and to operate the relay contacts. For practical purposes the flux density in the permanent-magnet air-gap may be considered constant over the range of movement of the coil, and the deflection is thus proportional to the input current. The relay is a d.c. measuring device and is also suitable for the comparison of the magnitudes of two currents. For this duty the coil is in two sections, connected to exert opposing forces. When it is used on a.c. circuits, rectifiers are incorporated.

Moving-Iron Instrument. A common form of indicating instrument. A fixed coil and a moving iron may be arranged for mutual attraction; alternatively, two irons may be similarly magnetized

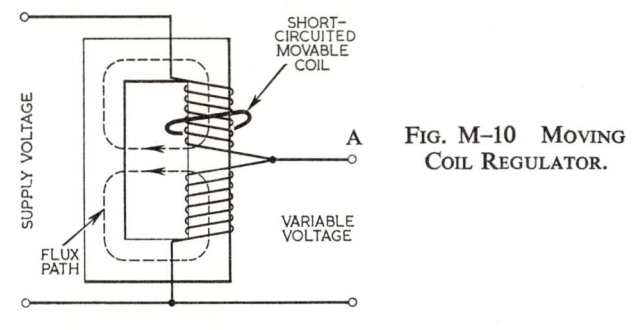

Fig. M-10 Moving Coil Regulator.

by the same coil and mutually repel each other: if now one iron is fixed and the other pivoted the movement will depend on the current. Both these *attraction* and *repulsion* types are in use. Moving-iron instruments of good design read voltage and current equally well on direct current and alternating current of power frequencies. Being relatively cheap they form the bulk of a.c.

Moving-Magnet Instrument

voltmeters and ammeters for a.c. power-system use. They are sometimes known as *soft-iron instruments*.

Moving-Magnet Instrument. A MOVING-IRON INSTRUMENT in which the movable component consists of a piece of permanently magnetized material.

Multi-break Circuit-Breaker, Switch. A circuit-breaker or switch in which there are two or more breaks in series in each pole or phase.

Multiple Earthing. See PROTECTIVE MULTIPLE EARTHING.

Multiple Tariff. A TARIFF in which different rates are charged for units supplied for various purposes.

Multi-point Heater. Water heater supplying hot water to a number of outlets. It may be a *pressure heater* or a *cistern heater*. Cf. NON-PRESSURE HEATER.

Multi-polar Machine. A machine in which the field magnet has more than two poles.

Multi-speed Induction Motor. Motor with two or four operating speeds, obtained by changing the number of poles in the stator, the squirrel-cage rotor being adaptable for any number of poles. It is also known as a *change-pole motor*.

Multivibrator. A useful form of electronic RELAXATION OSCILLATOR.

Mumetal. A ferromagnetic alloy, approximately 75% nickel, 5% copper and 2% chromium. Its important properties are high permeability and low hysteresis loss in relatively low magnetizing fields. The initial and maximum relative permeabilities are of the order of 30,000 and 100,000. See NICKEL ALLOYS.

Murray Loop Test. A cable fault localization test. The faulty conductor, which must be continuous, is looped to a sound conductor of the same length and size, and a resistance bridge and galvanometer are connected across the open ends of the loop (Fig. M–11). A battery or other d.c. source is connected between

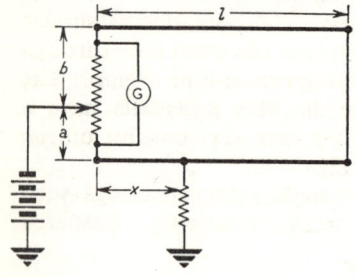

FIG. M–11. MURRAY LOOP TEST FOR LOCATING A FAULT IN A CABLE.

the bridge tap and earth. A Wheatstone bridge arrangement is formed by the circuit, two of the arms being given by the resistances a and b, and the other two by the parts of the cable loop on either side of the fault. Balance is obtained by adjustment of a and b, and the distance to the fault from the test end is given by

$$x = l \cdot 2a/(a + b),$$

where a is the arm connected to the faulty core and l is the route length of the cable. See also WERREN OVERLAP TEST, FISHER LOOP TEST.

Mush Winding. Winding for small a.c. machines in which the conductors are dropped individually into lined slots, the end connections then being insulated separately. It is also known as a *random winding*.

Mutual Coupling. Coupling of COUPLED CIRCUITS in which the two circuits have a partly common magnetic circuit.

Mutual Impedance. Alternative name for TRANSFER IMPEDANCE (q.v.).

Mutual Inductance. The ratio of the magnetic flux-linkage in an electric circuit to the current flowing in another circuit linked magnetically with the first. The two circuits have unit mutual inductance of one henry if (*a*) the energy stored in the common magnetic field is 1 J when the current in each circuit is 1 A, (*b*) a primary current changing at the rate of 1 A/sec induces an e.m.f. of 1 V in the secondary, or (*c*) the number of flux-linkages with the secondary totals 1 Wb-turn when the primary current is 1 A.

Mutual Surge Impedance. A form of SURGE IMPEDANCE involving two transmission lines.

N

n–p–n Transistor. TRANSISTOR consisting basically of two outer n regions separated by a thin p layer. The operation is similar to that of a p–n–p TRANSISTOR, except that electrons flow through the p-type base instead of holes through an n-type region. The main difference, practically, is that the bias potentials have a polarity opposite to that for the p–n–p type (i.e. positive on the collector and negative on the emitter).

n-Type Semiconductor. A quadrivalent SEMICONDUCTOR such as germanium, which includes a small, accurately controlled

Nano-

amount of a pentavalent *donor* element so that the crystal lattice contains free electrons to act as charge carriers.

Nano-. A decimal prefix indicating multiplication by 10^{-9} (abbreviation: n).

Narrow-Base Tower. A Lattice Tower constructed as a cantilever with its base embedded in the ground.

Natural Frequency. The frequency at which an oscillatory system will oscillate if provided with energy and then left free of restraint.

Natural Graphite Brush. A Brush similar to a Carbon Brush but with natural graphite (i.e. plumbago) instead of carbon. The excellent lubricating properties of graphite and its high thermal conductivity make these brushes suitable for the larger heavy-duty d.c. generators and motors and for high-speed applications such as turbo-alternator slip-rings.

Natural Load. A load impedance, normally nearly resistive, equal to the characteristic impedance of the transmission line that it terminates. The term is used in connection with energy transmission over long power lines. Telecommunication lines are always terminated on their natural load in order to avoid signal reflection. They are then said to be *matched*.

Needle Gap. A Spark Gap used for measuring voltages of a few kilovolts. It has a larger, and therefore more convenient, spacing than a sphere gap at these low voltages. Consistent results are, however, difficult to obtain.

Negative Booster. A Booster arranged to reduce the voltage supplied from another source.

Negative Feedback. Return of energy from the output of an equipment to the input with reverse polarity or in anti-phase. It will reduce the gain of an amplifier but neutralizes a large amount of distortion. It is also known as *degenerative feedback*.

Negative Feeder. In an electric-traction system, the Feeder connecting the track rails or negative conductor rail to the negative busbars at the substation or generating station. It is also called a *return feeder*.

Negative Feeder-Booster. A Booster arranged to reduce the potential difference between two points of an earth return.

Negative Glow. The luminous glow in a discharge tube that may occur round the cathode at low gas pressures.

Negative Phase Sequence. A sequence of phase voltages or currents that is opposite to the normal (or positive sequence) in a polyphase a.c. system. If the positive sequence of three phases is ABC, then ACB represents a negative sequence.

Negative Resistance. The property of a circuit parameter such that the voltage across it is $V = -IR$. Such properties arise in certain parts of the current/voltage characteristics of devices in which, although they are always dissipative as a whole, there is a region in which the voltage falls as the current rises.

Neon Tube. An evacuated glass tube or bulb, with two electrodes in an atmosphere of neon or other gas at low pressure. By common usage, the term now includes discharge tubes normally used for exterior illumination.

Neoprene. Trade name for the first commercial synthetic rubber, polychlorobutadiene, derived from acetylene and hydrochloric acid. It is employed as a cable sheath.

Neper. A dimensionless unit used to express the scalar ratio of two currents or two voltages. In its passage through a passive quadripole having dissipation, a signal voltage or current suffers attenuation. When the quadripole is matched, the attenuation is represented by the ratio of output voltage to input voltage $|V_2/V_1|$, or output current to input current $|I_2/I_1|$. It is convenient to express these in the exponential form:

$$|V_2/V_1| = |I_2/I_1| = \exp(-\alpha)$$

Then α is the attenuation in nepers.

Network. A complex circuit comprising a number of *branches*, connected at junctions or *nodes*, and forming closed loops or *meshes*. The term "network" is used both for an actual electrical system, such as a power-supply network, and for the diagram by which a system can be represented.

Network Analyser. Device that facilitates the calculation of the performance of electrical power systems and other electrical circuits. Network analysers form a distinct group of electrical circuit machines in the general classification of analogue calculators. Arrays of circuit units represent the parameters of generators, lines, transformers and loads of a system. The problems for which a network analyser may be employed concern the currents and voltages to be expected in the real circuits under normal operating conditions, and also during faults or other disturbances. The phenomena may concern power frequencies only, or they may include high-frequency transient or resonant conditions. It is also

Network Analysis

known as a *network calculator*, *a.c. calculating table*, *a.c. board*, or *network model*.

Network Analysis. The determination of the voltage, current, power dissipation, energy storage, etc., in an electric circuit. All methods of analysis are based on KIRCHHOFF'S LAWS which, being based on the continuity of current and the principle of energy conservation, are always applicable. The direct application of the laws may be laborious, however, and several network theorems have been devised to facilitate circuit analysis (see, for example, SUPERPOSITION, COMPENSATION THEOREM, RECIPROCITY THEOREM, STAR/MESH CONVERSION and NODE-VOLTAGE ANALYSIS).

Network Function. See SYSTEM FUNCTION.

Network Synthesis. The process of designing an electric network to have specified properties; a process that is the inverse of NETWORK ANALYSIS. The properties required of the network to be designed are expressed by a SYSTEM FUNCTION which describes what kind of response is to result from a given stimulus. The synthesis proceeds on a step-by-step basis, to yield eventually a finite number of network elements (resistors, inductors, capacitors, ideal transformers) which form a network having the required properties.

Neutral. A point, conductor or region in an electrical system or device possessing symmetry in some special defined sense, e.g., the position of brush axis of a machine such that the electromotive force appearing at the brush is a maximum or minimum (see also NEUTRAL POSITION), or a point on a supply system which normally has the same magnitude of potential difference from each of the line conductors. In the latter case the point is usually at earth potential.

Neutral Conductor. Conductor connected to the neutral point of a supply system and laid along with the supply conductors. The neutral points of load equipment may be connected to it. It is usually earthed at the supply end of the system. The use of a neutral conductor gives a choice of voltages for supplying a two-terminal load.

Neutral Inversion. A phenomenon resulting in the phase-to-neutral voltages of a supply system becoming unequal so that the neutral-point potential moves away from the geometrical centre of the supply-voltage triangle, and takes a position possibly outside the triangle (Fig. N–1) where all phase-to-neutral voltages are far above their normal values. Such a situation may arise if

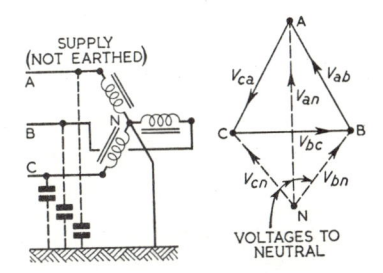

FIG. N-1. NEUTRAL INVERSION.

the impedances to neutral of the three phase conductors become widely unbalanced; this could occur in the normally balanced circuit where a capacitance is in parallel with an iron-cored inductance such as that of an unloaded power or voltage transformer, the only earth point on the system being that shown in the figure.

Neutral Position. A position of brushes in an electrical machine. In an a.c. machine it is that position obtained when the axes of the main stator and rotor windings coincide; in a commutator machine, it is the position that gives equal speeds in either direction of rotation when the machine is run at a constant load. In the latter case, it is approximately the position where the mutual induction between the field and armature windings is zero.

Neutral Zone. The part of the commutator of a d.c. machine in which, when the machine is running at no load, the voltage between adjacent bars is almost zero.

Neutron. Component of the nucleus of an atom, approximately equal in mass to the PROTON but having no associated charge. Neutrons resulting from nuclear fission have very high energies, e.g. 2 MeV, and may be used directly in a FAST REACTOR. In a THERMAL REACTOR they are slowed down by a MODERATOR, and are then known as *slow* or *thermal neutrons*.

Newton. The M.K.S. unit of mechanical force (abbreviation: N). It is that force which imparts to unit mass (1 kg) a unit acceleration (1 m/sec^2). It is also that force which, expended over the distance of one metre, does work amounting to one joule, so relating it directly to the usual electrical units.

Nickel. A silver white metallic element (symbol: Ni; atomic weight: 58·69; specific resistivity at 20°C: 10·9 $\mu\Omega$/cm^3). The principal uses of nickel are in alloy steels (to which it gives strength, valuable high-temperature properties and corrosion

Nickel Alloys

resistance); in electroplating and coinage; in battery manufacture; and in special ferromagnetic alloys.

NICKEL ALLOYS

Alloy	Composition	Application
ALCOMAX	Ni 10, Al 6, Co 25, Cu 2, Ti 2, Fe 55	Permanent magnets
ALNI	Ni 25, Al 12, Cu 4, Fe 59	Permanent magnets
ALNICO	Ni 18, Al 10, Co 12, Cu 6, Fe 54	Permanent magnets
ALUMEL	Ni 97, Al 3	Thermo-couples
Brightray	Ni 80, Cr 20	Heating elements
CHROMEL	Ni 80, Cr 20; etc.	Thermo-couples
CONSTANTAN	Ni 40, Cu 60	Resistors
Hypernik	Ni 50, Fe 50	High-μ
MUMETAL	Ni 76, Cu 5, Cr 2	High-μ
Nichrome	Ni 65, Cr 15, Fe 20	Heating elements
PERMALLOY	Ni 78·5, Fe 21·5	High-μ
Permalloy C	Ni 78·5, Mo 3·5, Fe 18	High-μ
Supermalloy	Ni 79, Mo 5, Mn 0·5, Fe 15	High-μ

Nickel Alloys. Mixtures of nickel with other elements, particularly iron. Nickel–iron alloys with 45–80% nickel have high permeability and low loss at small magnetic field intensities. Excellent magnetic properties are also obtained from alloys of nickel, iron and additional metals. The properties of the three *permanent-magnet* materials in the main table are (remanence B_r in Wb/m², coercivity H_c in AT/m, and $(BH)_{max}$ in J/m³):

	B_r	H_c	$(BH)_{max}$
Alcomax	1·2	40,000	30,000
Alni	0·6	40,000	10,000
Alnico	0·7	44,000	13,000

The *high-permeability* (high-μ) materials listed have initial and maximum relative permeabilities, coercivities in AT/m, saturation densities B_s in Wb/m², and resistivities ρ in $\mu\Omega$-cm, as follows:

Node-Voltage Analysis

	μ_i	μ_{max}	H_c	B_s	ρ
Hypernik	8,000	60	8	1·60	45
Mumetal	20,000	80	2·5	0·75	42
Permalloy	10,000	100	4	1·07	17
Permalloy C	21,000	75	4	0·85	55
Supermalloy	100,000	1,000	0·3	0·80	65

Nickel–Cadmium Cell. A STEEL–ALKALINE CELL in which the electrolyte is dilute potassium hydroxide, the positive plate nickel hydrate, and the negative plate cadmium with a small proportion of iron. The use of cadmium gives this cell a lower charging voltage than the NICKEL–IRON CELL and a reduced internal resistance, so that for those duties where the difference between charge and discharge voltage is important or where the ability to give high rates of discharge with a well-maintained voltage is essential, the nickel–cadmium cell is in more general use. It is employed for such duties as engine starting, switch closing and tripping, starting and lighting purposes on vehicles, for lighting and manoeuvring on trolleybuses, etc.

Nickel–Iron Cell. A STEEL–ALKALINE CELL in which the electrolyte is dilute potassium hydroxide, the positive plate nickel hydrate, and the negative plate iron. Its use is principally confined to such duties as traction, house-lighting, etc., for which duties it is particularly suited and where its rather higher internal resistance and charging voltage compared with a NICKEL–CADMIUM CELL are of little importance.

No-Lag Motor. Trade name for a compensated induction motor having stator and rotor similar to those of a SCHRAGE MOTOR but with the brush gear fixed.

Node. A point at which the amplitude of a waveform, such as a current or voltage wave, is zero.

Node-Voltage Analysis. A systematic method of solving for the voltages in a NETWORK. Each junction point is a NODE. One node is used as a reference so that the voltages of all the other independent nodes are found with respect to it. All sources are converted to equivalent current generators, and the total currents fed into the nodes are expressed by the simultaneous equations

$$I_1 = V_1 Y_{11} + V_2 Y_{12} + \ldots + V_n Y_{1n}$$
$$I_2 = V_1 Y_{21} + V_2 Y_{22} + \ldots + V_n Y_{2n}$$
$$\cdot \quad \cdot \quad \cdot \quad \cdot \quad \cdot \quad \cdot$$
$$I_n = V_1 Y_{n1} + V_2 Y_{n2} + \ldots + V_n Y_{nn}$$

Noise

Here $Y_{11}, Y_{22} \ldots Y_{nn}$ are the self-admittances of the nodes, i.e. the sum of all the admittances of the branches meeting at nodes $1, 2 \ldots n$. Terms of the form $Y_{12}, Y_{1n}, Y_{2n} \ldots$ represent the mutual admittances respectively between the nodes 1 and 2, 1 and n, 2 and $n \ldots$. In each case $Y_{pq} = Y_{qp}$, and if all the voltages of the independent nodes are taken as having the same polarity with respect to the reference node, then all the mutual admittances are taken as negative.

Noise. Unwanted electrical signals and their audible result. The term includes *transformer noise*, which is audible and has its origins in magnetostriction and in magnetic forces on the core laminations, and *circuit noise* which is random and may be due to electron movement (*thermal noise* or JOHNSON NOISE).

Nominal T and Π Networks. Approximate representations of a transmission line, for the purposes of analysis. The line can be considered to be passive QUADRIPOLE of T or Π formation. Such a quadripole can be devised to give characteristics identical with those of the line for any given frequency. Such T and Π sections would then be termed equivalent. For a power-transmission line not electrically long (e.g. not longer than 1/30 of a wavelength, corresponding to 200 km of overhead line worked at 50 Hz), a close approximation to the behaviour of the line is obtained from a quadripole made up of the total line impedance $Z = R + j\omega L$ in the series arm, and a total shunt admittance $Y = G + j\omega C$ comprising the total leakance and capacitive susceptance of the line. The sections so formed are called nominal because their lumped nature does not in fact exactly accord with the distributed parameters of the line, although the difference is small. In the T network, Y is concentrated symmetrically at the junction of two $\frac{1}{2}Z$s; while in the Π network, $\frac{1}{2}Y$ is placed at each end of the section.

Non-bleeding Cable. A pre-impregnated or drained impregnated cable. Under working conditions there is no danger of the impregnating material exuding.

Non-pressure Heater. Water heater having an open outlet. It is connected to a cold-water supply and is controlled by a stop valve on the inlet side (see Fig. N–2). It is also known as a *displacement heater*.

Non-reactive Load. A load in which the current and voltage are in phase at the terminals.

Nuclear Fusion

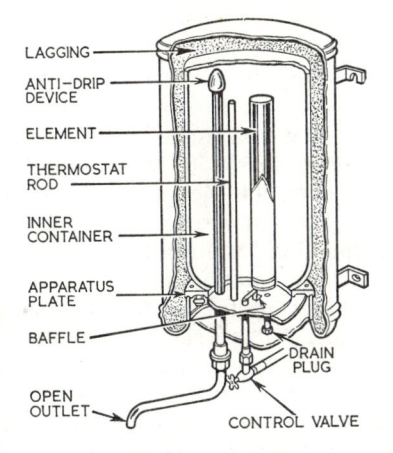

FIG. N–2. BASIC PARTS OF A NON-PRESSURE OR DISPLACEMENT WATER HEATER.

Normal Bend. A conduit fitting, having a longer radius than an ELBOW, to connect two lengths of conduit at right-angles. A similar fitting for connecting at $1\frac{1}{2}$ right-angles is a *half-normal bend*.

Norton Theorem. See HELMHOLTZ–NORTON THEOREM.

Nose Suspension. A method of mounting a traction motor on an electric truck. One side of the motor is supported on the axle by means of suspension bearings and the other on the framework by means of a lug projecting from the motor case.

Notching Controller. A switch actuated mechanically by the mechanism of a tap changer to ensure that the tap changer, having once been set in motion, continues its movement until the change in tapping is complete. It is also known as a *sequence switch*.

Nuclear Fission. Term applied to a particular type of nuclear reaction in which a heavy nucleus is split into two nuclei of medium weight. Some elements undergo fission spontaneously, but the most important reactions are those induced by neutrons. In each nuclear fission a large amount of energy is released and most appears as kinetic energy of the fission fragments. The great importance of fission, however, is due not only to the large amount of energy that is released but also to the fact that more than one neutron is emitted per fission. Hence a chain-reaction can be initiated. The process is initiated and controlled in a NUCLEAR REACTOR.

Nuclear Fusion. Combination or fusion of light elements

Nuclear Reactor

into heavier with the release of energy. This may be achieved in a *thermonuclear reactor* or PLASMA.

Nuclear Reactor. Plant for the controlled production of a neutron chain-reaction producing NUCLEAR FISSION. Bombardment of a uranium-235 nucleus by a neutron, under appropriate conditions in a nuclear reactor, results in the fission of the nucleus into two parts and the release of about 200 MeV ($3 \cdot 2 \times 10^{-11}$ J) of heat energy, and the production of two or three more neutrons which can immediately cause further fissions in adjacent nuclei of the uranium, thus setting up a chain reaction and giving rise to a continuous output of heat. The most useful fuel for a nuclear reactor is natural uranium, comprising 0·7% fissile uranium-235, the rest being uranium-238. Uranium-235 is the only fissile material occurring in nature, but two other fissile materials, plutonium and uranium-233, can be made artificially by neutron bombardment of uranium-238 and thorium, respectively. A nuclear reactor is sometimes known as an *atomic pile*. See also FAST REACTOR, THERMAL REACTOR.

Nucleon. Constituent of an atomic nucleus; principally a PROTON or a NEUTRON.

Null Detector. Instrument for comparing, by equating them, an unknown quantity with a known, e.g. a galvanometer in a bridge circuit.

Null Measurement. Method of making an electrical measurement in which the quantity to be measured, which must be steady, is compared with a known quantity using an indicating instrument as a NULL DETECTOR. It is more accurate than *deflectional* methods and so is employed for high precision and for calibrating pointer instruments for commercial use.

Nylon. A thermoplastic material with a wide range of uses. It is chiefly used in the electrical industry as cable sheathing, and for the production of small moulded components. It has great mechanical strength.

Nyquist Criterion. A means whereby the degree of stability or instability of a feedback system may be assessed from its complexor locus diagram.

O

O-Class Insulation. A class of insulating materials now designated Y-CLASS INSULATION (q.v.).

Oersted. The C.G.S. unit of magnetizing force, equal to $4\pi/10$ times the ampere-turns per centimetre.

Off-Peak. During a period when the demand on the power supply system is not at a maximum, e.g. at night. See PEAK LOAD.

Off-Peak Tariff. A TARIFF applied to units supplied for storage heaters or other loads that are automatically switched on during selected off-peak periods only.

Ohm. The M.K.S. and practical unit of resistance (abbreviation: Ω). A conducting path has a resistance of 1 Ω when the passage of a current of 1 A requires a potential difference across the path of 1 V.

Ohmic Resistance. The resistance offered by a conductor to the free flow of a direct current, undisturbed by such influences as eddy currents, thermo-e.m.f.s, or pinch, skin and proximity effects.

Ohmmeter. Instrument containing voltage and current coils to read the quotient of the voltage across an unknown resistance and the current through it. A major application, incorporated in an instrument with a hand-driven magneto generator, is the measurement of insulation resistance.

Ohm's Law. The current I in a conductor is proportional to the potential difference V across its ends. The usual form of the law is $V/I = R$, where R is the resistance (unit, ohm).

Oil. See INSULATING OIL.

Oil Circuit-Breaker. Circuit-breaker in which the arc is drawn in oil and its high temperature causes the formation of a gas bubble consisting mainly of hydrogen (see Fig. O–1). The turbulence of the gas and the cooling effect of the surrounding oil disperse and de-ionize the arc path at a current zero and restore its dielectric strength. Arc-control devices are used to confine the arc and improve performance. The main types of oil circuit-breaker are the BULK-OIL CIRCUIT-BREAKER and the SMALL-OIL-VOLUME CIRCUIT-BREAKER. IMPULSE CIRCUIT-BREAKERS belong to the second category.

Oil Conservator. A chamber fitted above the main tank of an oil-filled transformer. It normally has a capacity of about 10% of that of the main tank. It is connected to the tank by a small-diameter pipe so that oil may expand freely into it, and is intended to ensure that the oil is maintained in good condition, firstly by excluding moisture and secondly by reducing the area of the oil in contact with the air.

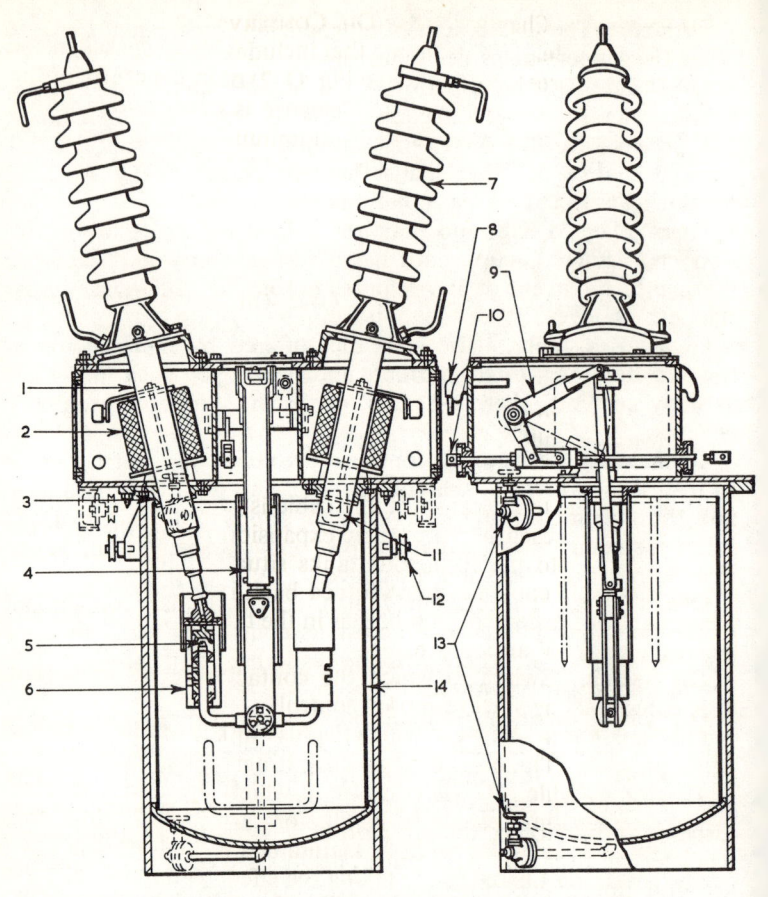

FIG. O–1. SECTIONAL DRAWING OF ONE PHASE OF 66 kV FRAME-MOUNTED OIL CIRCUIT-BREAKER.

1. Terminal bushings.
2. Current transformer cores and secondaries.
3. Oil sealing gland.
4. Guides, opening springs and dashpots.
5. Moving contacts.
6. Arc-control pots and contacts.
7. Porcelain weather shield.
8. Vent.
9. Phase-operating mechanism.
10. Flexible coupling.
11. Oil level indicator.
12. Rope pulleys for tank-lowering gear.
13. Filtering and sampling cocks.
14. Vulcanized fibre insulating linings.

(*The General Electric Co. Ltd.*)

On-Load Tap Changing

Oil Expansion Chamber. See OIL CONSERVATOR.

Oil-Filled Cable. Cable design that includes metal spirals in the interstices of a three-core cable (see Fig. O–2) or in the centre of the strand of a single-core cable. The dielectric is impregnated with a thin oil after a reinforced-lead or aluminium sheath has been

FIG. O–2. OIL-FILLED THREE-CORE CABLE.

applied. As the temperature of the cable is increased by loading, the excess oil caused by volumetric expansion passes through the metal spirals into membraneous tanks situated along the route. When the cable cools the oil is forced back into the cable by the movement of the flexible membranes in the tanks. Oil-filled cables are used at 33 kV and above.

Oil Switch. Switch in which the contacts are oil immersed. It is commonly applied to motor control.

Oil-Tank Fuse. A fuse in which the fuse-link is totally immersed in an oil tank. The oil acts as an arc-extinguishing medium.

Oilostatic Cable. Oil-filled cable in which the sheath is eliminated and the three insulated conductors are pulled into a steel pipe which is then filled with oil maintained at high pressure.

On-Load Tap Changing. Varying the effective number of turns in the primary or secondary windings of a transformer while the transformer is on-load. There must be no interruption in the winding circuit while the selector switches operate to change the active tappings. This means that, for a short period, there must be simultaneous connection to two adjacent tappings. Means must therefore be provided to limit the short-circuit current in the turns between the bridged tapping points.

The methods in use are illustrated in Fig. O–3. They are:
(a) *Parallel Windings.* With the tapping section (or all) of the winding arranged as two parallel paths, each provided with

Open and Closed Test

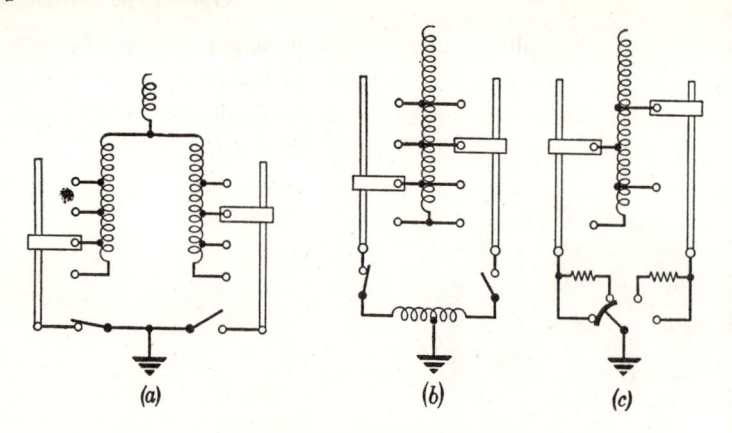

FIG. O-3. THREE METHODS OF TAP-CHANGING.
They involve (a) parallel windings, (b) an inductor bridge, and (c) a resistor bridge.

tappings, the circulating current is limited by the inherent leakage reactance between the parallel halves. (b) *Inductor Bridge*. An inductive reactor is used to make the temporary bridge between successive tappings. Normally the two sides of the reactor convey equal currents in opposite directions, so that its effective impedance is low. (c) *Resistor Bridge*. A pair of resistors is connected to be in series as regards the circulating current between tappings. It is not permissible to operate with the tap-changer device permanently in the central switch position.

Open and Closed Test. A modification of the MURRAY LOOP TEST for locating faults in cables, designed for use when the insulation of the return conductor has a low value.

Open- and Short-Circuit Characteristics. Relations from which all the essential details of the circuit behaviour of an a.c. device can be, in principle, derived.

Consider a passive QUADRIPOLE in which input and output voltage and current, $V_1 I_1$ and $V_2 I_2$, are related by the general parameters $ABCD$ so that

and
$$V_1 = V_2 A + I_2 B$$
$$I_1 = V_2 C + I_2 D$$

If the receiving end is open-circuited, then $I_2 = 0$, whence $V_1 = V_2 A$ and $I_1 = V_2 C$. Thus, if the input voltage and current,

Open-Type Switchgear

and the output voltage, are measured in magnitude and phase, then
$$A = V_1/V_2 \quad \text{and} \quad C = I_1/V_2.$$
Again, if the output terminals are short-circuited, $V_2 = 0$, giving
$$B = V_1/I_2 \quad \text{and} \quad D = I_1/I_2.$$
As the parameters $ABCD$ are now known, the behaviour of the quadripole can be predicted in network terms. Examples of *o.c.c.* and *s.c.c.* are given in Fig. O–4.

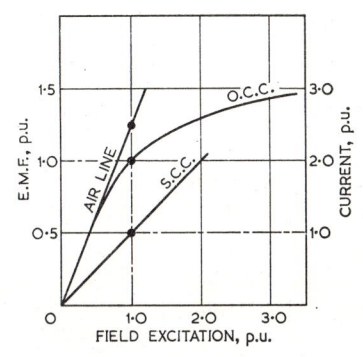

Fig. O–4. Open-Circuit Characteristic, Short-Circuit Characteristic and Air-Gap Excitation of a Synchronous Machine

Open Arc. A carbon arc which is freely exposed to the external atmosphere. It may however be partially enclosed to eliminate draughts or diffuse the light.

Open-Circuit. A break or electrical discontinuity in a circuit through which current can normally flow. The term is applied to a generator or transformer when a voltage difference exists between its terminals which is rendered ineffective by the lack of a complete connecting circuit.

Open-Circuit Characteristic. See Magnetization Characteristic, Open- and Short-Circuit Characteristics.

Open Machine. Machine in which no restriction is placed on the ingress of ventilating air and in which the commutator and other moving parts are easily accessible to accidental contact and ingress of dirt.

Open-Type Switchgear. Switchgear in which the various items are mounted on steel or concrete structures and connected with bare conductors. The use of air as the main insulation reduces the cost, provided that the site is cheap. For this reason a rural or suburban site is normally chosen. Voltages range from 11 kV to the highest in use.

Optical Pyrometer

Optical Pyrometer. A pyrometer which compares the intensity of visible radiation from a body with that of the same colour and wavelength derived from a calibrated comparison source, usually a filament lamp.

Orbital Accelerator. A direct PARTICLE ACCELERATOR having either a constant magnetic field in which the particles move in orbits consisting of a series of arcs of circles of discrete and increasing radii (CYCLOTRON, SYNCHROCYCLOTRON or MICROTRON), or a changing magnetic field with nearly constant orbit radius (BETATRON, SYNCHROTRON).

Oscillating Neutral. An insulated star-point that oscillates about the neutral potential at third-harmonic frequency. It may occur when the load comprises iron-core devices, and the insulated star-point of the generator is earthed.

Oscillograph. Literally an instrument for writing or recording oscillations. The term is sometimes used for an OSCILLOSCOPE. A mechanical form is the DUDDELL OSCILLOGRAPH.

Oscilloscope. Instrument for recording oscillations in a visible form. Its most common form incorporates a CATHODE-RAY TUBE.

Output Coefficient. A coefficient expressing the volt–ampere rating of an electrical machine per unit volume and per unit speed.

Overcompensation. The result of a COMPENSATING WINDING having a greater effect than the armature reaction which it opposes. A cumulative action is obtained, tending to make output voltage rise as load current is increased.

Overcompounded. Term applied to a compound-wound generator in which the series winding is so designed that the voltage increases with the load.

Overcurrent Protection. Protection of a system against excessive currents. The FUSE forms the basis of most small, simple distribution-system protection, combining overcurrent protection and fault isolation. However, its general application on more extensive networks is limited by the difficulty of achieving satisfactory discrimination between fuses connected in series and the inconvenience of replacement after operation. British practice favours the *definite-minimum-time relay*. There are various types of construction, mostly employing an eddy-current disc in which torque is developed by the interaction of fluxes. Movement of the disc is controlled by an eddy-current brake and a return spring. Essential characteristics are: (*a*) a standard inverse time/current characteristic with a definite minimum operating time at large

Overhead

multiples of setting current; (*b*) adjustment for current setting, usually by tapped operating coil; (*c*) adjustment for time setting, usually by travel of the disc and moving contact. An overcurrent relay is frequently required to operate only for fault power flow in one direction. See also OVERLOAD RELAY.

Overhang. The part of a winding also known as the END WINDING (q.v.).

Overhead Distribution. The supply of power from generating stations or main substations by means of tower-supported transmission lines to load centres or secondary substations (*primary distribution*), or the supply of power by similar means from secondary substations to consumers' premises (*secondary distribution*).

Overhead Line. Cable employed in overhead distribution of power. In the early years of overhead lines, conductors were invariably of hard-drawn (*h.d.*) copper used at moderate voltages with short-span wood-pole supports. Economic trends called for tighter sags and longer spans. Alternative and stronger conductor materials were introduced for this purpose: cadmium–copper alloy, steel-cored copper, copper-cored steel, and finally aluminium alloy and steel-cored aluminium (Fig. O–5).

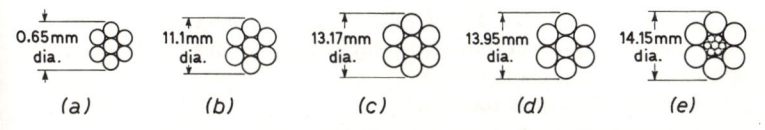

FIG. O–5. COMPARATIVE SIZES OF CONDUCTORS.
The conductors of the dimensions shown have approximately equal conductances.
(*a*) Copper 7/3·55 mm, (*b*) Cadmium copper 7/3·70, (*c*) Aluminium 7/4·39 mm, (*d*) Aluminium alloy (Silmalec) 7/4·65 mm, (*e*) Steel-cored aluminium 6/4·72 mm aluminium + 7/1·57 mm steel.

Overlap Angle. The time angle during which successive anodes conduct simultaneously in a multi-anode rectifier. It results from the effect of anode-circuit inductance, which prevents the currents conducted by anodes from being instantaneously commutated.

Overlap Span. Alternative name for SECTION GAP (q.v.).

Overload. A load on a machine or system which is greater than it is designed to withstand continuously. It is expressed numerically as the amount in excess of the rated load.

Overload Protection

Overload Protection. See OVERCURRENT PROTECTION, OVERLOAD RELAY.

Overload Relay. Relay designed primarily to protect a motor or other equipment from damage due to currents in excess of normal. In the case of motors, the relays should generally incorporate some form of time delay to prevent their operation during normal starting current peaks and also during short-time overloads insufficient to damage the motor. Upon operation they open the starter usually by means of an associated normally closed contact breaking the coil circuit of the undervoltage release. The relays can be divided into two main groups, THERMAL OVERLOAD RELAYS and MAGNETIC OVERLOAD RELAYS.

Overvoltage. (1) An increase in system voltage above the normal or declared values. It may be transient or sustained. A *transient* overvoltage may cause damage in the system through electrical breakdown but is unlikely to cause breakdown through excess current resulting from the overvoltage. A *sustained* overvoltage may of course lead to breakdown in the system through the excess current causing overheating.

(2) In electrochemistry, the difference between the electrode potential required to liberate a gas, usually hydrogen, at a specific electrode, and the observed, or calculated potential in the same reaction at the same current density on an electrode free from polarization. The term is also applied to other electrode processes, such as metal deposition.

Overvoltage Test. Test applied to transformers, etc., to prove the quality of the internal insulation. It may be an induced-voltage test or a separate source test.

Oyster Fitting. A robust light fitting designed for attachment to bulkheads, etc., where space is restricted, and emitting light simultaneously on both sides of the bulkhead.

P

pH Meter. A d.c. valve voltmeter arranged to measure the electromotive force produced in a HALF CELL, the electrolyte of which is the solution on test, and hence give indication of the pH VALUE of the solution.

pH Value. A term used to indicate the concentration of hydrogen ions in a solution (i.e. the acidity). It is the common logarithm of the reciprocal of the concentration.

p–n–p Transistor

p–n Junction. Junction between a *p*-TYPE SEMICONDUCTOR and an *n*-TYPE SEMICONDUCTOR, having rectifying properties. If a voltage is applied as shown in Fig. P–1 (*a*), electrons in the *n* side will be attracted across the junction to the positively biased terminal and holes in the *p* side to the negatively biased terminal, and

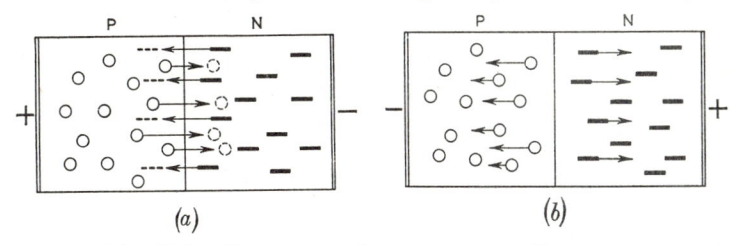

FIG. P–1. RECTIFYING ACTION OF A *p–n* JUNCTION.
(*a*) Forward bias; (*b*) Inverse bias.

a large current will flow. Now if the polarity is reversed, (*b*), both electrons and holes will be attracted away from the junction until the electric field produced by their displacement equals the applied field. Only a small current can flow, due to free electrons and holes being produced near the junction by thermal agitation.

p–n–p Transistor. TRANSISTOR consisting basically of two *p*-type areas with a thin *n*-type layer between them (Fig. P–2).

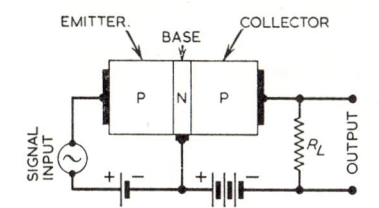

FIG. P–2. BASIC *p–n–p* TRANSISTOR.

The first *p* region is the *emitter*, the *n* region the *base*, and the second *p* region the *collector*. There are thus two *p–n* rectifying junctions, the emitter junction being biased in the forward direction and the collector junction in the reverse direction. The resistivity of the emitter *p* region is made very low compared with the base *n* region, and so the forward current across the emitter-base junction will consist mainly of a flow of holes from the emitter to the base. Once these holes are in the base they diffuse away from the emitter junction and most of them eventually come within

p-Type Semiconductor

the field existing across the collector junction, when they are immediately attracted into the collector region by the negative voltage gradient. Thus, the hole flow produced across the emitter junction by a low voltage causes a current to flow through the high-resistance collector load R_L and produces a high voltage output.

_p_-Type Semiconductor. A quadrivalent SEMICONDUCTOR which includes a suitable trivalent *acceptor* element so that the crystal lattice contains a number of holes, or positive charge carriers.

Π-Network, Pi-Network. A network containing a shunt arm, a series arm and a further shunt arm equal to the first (see Fig. P–3).

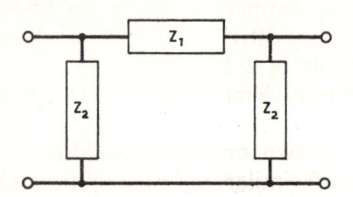

FIG. P–3. A SYMMETRICAL UNBALANCED Π-NETWORK.

Palmer Limit Switch. A special switch for use with d.c. series motors on cranes, where the motor may often be running at over double normal speed and quick stopping is necessary. It has two poles normally closed and two normally open, the latter closing immediately the former have opened. In use it is connected with armature and field in such a way that the main circuit is opened and then a dynamic braking resistance connected across the armature, the current in the braking resistance being also led round the field so as to maintain it in the same direction as immediately prior to the operation of the switch.

Pantograph. A sliding current-collector used in electric traction with overhead wires. It consists of a bow-shaped contact strip mounted on a lozenge-shaped or cross-arm frame which is hinged to provide vertical motion for the contact strip.

Paper. Cellulose fibres felted to form a mechanically strong sheet.

Cable papers have a density normally from 0·7 to 0·9 g/cm³ dry. To improve the dielectric strength of a cable paper it is impregnated with a suitable insulating oil so that approximately one-half of the cable dielectric is occupied by the impregnant. The permittivity is approximately 2·25, but the power factor varies somewhat with

Parallel-Plate Capacitor

temperature, between 0·002 at 20°C and 0·0027 at 100°C for wood-pulp paper, and somewhat less for washed wood-pulp paper.

Capacitor paper is usually a kraft paper of the order of 10 to 15 μm thick impregnated with mineral oil or petroleum jelly. **The effective permittivity at 50 Hz of an oil-impregnated paper is of the order of 4·0.** The loss angle at 50 Hz is 0·0015 to 0·003; the insulation resistance at 20°C is 2×10^4 ohm-farads; the approximate working stress is 15 to 20 r.m.s. V/μm; and the working temperature range up to 90°C.

Paraffin Wax. A wax-like substance left after the distillation of petroleum, employed for the impregnation of some capacitors, radio coils, etc., where the operating temperatures are not high. It has a melting point of 50–60°C, resistivity at 20°C of 10^9–10^{13} MΩ per cm cube, and permittivity of 2·2.

Parallel-Generator Theorem. Alternative name for the MILLMAN THEOREM (q.v.).

Parallel Operation. Operation of two or more machines or transformers with their terminals of similar polarity connected together. This implies equality of terminal voltage. This condition imposes restrictions on the characteristics of the machines, and affects stability and load sharing.

The conditions that must be observed for the parallel connection of *transformers* on both primary and secondary sides are that they should have the same voltage-ratio, the same polarity, the same phase-sequence, and zero relative phase displacement. *D.C. generators* are not often worked in parallel with each other, but the case of a d.c. generator paralleled with a battery is very common. When two *synchronous generators* are connected in parallel, they have an inherent tendency to remain in step, on account of the changes produced in their armature currents by a divergence in phase, and the development of synchronizing power.

Parallel-Plate Capacitor. A capacitor consisting basically of two equal parallel plates separated by a dielectric (Fig. P–4). If the area of each plate is A, their separation s, and the charge 1C,

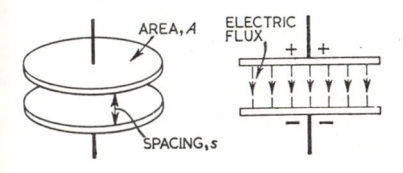

FIG. P–4. A BASIC PARALLEL-PLATE CAPACITOR.

Parallel-T Bridge

then ignoring edge effects, the dielectric flux density at any point between the plates is $D = 1/A$, so that the electric field intensity $E = D/\epsilon = 1/A\epsilon$. Because the field is uniform, integration of E yields $Es = V = s/A\epsilon$. Hence the capacitance is

$$C = 1/V = \epsilon A/s \text{ farads.}$$

Parallel-T Bridge. A six-arm A.C. BRIDGE for use at radio frequencies. It has the advantage that one end of the supply, the detector, unknown impedance Z_3 and standard comparison arm Z_6 (see Fig. P–5) are all at earth potential. Star/delta conver-

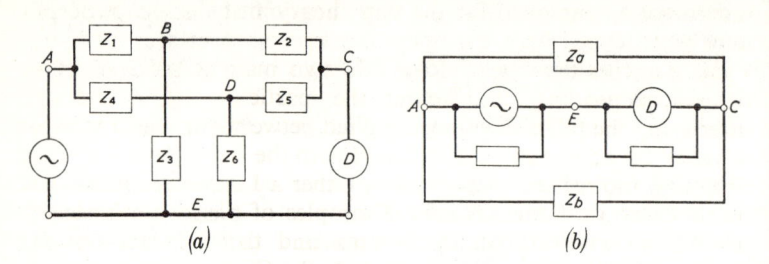

FIG. P–5. A PARALLEL-T BRIDGE AND ITS STAR/DELTA CONVERSION.

sion gives the form shown in (b), from which it is seen that balance is obtained with $Z_a = -Z_b$, where

$$Z_a = Z_1 + Z_2 + Z_1 Z_2/Z_3,$$
$$Z_b = Z_4 + Z_5 + Z_4 Z_5/Z_6.$$

Suppose Z_1 and Z_2 are pure reactances of like sign and Z_3 is a resistor: then

$$Z_1 Z_2/Z_3 = (-jX_1)(-jX_2)/R_3 = -X_1 X_2/R_3$$

equivalent to a negative resistance, enabling a real resistance in Z_b to be balanced out. In a typical arrangement Z_1 and Z_2 are equal capacitors, Z_4 is a capacitor, Z_5 is a resistor, and Z_3 is a capacitor across which the test impedance can be connected.

Paramagnetism. Phenomenon exhibited by materials having a constant permeability greater than that of a vacuum, i.e. greater than unity.

Parametric Amplifier. A reactance amplifier working on the principle that if, in an oscillatory LC circuit, either the capacitance C or the inductance L is varied periodically at appropriate frequency and in appropriate phase, electrical oscillations can be maintained in the circuit.

Particle Accelerator. A machine for accelerating electrons or ions to high energies. The lowest energy accelerators can be used to obtain information on the simple characteristics of the nucleus; they are also used considerably in medical research and in engineering, and for food sterilization. At higher energies, the velocities of the accelerated particles are so great that they may be considered as interacting with individual nucleons in the nucleus. They may therefore be used to investigate the forces acting between nucleons. At energies above 1 BeV, heavy mesons and hyperons may be produced.

Accelerators may be divided into two main classes, *direct* or *indirect*, depending on whether the particles are accelerated directly by the full d.c. voltage applied between two electrodes, or whether the acceleration is obtained with the aid of an r.f. electric field. An indirect machine may be either a LINEAR ACCELERATOR or an ORBITAL ACCELERATOR. Examples of direct machines are the COCKROFT-WALTON MULTIPLIER and the VAN DE GRAAFF GENERATOR. Orbital machines include the CYCLOTRON, SYNCHROCYCLOTRON, MICROTRON, BETATRON and SYNCHROTRON.

Paschen's Law. In a semi-evacuated space, containing two plane-parallel electrodes, the sparking, breakdown or disruptive voltage depends upon the number of gas atoms in the interelectrode space.

Pasted Plate. A plate of a LEAD-ACID CELL which is made mechanically by the application of lead-oxide paste on to a grid. It is also known as a *Faure plate*.

Paxolin. Trade name for a type of SYNTHETIC-RESIN-BONDED PAPER (q.v.).

Peak Factor. The ratio of the PEAK VALUE of an alternating or pulsating wave to its ROOT-MEAN-SQUARE VALUE. A sine wave has a peak factor of $\sqrt{2}$.

Peak Load. The maximum demand on a power supply system. To meet it, the less efficient generating stations are employed in addition to those supplying the BASE LOAD.

Peak Value. The maximum instantaneous positive or negative value of an alternating, oscillating or vibrating quantity. It is sometimes referred to as the *amplitude*.

Pearl Lamp. An electric lamp having its bulb etched (or *frosted*) on the inside to diffuse the light.

Pedersen Potentiometer

Pedersen Potentiometer. A quadrature-type A.C. POTENTIO-
METER. Two sets of slide-wires and resistance coils are fed in
quadrature through inductive and capacitive reactances respectively to secure a 90° displacement (see Fig. P–6). Separate
measurements are taken of each coordinate of voltage, and provision is made for reversal so that voltages in any quadrant can be
balanced.

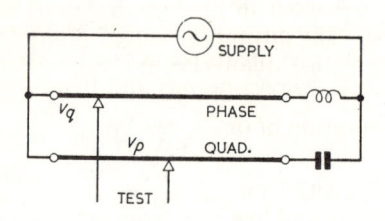

FIG. P–6. PRINCIPLE OF PEDER-
SEN A.C. POTENTIOMETER.

Peltier Effect. A thermoelectric effect. When an electric
current is maintained across the junction of two dissimilar metals,
the junction temperature changes unless heat is supplied or
removed by external means. The rate at which heat must be
supplied to the juntion to keep its temperature constant is
proportional to the current, and changes sign when the current
direction is reversed. If the current flowing across a junction
causes an absorption of πi ergs per second, π is known as the
Peltier coefficient.

Pelton Wheel. An impulse-type WATER TURBINE developed
from the simple water wheel. It consists of a series of buckets
placed round the perimeter of a disc, two roughly hemispherical
buckets being employed at each position with a cutaway portion
in the centre to prevent obstruction to the jet of water by the
succeeding pair of buckets. It is fundamentally suitable for dealing
with a high water velocity rather than with the large quantities of
water which must of necessity be employed to produce the same
power under low-head conditions. Two or more jets may be
employed on a single runner, or two runners may be coupled
together on the same shaft, each with one or two jets.

Pentode. An electron tube or valve having five electrodes.

Pentode Ignitron. An IGNITRON for high-voltage applications.
It has three grids in addition to its anode and cathode. It may be
used for high-voltage rectification and inversion, or as a high-
voltage switch for capacitor switching in special pulse applications.

Per-Unit System. A method of designating parametric constants and operating values, in common use in the analysis of electrical machines and supply systems. The various values can be expressed by the appropriate numbers of ohms, volts, amperes, etc. But these do not sufficiently represent the comparative importance of the factors involved. A better indication is given by expressing the quantities in percentage values to a common base. In the per-unit system the quantities are expressed in decimal fractions of unity instead of in parts per 100. The advantage is that multiplications or divisions of per-unit (p.u.) quantities can be carried out straightforwardly.

Period, Periodic Time. The duration of one cycle of a sustained oscillation or alternation.

Permalloy. An alloy of nickel and iron containing over 78% of nickel. It has a high permeability and low coercivity, and hence low hysteresis loss. See NICKEL ALLOYS.

Permanent Magnet. Magnet made from a material which, when once magnetized, retains more or less permanently the greater part of its magnetic properties. Powerful magnets may be constructed of barium ferrite.

Permeability. The ratio, in any material medium, of magnetic flux density to magnetic field intensity, compared with the same ratio for free space. It is preferably known as *relative permeability*. See ABSOLUTE PERMEABILITY.

Permeameter. Instrument for measurement of the magnetic characteristics of ferromagnetic materials (see YOKE PERMEAMETER).

Permeance. The magnetic flux per ampere-turn of total magnetomotive force in the path of a magnetic field. It is the reciprocal of RELUCTANCE.

Permeance Coefficient. The ratio of area to length of a non-magnetic path. If permeance is Λ, permeance coefficient λ, and absolute permeability of free space μ_0, then

$$\Lambda = \mu_0 \lambda.$$

Permit To Work. A form of declaration issued to a person in charge of work on or near electrical equipment. It identifies which equipment is isolated and earthed, and which it is safe to work on or near.

Permittivity. The ratio, in an insulating material, of the electric flux density to the electric field intensity, compared with the same

Perspex

ratio for free space. It is alternatively known as *relative permittivity, specific inductive capacity* or *dielectric constant*. See ABSOLUTE PERMITTIVITY.

Perspex. Trade name for a POLYMETHYL METHACRYLATE resin. The outstanding characteristics of Perspex are its stability, relatively high softening point and clarity.

Petersen Coil. See ARC-SUPPRESSION COIL.

Phase Advancer. A machine that provides an e.m.f. for injection into the rotor circuit of a slip-ring induction motor to provide magnetizing volt-amperes, so that the power factor at the stator terminals is improved. Four basic types of machine may be employed (see LEBLANC ADVANCER, WALKER ADVANCER, SCHERBIUS ADVANCER and FREQUENCY CONVERTER).

The KAPP VIBRATOR is a distinct type of phase advancer which produces the equivalent of a very large capacitor in the rotor circuit.

Phase Angle. The angle by which a sinusoidally varying quantity is displaced in time from another of the same frequency.

Phase-Change Coefficient. The change of phase angle of a sine signal passed through a quadripole; or the change per unit length of an electromagnetic wave in its progress along a transmission line, through a material medium, or through free space (see also PROPAGATION COEFFICIENT).

Phase Changer, Phase Converter. A machine for converting alternating current of a certain number of phases to alternating current of a different number.

Phase-Equalizer. Alternative name for ABSORPTION INDUCTOR (q.v.).

Phase Modifier. A synchronous machine arranged for control of its excitation to take leading or lagging reactive volt–amperes so that a load may have its effective power factor modified; in particular a synchronous condenser connected at the end of a transmission line for this purpose.

Phase Modulation. The imposition of a signal on to an alternating carrier wave in such a way that the instantaneous phase of the carrier varies by an amount proportional to the amplitude of the modulating signal.

Phase-Rotation Indicator. An instrument to indicate phase rotation comprising an iron base with three short iron poles each carrying a coil connection to one lead, with a common neutral point which is not brought out. Pivoted above the magnetic

Phase-Shifting Transformer

pole is a light iron disc which will revolve when the coils are connected to a three-phase supply. The direction of rotation will depend on, and indicate, the phase sequence of the supply.

Phase-Sequence Components. Components into which the current values are resolved in analysis of an unbalanced system by method of SYMMETRICAL COMPONENTS.

Phase-Shift Distortion. A change in waveform, where the input components have been altered in their relative phase because of a non-linear phase characteristic which does not produce a phase shift that is an integral multiple of 2π radians. This is also known as *delay*.

Phase-Shifting Transformer. Transformer employed where it is necessary to alter phase. Two types have been developed (see Fig. P–7). The first is based on a 3/12- or 3/24-phase transformer.

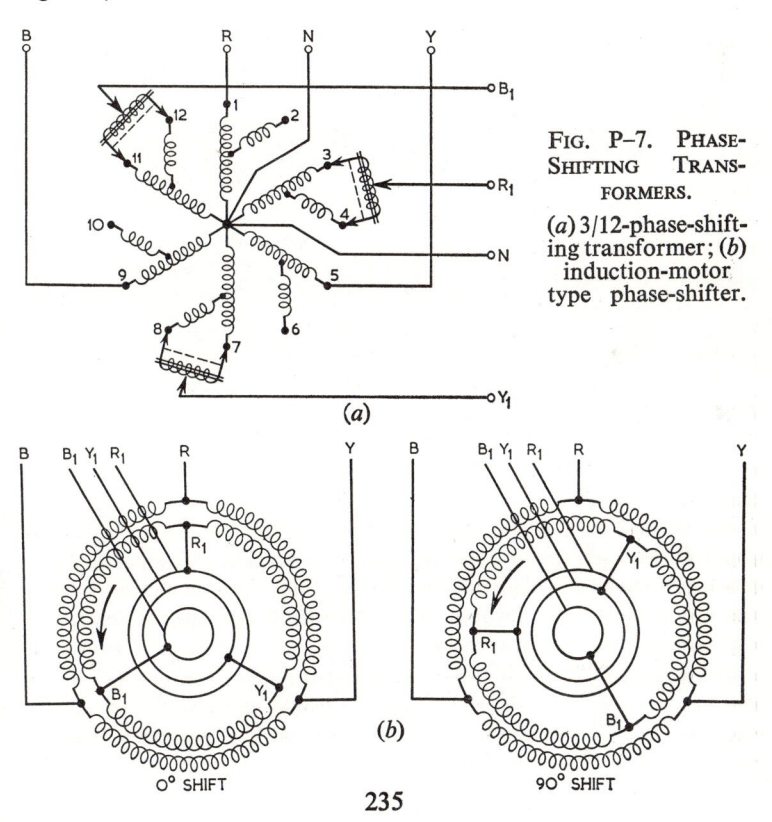

FIG. P–7. PHASE-SHIFTING TRANSFORMERS.

(*a*) 3/12-phase-shifting transformer; (*b*) induction-motor type phase-shifter.

Phase Voltage

This arrangement gives voltage spaced every 30° for a 3/12-phase or every 15° for a 3/24-phase transformer. To obtain fine adjustment of the phase angle between the 30° and 15° steps, variable-ratio auto-transformers of the VARIAC type can be connected between adjacent phases as in (*a*). The second type is virtually a three-phase slip-ring induction motor in which the rotor is turned to any desired fixed position by hand. When the primary (stator) winding is energized from a three-phase supply, the voltages induced in the secondary (rotor) winding will remain substantially constant in value but their phase relationship with respect to the primary voltages will depend on the relative positions of the windings. This is shown in (*b*) for two positions of the rotor.

Phase Voltage. See STAR VOLTAGE.

Phasor. A term describing the representation of a steady-state sine-varying quantity (such as an alternating current or voltage) as a complex number (see COMPLEXOR).

Phenolic Resin. Thermosetting resin produced by the controlled interaction of phenol and formaldehyde. It is used in varnish form to impregnate the filler materials in the majority of thermosetting laminates employed in the electrical industry. Phenolic materials play an important part in insulators for electrical generation and distribution, particularly in switch and control gear.

Phosphor. Material having the property of emitting visible light when bombarded with swiftly moving electrons or irradiated with certain forms of electromagnetic radiation. Suitably prepared zinc-sulphide phosphors are outstanding in exhibiting ELECTROLUMINESCENCE.

Photoconductivity. Property possessed by certain materials, notably selenium, whereby their conductivity increases with the intensity of the light falling upon them.

Photoelectric Cell, Photocell. A light-sensitive device comprising two electrodes in a glass bulb, evacuated or filled with an inert gas. One of these electrodes, the cathode, has a surface coating of an intermetallic compound of one of the alkali metals, the layer being of molecular dimensions and giving off electrons by PHOTO-EMISSION when illuminated. The other electrode, the anode, which is maintained at a potential positive to the cathode, serves to collect the emission from the cathode surface. The current in the external circuit increases with the illumination but is not necessarily proportional to it. The device is also known as an *emission cell*.

Photoelectric Effect. A change produced, by the action of light, in the electrical properties of a body.

Photoelectric Relay. A combination of a photoelectric device operating a conventional sensitive relay, either directly or through an amplifier.

Photoemission. The emission of electrons from a surface due to the incidence of light. When a metal surface is irradiated by light, low-energy electrons are emitted provided the photon energy, which is absorbed and imparted to the lattice electrons, is at least equal to the work-function for the metal. Each photon emits only one electron and therefore the number of electrons emitted per second (photo-current) is proportional to the intensity of the light (number of photons per second).

Photometry. The science of light measurements. The photometric measurements most commonly made are those of the light output of a source, the illumination of a surface, and the brightness of an illuminated or self-luminous surface. The total light output of a source is measured in lumens (lm); the directional luminous intensity is measured in candelas (cd), formerly "candles".

Photosensitive Transistor. A TRANSISTOR which uses the fact that a $p-n$ junction is photosensitive. Conventional transistor structures are used, arranged to receive incident light on the emitter junction.

Pi-Network. See Π-NETWORK on page 228.

Piezo Crystal. A crystal which exhibits the phenomenon of PIEZOELECTRICITY. Many do to some degree, but only about ten are suitable for practical applications. Four have been exploited commercially: ROCHELLE SALT, QUARTZ, AMMONIUM DIHYDROGEN PHOSPHATE and lithium sulphate. The last is used in ultrasonic flaw detectors.

Piezoelectricity. The production of an electric charge on the surfaces of a PIEZO CRYSTAL when it is subjected to mechanical strain or pressure. The converse effect is that if an electrical charge is applied to two surfaces the crystal will mechanically deform. In both cases there is a linear relationship between the magnitude of electrical and mechanical factors.

Pilot Controller. A multi-way switch for controlling the operation of a set of contactors. It is sometimes called a *master controller*.

Pilot Wire. A conductor in a power system used for auxiliary purposes such as measurements, communications or protection.

Pilot-Wire Protection

Pilot-Wire Protection. A fast-acting form of network protection. Used in conjunction with overcurrent protection it reduces operating times and alleviates difficulties of grading. The basic principle of pilot protection is the algebraic comparison via pilot wires of some quantity which is equal at the two ends for external or through-fault conditions; the relays must then stabilize for equality or near-equality and trip if there is wide disparity. In a *balanced-voltage* system, a voltage, varying as the current-transformer primary current, is produced at each line end and balanced on the pilot wires, which thus carry no current (ideally) for a through fault.

Pin Insulator. An insulator in an overhead-line system, supported from the cross-arm by a pin on which it is rigidly mounted. It is usually below the phase-conductor that it supports. It is suitable up to 33 kV, and has been used for 66 kV (see Fig. P-8).

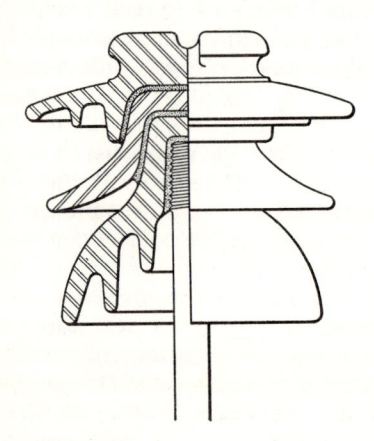

FIG. P-8. PIN INSULATOR.

Pin Winding. A hand-applied CONCENTRIC WINDING which, although obsolete for the normal stator, is necessary in certain types of machine.

Pinch. In a lamp, the airtight seal round the wires connecting the cap contacts to the filament or electrode.

Pinch Effect. The mechanical force that tends to urge the current in a wire to flow along the axis of the wire. It arises because the current produces a self-magnetic field in which the current itself lies, and the interaction law operates to confine the current to a central filament.

Pipe-Ventilated Machine. Machine which is completely enclosed except for inlet and outlet pipes through which ventilating air may be drawn by means of fan blades mounted on the shaft or by an external fan. A refinement, used on large machines, employs a closed air (or other gas) circuit, the warm air from the outlet being passed through a separate cooler before being returned to the inlet.

Plain-Break Circuit-Breaker. An OIL CIRCUIT-BREAKER having no arc-control device.

Planté Plate. A plate of a LEAD-ACID CELL, prepared by electrolytic conversion. It is also known as a *formed plate*.

Plasma. A gas, or mixture of gas and vapour, at very high temperature. It consists of ionized atoms and free electrons, and is so ubiquitous in nature that it is sometimes referred to as a fourth state of matter, along with solids, liquids and gases. The everyday use of plasmas is at present confined to certain types of discharge tube used in lighting. Attempts are being made to realize arrangements in which a gas is kept hot enough for long enough for thermonuclear fusion to occur in heavy hydrogen with a useful output of power.

Plastics. Organic substances possessing useful combinations of electrical and mechanical properties. They are plastic under pressure and heat, and can be moulded, extruded, etc. Plastics fall, broadly, into two groups—thermosetting and thermoplastic—with the thermosetting or heat-hardening group being structurally rather more complicated. For, whereas both at some state in their production are capable of being softened by heat or dissolved in suitable solvents, the thermoplastics retain this characteristic indefinitely, while thermosetting materials, when subjected to further heat, undergo chemical changes which render them permanently rigid, infusible and insoluble. Materials of both groups are in constant use in the electrical industry. For reference to entries on individual materials or groups of plastics, see under THERMOPLASTICS and THERMOSETTING PLASTICS.

Plate Electrode. An EARTH ELECTRODE, generally of cast iron, mild steel, or copper, not larger than 1 to 2 m square. Plates are now rarely used, except at power stations and substations where prospective fault currents are high and the cost of installing them is small in relation to total costs.

Plate Frame. Of a nickel–iron accumulator, the nickel-plated steel framework that supports the perforated steel tubes containing the active materials of the electrode.

Plugging. A method of braking an electric motor by reversal of the supply connections. It is sometimes known as *counter-current braking* or *reverse-current braking*.

Plus Tapping. A tapping on a winding, having its position such that more turns are included in the active part of the winding than are required for the service voltage or current ratio.

Point-Contact Transistor. TRANSISTOR in which the base is a crystal of *p*-type or *n*-type semiconductor and the emitter and collector are, in effect, small areas of material of opposite type formed round the points of two closely spaced cat's whiskers. It is now largely superseded by the JUNCTION TRANSISTOR.

Polarization. (1) Orientation of the positive and negative molecular charges in a dielectric material situated in an electric field (see POLARIZATION CURRENT).

(2) The formation of a film of gas or a strong ionic concentration at an electrode of a primary cell, causing a reduction in the electromotive force.

(3) The direction of the planes of the electric and magnetic components of the field of an electromagnetic wave.

Polarization Current. Movement of electric charges within the structure of a solid dielectric placed in an electric field. The field will act on the molecules of the dielectric to "stretch" or "rotate" them, and so to orient the positive and negative charges in opposite directions. The dielectric is said to be polarized, and in becoming so there is a true movement of electric charges within its structure. The movement will occur momentarily when a direct-voltage potential difference is applied; or continuously when the p.d. changes or alternates. Polarization current is additional to DISPLACEMENT CURRENT, and provides a mechanism for explaining the relative permittivity of a dielectric material.

Pole. (1) Each of the lines or terminals of a circuit or equipment between which a large voltage exists.

(2) Each of the points or regions of a magnet to which the lines of force converge.

(3) The part of the magnetic circuit of a machine between the yoke and the air-gap (otherwise known as the *pole piece*).

(4) The extremity of each electrode between which an arc burns.

(5) The terminal or accessible part of an electrode in a voltaic cell.

(6) An upright beam inserted in the ground and supporting the phase conductors of an overhead line.

Pole-Change Motor. An INDUCTION MOTOR in which the speed is varied by changing the number of primary (stator) poles.

Pole Core. The part of a POLE PIECE which is surrounded by the excited coils. It is also known as the *pole shank*.

Pole Horn. The part of a POLE PIECE that projects circumferentially beyond the excited coils.

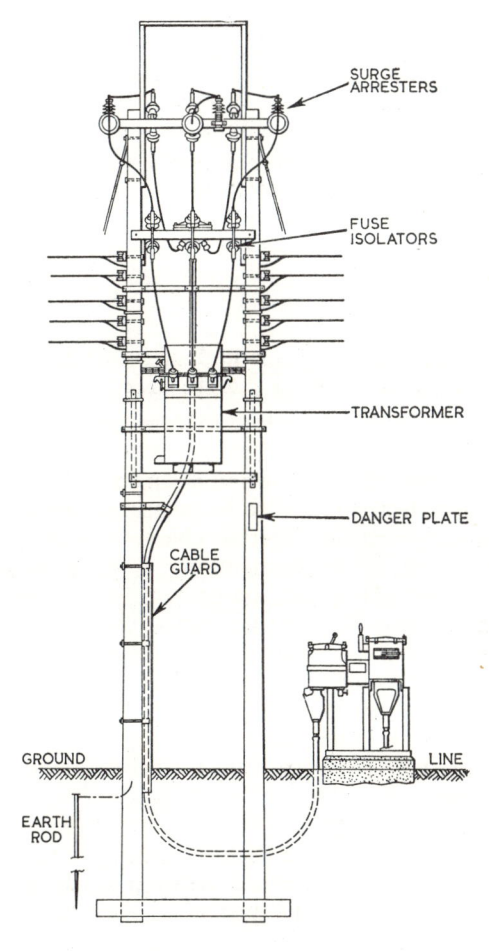

FIG. P–9. TYPICAL H-TYPE TRANSFORMER SUBSTATION ON CONCRETE POLES.

Pole Modulation

Pole Modulation. A method of winding a POLE-CHANGE MOTOR which gives a speed ratio within the range 1·5 : 1. It involves tapped stator windings, and a change in speed is obtained by reconnection of the winding tappings by means of a simple drum controller.

Pole-Mounted Transformer. A transformer substation mounted directly on a pole of the overhead transmission line. Transformers up to 100 kVA three-phase have been successfully mounted on single poles and up to 200 kVA on H-poles (see Fig. P–9). The transformer either is supended from a cross arm on the pole by means of hangers welded to the tank, or stands on a platform consisting of two brackets in the case of a single pole or two horizontal channels on either side of an H-pole.

Pole Piece. The part of the magnetic circuit of a machine between the yoke and the air-gap. It is alternatively known as the *pole* or *magnet pole*.

Pole Shoe. The separable part of a POLE PIECE that faces the armature of a machine.

Pollack Construction. A method of fitting long commutator bars to eliminate any lifting due to centrifugal force. In this arrangement the commutator bars A (Fig. P–10) with the uncured

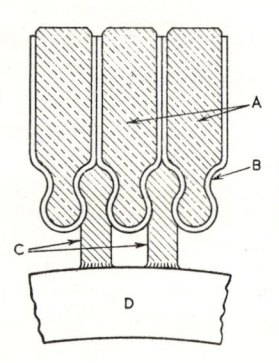

FIG. P–10. POLLACK CONSTRUCTION.
Centrifugal stresses tend to lift the centres of long commutator bars. This difficulty may be overcome by the Pollack construction.

insulation B and the supporting steel strips C extending along most of the bar, are assembled, and the structure is then consolidated. Some discs D are then fitted, to which the strips C are welded, so retaining them in position against centrifugal force and locking the segments against radial movement while still allowing slight relative axial movement by sliding on the insulation.

Polyester Resin. A synthetic thermosetting resin which may be employed in ENCAPSULATION. It is also used with various fillers and in the production of reinforced plastics. Polyester/glass-fibre fabrications are used for cable ducts.

Polyethylene. A thermoplastic polymer of ethylene, also known as *polythene*. Its electrical properties are outstanding, and in combination with its mechanical properties make it an excellent insulator. Two major advantages are its extremely low permittivity and power factor. Its breakdown voltage is high, and its volume resistivity considerably higher than those of other important insulating materials. The properties of polyethylene make it an ideal material for wire- and cable-covering.

Polymethyl Methacrylate. A rigid glass-clear transparent resin of good optical, mechanical and electrical properties. An insulating material, its power factor is not as low as that of POLYETHYLENE or POLYSTYRENE, but nevertheless it is of some value for high-frequency purposes. Its low softening point (about 60°C) limits its applications to situations where operating temperatures are not high.

Polyphase System. An a.c. network to which are supplied two or more electromotive forces of the same frequency but displaced in time phase. A symmetrical M-phase system of e.m.f.s has M e.m.f.s equal in magnitude, waveform and frequency, and separated in time phase by $1/M$ of a period. Such e.m.f.s are normally generated in an electromagnetic machine having M identical phase windings displaced from each other by $2/M$ of a pole-pitch. The voltages are then

$$v_A = v_m \sin \omega t,$$
$$v_B = v_m \sin(\omega t - 2\pi/M),$$
$$v_C = v_m \sin(\omega t - 4\pi/M),$$
$$\cdots\cdots\cdots\cdots$$
$$v_M = v_m \sin[\omega t - (M-1)2\pi/M],$$

where v_m is the peak voltage of each phase $A, B, C \ldots M$. The sum of the voltages at every instant is zero:

$$v_A + v_B + v_C + \ldots + v_M = 0.$$

Polystyrene. A thermoplastic material whose electrical properties make it suitable for moulded high-frequency insulating components. Insulating film is produced from it, and as a moulding material it is used for coil formers, transformer bobbins, etc.

Polytetrafluoroethylene

Polytetrafluoroethylene. A thermoplastic material finding increasing use in the electrical industry. An excellent electrical insulator, it is serviceable over a very wide range of temperatures. Its uses range from cable and wire insulation to stand-off insulators and accumulator components. It is sometimes known as *p.t.f.e.*

Polyvinyl Chloride. A thermoplastic material, available in a wide range of bright colours, either opaque or transparent. It has low water absorption and is resistant to corrosion and mechanical abrasion. It is used considerably in the electrical and associated industries for extruded cable insulation and sheathing. It is abbreviated *p.v.c.*

Porcelain. Dielectric material made from china clay with quartz and feldspar. It is used in the manufacture of high-voltage insulators.

Porter-Bentley Discriminator. A crane LOAD DISCRIMINATOR employing a series diverter rheostat which is mechanically operated by a deflection proportional to the torque being exerted.

Positive Booster. A BOOSTER arranged to increase the voltage supplied from another source.

Positive Feedback. Return of energy from the output of an equipment to the input, in phase with the input or of the same polarity. It may be used for increasing the gain of an amplifier or for generating oscillations. It is also known as *regenerative feedback*.

Positive Phase Sequence. That phase sequence which corresponds to the direction of rotation of a polyphase machine.

Positron. A "positive electron", i.e. an elementary particle having a mass similar to that of the ELECTRON but carrying an equivalent positive charge.

Post Insulator. An insulator built in the form of a post. It is used for supporting busbars in an h.v. outdoor substation, and is suitable for use in heavy-industrial districts with contaminated atmospheres.

Post Office Bridge. A self-contained resistance bridge developed originally by the Post Office. It is a form of WHEATSTONE BRIDGE.

Potential. At a point, the POTENTIAL DIFFERENCE between that point and earth, the latter being considered at zero potential. The unit of potential is $1 \text{ J}/1 \text{ C} = 1 \text{ V}$.

Potential Difference. The difference in the electrical states at two points, causing a movement of electricity from one point to the other. The practical unit is the volt. In electric-circuit

Potentiometer

analysis the terms *potential difference*, *voltage drop* and *electromotive force* are often used interchangeably. The term potential difference is, however, only applicable to circuits in which the currents do not vary with time; for the potential difference between two points is the work done in carrying unit charge from one point to the other, and the path is immaterial. When the currents are time-dependent, changing magnetic fields appear in which the path chosen affects the result and the original definition of potential becomes meaningless. Established custom ignores this error, and carries over the concept from the static electric conservative field in which it originated.

Potential Divider. See VOLTAGE DIVIDER.

Potential Gradient. The potential difference per unit length in the direction in which it is a maximum. It is measured in volts per unit length.

Potential Transformer. See VOLTAGE TRANSFORMER.

Potentiometer. Part of an instrument for comparing voltages, requiring in addition a cell or battery to supply a steady direct current, a STANDARD CELL and a null-detecting galvanometer; it will then measure an unknown direct electromotive force or potential difference within its range, there being no current taken from the test source at balance. Basically it comprises a length of uniform resistance wire, stretched against a scale calibrated in length units. The position of a sliding contact is read against the scale. The elementary circuit is shown in Fig. P–11, where a low-voltage supply feeds the slide-wire. From one end, connections are taken to the positive of the standard cell or of the voltage to be measured. The negative leads then go to the slide-wire contact

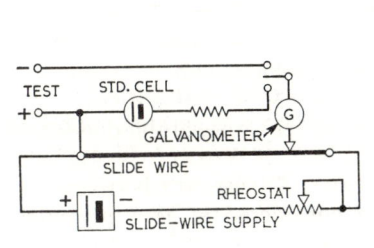

FIG. P–11. PRINCIPLE OF AN ELEMENTARY D.C. POTENTIOMETER.

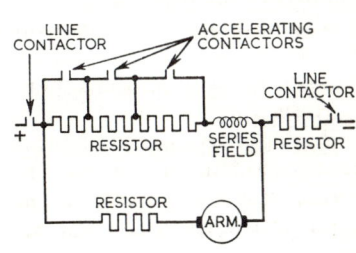

FIG. P–12. TYPICAL MAIN CONNECTIONS GIVING POTENTIOMETER CONTROL.

Potentiometer Control

through the centre-zero galvanometer which indicates the presence of a current in the slide-wire contact lead. See also A.C. POTENTIOMETER.

Potentiometer Control. Form of electric braking frequently employed on d.c. motors in crane control. The series motor is connected generally as in Fig. P–12. If the load is too small for overhauling to occur, the motor exerts a lowering torque, but with overhauling loads the motor acts as a generator, the current flowing around the dynamic braking loop circuit. The regenerated current exerts a braking force on the armature, thereby limiting the speed. Since the amount of resistance in the dynamic braking circuit is varied on each step of the controller, speed variation also is obtained.

Potentiometer-Type Field Rheostat. A FIELD REGULATOR in which the resistor may be connected across the source of supply, means being provided for the field winding to be connected between various points on the resistor.

Potier Reactance. Reactance of a synchronous machine as determined by a specific graphical construction which provides also the armature reaction and regulation from the open-circuit, short-circuit and zero-power-factor charactersitics.

Potting. See ENCAPSULATION.

Power. The rate at which work is done or energy is converted from one form to another. Fig. P–13 shows in tabular form the units of power that have been in common use up to the advent of the SYSTÈME INTERNATIONAL. The basic unit of ENERGY is the joule, and the joule per second, or WATT, is the basic unit of power. The watt, with its multiples and sub-multiples, forms the main column of the diagram. To the left are the older thermal units and to the right the British mechanical units, which will be encountered for some time to come. The numerical values marked show the relation between the units above and below.

Power Angle. Alternative name for LOAD ANGLE (q.v.).

Power Chart. A chart for a synchronous generator, motor or condenser from which its performance in relation to stability, permissible loading and excitation can be readily deduced.

Power Component. Alternative name for ACTIVE COMPONENT (q.v.).

Power Factor. The ratio of average power dissipation to the apparent power (i.e. the product of r.m.s. voltage and current)

Power-Factor Correction

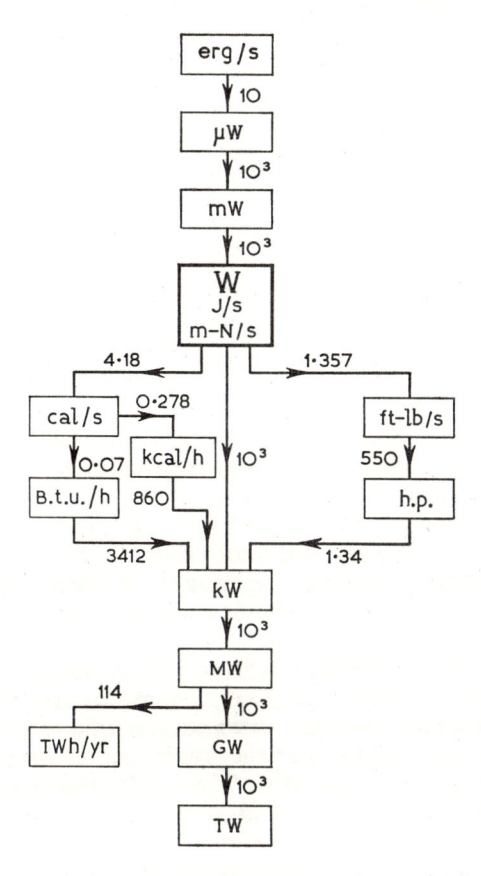

FIG. P-13. THE RELATION BETWEEN S.I. AND IMPERIAL UNITS OF POWER.

in an a.c. network or part thereof. If both voltage and current are sine functions and under steady-state conditions have a phase-displacment ϕ, then the power is $P = VI \cos \phi$, the volt-ampere product is VI, and the power factor is

$$P/VI = \cos \phi.$$

Power-Factor Correction. Use of equipment to restore a lagging power factor to near unity. SYNCHRONOUS CONDENSERS or static capacitors may be employed, or a compensated motor used.

Power-Factor Meter

Where the load is several thousand kVAr, the use of synchronous condensers is generally more economical. As the amount of correction required varies with the load, means must be provided for automatic variation. With the synchronous condenser this can be easily achieved by varying the field current, but with static capacitors it involves the switching of the units of the capacitor bank. See also POWER-SYSTEM CAPACITOR.

Power-Factor Meter. An instrument which measures the difference in phase between two electrical quantities of the same frequency, normally the voltage and the current in a circuit. It is sometimes called a *phase meter*.

Power Selsyn. SELSYN designed to ensure synchronous rotation of two or more mechanisms, the relative angular displacement varying with the transmitted torque. In operation it is similar to a general-purpose Selsyn, but is transmits much greater torques. It is generally built into the frame of a standard slip-ring induction motor.

Power Signalling. The operation of railway points and signals by electric or electro-pneumatic power.

Power-System Capacitor. Capacitor included in a power system to improve the characteristics of that system. Capacitors possess negative reactance and may be connected in series with the phases of an a.c. circuit to compensate the inductive reactance of the system. Capacitors so used are termed SERIES CAPACITORS. When connected across the phases parallel to an inductive load, capacitors draw leading reactive power which compensates the lagging reactive power drawn by the load, thus improving the overall power factor. Capacitors so connected are termed SHUNT CAPACITORS.

Precipitator. See ELECTROSTATIC PRECIPITATOR.

Precision Instrument. An instrument designed and constructed for testing work requiring high accuracy. It usually has a knife-edge pointer and finely divided mirror scale.

Preece's Law. A law relating the diameter of a wire to its fusing current and temperature. If the diameter is d, current I and melting point T, then $T = aI^2/d^3$ or $I = kd^{3/2}$. Thus if k is known for a wire of given diameter, the fusing current may be calculated.

Prefocus Lamp. Lamp in which the distance between the filament and the lampcap is set very accurately during manufacture.

Pre-impregnated Cable. A paper-insulated cable, the paper tapes of which are impregnated before assembly. This is in contrast to a *mass-impregnated* cable, which is impregnated after assembly.

Prepayment Meter. A domestic form of energy meter incorporating a coin mechanism, whereby the provision of energy is contingent upon the prior insertion of a suitable coin. The prime essential is the counting of energy units (kWh) against coin value in accordance with a preset tariff. An additional requirement may be the collection of money-value on a time basis to obtain a hire-purchase or rental facility. Fig. P–14 shows the essential mechanism schematically. Changes of cost/unit can be made without altering the gearing between the meter and the switch-trip position.

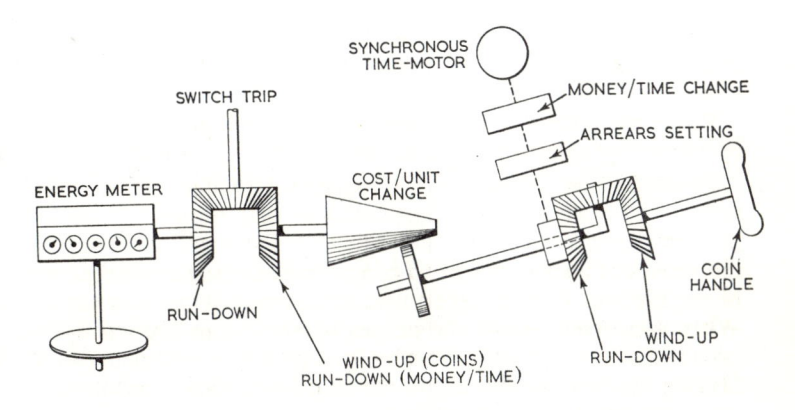

FIG. P–14. MECHANISM OF A PREPAYMENT METER.

The load between the two remains constant, although the gear-ratio between the differential and the coin handle is changed; but this altered loading is readily overcome by manual power. If the change compound is inside the meter case, then the gear connection to the coin handle is moved proportionally to money values, and the two-part tariff, hire-purchase or rental charges can be features readily incorporated.

Pressel Switch. A switch designed for attachment to a flexible cord. It is alternatively called a *pendant switch*, *pear switch* or *suspension switch*.

Pressurized-Water Reactor

Pressurized-Water Reactor. A nuclear reactor that uses natural water as the coolant (and possibly the moderator also) at temperatures of 200–300°C. To prevent boiling, this coolant must be pressurized to **10–14 MN/m^2 (1,500–2,000 lb/in^2.).**

Preventive Reactor, Resistor. A reactor or resistor connected across a tap-selector switch to reduce the circulating current due to the short-circuiting of adjacent tappings as the moving contact travels from one fixed contact to the next.

Price's Guard Wire. In insulation testing, a conductor placed round the edge of the insulating material to prevent the flow of earth leakage current through the measuring instrument.

Primary Cell. An electrolytic cell, in which two electrodes of different conducting materials associated with an electrolyte generate an electrochemical e.m.f. When the terminals are joined through a suitable resistance or other load, a current will flow in the circuit, and a chemical change takes place which is effectively irreversible. Cells are known as *wet* or *dry* depending on whether the electrolyte is a liquid or a paste. Wet cells are the DANIELL CELL and LECLANCHÉ CELL. A dry cell is the KALIUM CELL. The e.m.f.s vary from about 1 to 2 V per cell.

Primary Distribution. High-voltage distribution from generating stations or main substations to load centres or secondary substations (cf. SECONDARY DISTRIBUTION).

Primary Impedance Starting. A method of reducing the starting current of a motor, employing an impedance in series with the primary winding of the motor. The impedance will usually be a resistance but may be a reactance if it is desired to keep losses in the control cubicle to a minimum. This starting scheme gives the lowest torque per ampere of any, but gives smooth starting with a single step; as the motor accelerates and its current falls, the voltage drop in the series impedance falls also, so that more voltage is applied to the winding.

The method is used only when low values of torque are required, but is of value when starting machinery handling delicate materials since the torque is easily reduced to the desired value. It is also used when starting a motor which has to take over from a barring motor, the torque being reduced to avoid "snatch" at the instant of take-over. For large high-voltage motors, the starting impedance is usually a reactance connected in the star point of the motor as shown in Fig. P–15.

Protective Box

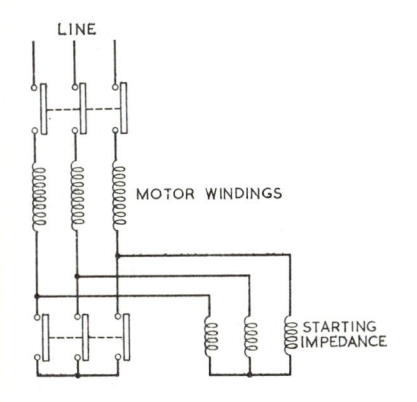

FIG. P–15. ELEMENTARY DIAGRAM OF CONNECTIONS FOR PRIMARY IMPEDANCE STARTING OF A LARGE HIGH-VOLTAGE SQUIRREL-CAGE MOTOR.

Primary Winding. The winding of a transformer to which the electromotive force is applied.

Printed Circuit. A circuit, including its connections and also certain of its components, produced by an etching process on metal foil deposited on an insulating sheet material as base.

Projection Welding. A development of SPOT WELDING, in which the current is concentrated at the desired points by projections on one of the components. Flat electrodes are used and a number of welds may be made simultaneously.

Proofed Tape. A cotton-cloth tape coated with a rubber compound which is wrapped round the insulation of rubber-insulated cables.

Propagation Coefficient. A combination of the ATTENUATION COEFFICIENT, α, and the PHASE-CHANGE COEFFICIENT, β, in the form $y = \alpha + j\beta$.

Propeller Turbine. A WATER TURBINE used for low heads of water. It is a development of the *reaction* or FRANCIS TURBINE. The runner takes the general form of a ship's propeller and comprises a number of blades attached to a heavy hub which revolves in a spiral casing. See also KAPLAN TURBINE.

Proportional Counter. Ionization counter operating at a voltage sufficiently low for the pulse of current to be small and proportional to the incident radiation.

Prospective Current. The current that would flow on the making of a circuit when the circuit is equipped for the insertion of the fuse but the fuse is replaced by a link of negligible impedance.

Protective Box. Box used to enclose lead joint boxes and

Protective Gap

sleeves, and afford mechanical protection to a joint. It may be of cast iron, earthenware or concrete and is usually filled with bitumen. Arrangements are made for through bonding of the wire or tape armouring of the cable. A cast-iron box will itself perform this duty but bonding clamps and straps must be used with other types of protective box.

Protective Gap. A form of SURGE DIVERTER consisting of two electrodes separated by an air space. It is connected between the conductor and earth to divert to earth part of any surge which may occur.

Protective Multiple Earthing. Method of earthing used in low-voltage distribution using the neutral for the return path to the transformer. The neutral conductor is itself earthed at intervals along the line, usually at consumers' premises and at the end of each distributor.

Protector Tube. Alternative name for EXPULSION GAP (q.v.).

Proton. Component of the nucleus of an atom. It has a positive charge equal in magnitude to that of an ELECTRON. The number of protons in the nucleus gives the ATOMIC NUMBER of the atom.

Proton Resonance. Resonance associated with the orientation of proton spins. Its frequency may be used in certain circumstances to measure the strength of a magnetic field to a very great accuracy.

Proton Synchrotron. See SYNCHROTRON.

Proximity Effect. Effect of one current-carrying conductor on another. When two or more neighbouring conductors carry currents (as in a coil), the current distribution in any one conductor is affected by the magnetic field produced by the adjacent conductors. The current density tends to be greatest in those parts of a conductor encircled by the smallest flux.

Proximity Switch. Switch which is operated without any physical contact. It incorporates a sealed head containing a sensor unit which detects the presence of magnetic materials and sends an impulse to operate a relay, etc.

Pulsactor. A saturating reactor employed with a capacitor and transformer to supply positive and negative pulses for the control of an ignitron.

Pulsating Current. A unidirectional current, the magnitude of which varies rhythmically with time.

Pulse. A stimulus or signal of short duration; it is generally of rectangular form.

Pulse Modulation. MODULATION in which the carrier is a series of pulses, some characteristic of which is varied in accordance with the modulating signal. The three types of pulse modulation are:

(a) *Pulse-amplitude modulation*, when the modulating signal is sampled at regular intervals; the amplitude is proportional to that of the modulating signal.

(b) *Pulse-width modulation*, when the amplitude and mean repetition frequency are maintained constant but the width of the pulses is varied in accordance with the amplitude of the modulating signal.

(c) *Pulse-position modulation*, where the amplitude, width and mean repetition frequency are maintained constant but the position of the pulse is varied in accordance with the modulation. In this case timing pulses are needed so that the variations in position may be determined.

Pulse Reflection Test. A method of localizing a fault in a cable, of particular application to telecommunication circuits.

Pulse-Type Regulator. An AUTOMATIC VOLTAGE REGULATOR in which the excitation current is fed into the exciter field in a series of pulses, which can vary in magnitude and/or frequency. It is generally used for medium-sized machines of capacity up to about 6,000 kW. Fig. P-16 illustrates schematically a typical pulse-type regulator. The voltage-sensitive device (I, V) is a moving-iron device positioned in a solenoid, and controls the position of a contact X with respect to a complementary contact W. W is made to oscillate with a constant amplitude and frequency by a motor-driven cam T. If the voltage under control rises, the contact X lifts, and hence the ratio of "contacts closed" to "contacts open" increases. If the voltage falls, this ratio will decrease. The small contacts energize relay N to operate contact M having larger thermal capacity to deal with the currents flowing in exciter-field rheostat C. Thus, as the moving iron I moves from the lowest to the highest position, the ratio of "contacts closed" to "contacts open" will vary smoothly from a maximum to a minimum, and the equivalent resistance of rheostat C will vary smoothly from a minimum to a maximum value.

Pulverized-Fuel Boiler. Boiler employed with generating plant in which coal is reduced to a fine powder and projected into the combustion chamber by means of a current of hot (primary) air. A further quantity of preheated (secondary) air to make up the required amount of combustion is also blown in separately.

Pump Line

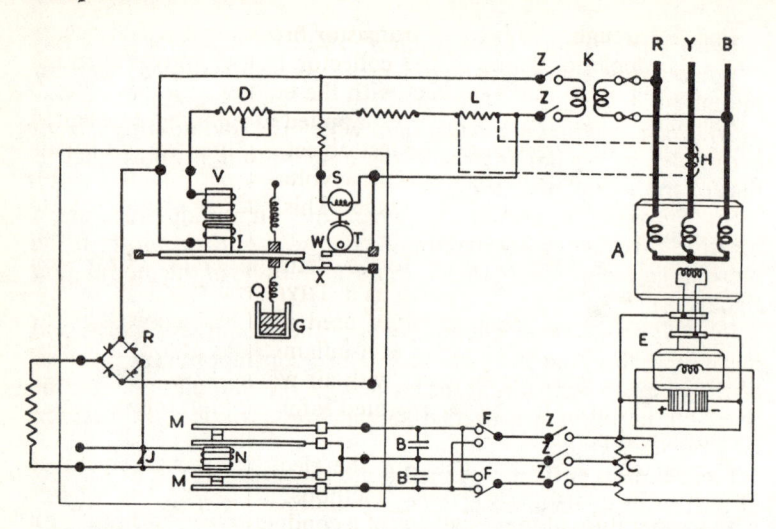

Fig. P–16. Pulse Type of Voltage Regulator Applied to a Three-Phase Alternator in Parallel with Others Similarly Connected.

A. Alternator.
B. Spark suppression capacitors.
C. Exciter shunt field rheostat.
D. Regulator control coil rheostat.
E. Exciter.
F. Reversing switch for contacts M.
G. Dashpot.
H. Current transformer.
I. Moving iron.
J. Surge suppressor.
K. Voltage transformer.
L. Compounding resistance.
M. Regulator main contacts.
N. Relay magnet.
Q. Main control arm.
R. Rectifier for supply to relay N.
S. Shaded-pole induction motor.
T. Motor-driven cam.
V. Fixed iron.
W, X. Contacts controlling relay switch.
Z. Isolating switch.

The ensuing turbulence in a high-temperature chamber facilitates thorough combustion.

Pump Line. In a railway train, a cable (or Train Line) extending throughout the length of the train, for the control of auxiliary apparatus such as air-compressors.

Pumped Storage. A method of increasing a power system load during periods of low electrical demand, by pumping water into storage reservoirs. This water may later be used for generation during periods of peak load.

Punch-Through. A form of transistor breakdown which occurs when the depletion layer of the collector barrier moves through the base region to make contact with the emitter.

Puncture-Withstand Test. A test applied to equipment designed to operate at high voltages. The test object is immersed in insulating oil and subjected to a specified voltage, for the minimum time necessary to measure the voltage. This test must be withstood without flashover or puncture.

Pylistor. Suggested name for a semiconductor controlled rectifier. Now preferably known as a **THYRISTOR** (q.v.).

Pyrochlor. A fire-resistant fluid used as a filling medium for transformers. As well as being non-inflammable, it will not produce explosive gases or mixtures with air when heated or exposed to an electric arc. It may be operated safely at high temperatures for sustained periods.

Pyrometer. An instrument for the electrical measurement of temperature. There are several methods, including the change with temperature of the resistance of a conductor or the application of the thermoelectric effect (see **OPTICAL PYROMETER**, **RADIATION PYROMETER**).

Q

Q-Factor. A figure of merit for a resonant device such as an LC circuit, a cavity resonator, a piezo crystal or, by analogy, any related device in the mechanical field. Such systems resonate by cyclic interchange of stored energy, accompanied by energy dissipation due, for example, to resistance, electromagnetic radiation or friction. If ω is the angular frequency of resonance,

$$Q = 2\pi \frac{\text{energy stored}}{\text{energy loss per cycle}}$$
$$= \omega \frac{\text{energy stored}}{\text{average power dissipation}}$$

For an inductive circuit, $Q = \omega L/R$; for a capacitive circuit, $Q = 1/\omega CR$.

Quad Cable. Cable containing four separately insulated conductors twisted together. In a *multiple twin quad*, the conductors are in two twisted pairs, with the two pairs twisted together. In a *star quad*, the four conductors are twisted about a common axis.

Quadrant Electrometer. A basic form of ELECTROMETER (q.v.).

Quadrature Axis. In an electromagnetic machine, the axis electrically at right-angles to the *direct axis*. It is also known as the *cross axis*. See Fig. D–5.

Quadrature Component. Alternative name for REACTIVE COMPONENT (q.v.).

Quadripole. A passive electrical network with one input pair and one output pair of terminals, also called a *four-terminal network*. A quadripole may be designed with dissipative elements as an attenuator, or with reactive elements as a filter. It may also be arranged as an equalizing or balancing network in telecommunications, or constructed to have characteristics resembling those of a transmission line. Power transmission lines may be analysed for steady-state operation by a quadripole representation, and the networks are often incorporated in a NETWORK ANALYSER to represent generators or transformers.

Quantity of Electricity. The product of the current flowing in a circuit and the time for which it flows. The practical units are the coulomb and the ampere-hour.

Quantum Theory. Postulation by Planck that the energy of a full radiating body is divided into a finite number of discrete units of energy known as *quanta* which are distributed at random among the individual oscillators making up the radiator. A quantum or photon of electromagnetic radiation always has associated with it an energy E proportional to its frequency f according to the expression.

$$E = hf$$

where $h = 6 \cdot 625 \times 10^{-34}$ J–s is *Planck's constant*.

Quarter-Phase System. A two-phase system, so-called because the displacement of one phase with respect to the other is one-quarter of a period. This asymmetric combination results in voltages in quadrature time phase that can generate a rotating field. A symmetrical two-phase arrangement would comprise equal and opposite voltages.

Quartz. A piezoelectric crystal. Plates cut from quartz are commonly employed as circuit elements in oscillators designed to provide the stable frequency required in most radio transmitters.

Quench Voltage. Voltage at which a discharge ceases when the applied voltage between two electrodes is reduced. It is usually lower than the voltage at which the discharge was initiated.

Quick Make-and-Break Switch. A switch that operates rapidly to make or break a circuit, the rate of motion of the blade contacts at the critical moment of contact or separation being independent of action of the operator. It is also known as a *snap switch*.

Quickstart Tube. A fluorescent tube, the exterior of which is coated by a silicone treatment.

Quill Drive. A form of drive employed with electric traction. It consists of a hollow shaft surrounding the driving axle and having sufficient clearance to permit the necessary relative movement between the spring-borne quill and the axle. The quill carries a gearwheel engaging with a pinion on the motor shaft. A twin- or double-armature motor is frequently employed, and to secure flexibility the pinions may be spring cushioned. The chief feature of the drive is the method of connecting the quill to the driving axle. This is accomplished by an arrangement of circumferential springs acting between a spider on the quill and a special spoke arrangement on the driving wheel. It is also known as *spring drive*.

R

R.M.S. Value. See ROOT-MEAN-SQUARE VALUE.

R Unit. Alternatiye name for RÖNTGEN (q.v.).

Rad. Unit of absorbed dosage. It is the amount of X-radiation that results in the production of 100 ergs/g in the absorbing medium.

Radar. Technique using radio waves for the detection of objects and the determination of their direction and distance from a reference point. The process uses waves of very short wavelengths, propagated as a narrow beam by means of a highly directional aerial and reflected off the objects.

Radial Feeder. See INDEPENDENT FEEDER.

Radial System. Distribution system in which single supplies are given as individual lines from a central substation, or as spurs from a main distributor. The system is cheaper but less reliable than other systems such as a RING SYSTEM. The circuit utilization is high, and capital costs are minimum.

Radiant-Arc Furnace. A modification of the HÉROULT ARC FURNACE. The electrodes are so arranged that the arc is deflected downwards away from the roof on to the furnace charge. In

Radiation Heating

Fig. R-1, two electrodes are shown entering from the side, with a third electrode vertical. They are connected to a three-phase supply so that the vertical electrode forms the common return carrying some 40% more current than the side electrodes. The arc is deflected by the electromagnetic forces to spread over the surface

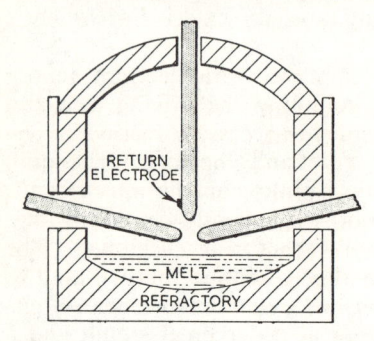

FIG. R-1. PRINCIPLE OF THE RADIANT-ARC STEEL FURANCE.

of the melt. Furnaces of this type range in capacity from 2 to 12 tons or more, and high currents are required: for example a 3-ton furnace requires about 750 kVA.

Radiation Heating. Heating by radiation, i.e. heating which warms a person or object in the path of the radiation, rather than the room or surrounding atmosphere. Unlike CONVECTION HEATING, it is very suitable for intermittent use. It may be *high temperature*, which effectively warms an individual, or *low temperature*, which gives a more uniform heat distribution from panels on the ceilings, walls or floor of a building.

Radiation Pyrometer. A pyrometer in which radiation energy of all wavelengths is focused by means of a quartz lens or concave mirror on to the hot junctions of a small, compact thermopile, the e.m.f. being measured potentiometrically or by means of an indicating instrument calibrated to read temperature direct.

Radioactivity. The property possessed by unstable atomic nuclei, of emitting radiations. These nuclei disintegrate, with the emission of ALPHA PARTICLES, BETA PARTICLES and GAMMA RADIATION.

Radio-Frequency Heating. See HIGH-FREQUENCY HEATING, DIELECTRIC HEATING, INDUCTION HEATING.

Radiograph. An image formed on a sensitive film or plate by the action of X-rays or gamma rays which have passed through an object. It is sometimes known as a *skiagraph* or *röntgenogram*.

Radiotherapy. The electromedical treatment of diseases, employing radiation. *Ultra-violet therapy*, or *actino-therapy*, uses radiation of wavelength 2,500–4,000 Å; *radiant-heat therapy*, 4,000–7,000 Å; *infra-red therapy*, 7,000 Å and above.

Ramp Function. A function of time, growing uniformly from zero at $t = 0$ and given at any instant t thereafter by

$$f(t) = kt.$$

The ramp function is of use in transient analysis, as it can represent the effect of a steep linear wavefront, or the initial part of a slowly rising sine function.

Random Winding. Alternative name for MUSH WINDING (q.v.).

Rate of Rise of Restriking Voltage. The rate at which the voltage across the contacts of an interrupting device rises (see RESTRIKING VOLTAGE). Across the contacts while an arc exists, the voltage is small and nominally zero. At a current zero the voltage rises from this nominally zero value to the RECOVERY VOLTAGE, usually through a high-frequency damped oscillation. The arc path with its residual ionization permits a very small current to flow initially, and if the rate of rise of restriking voltage is fast enough, the residual ionization is increased and current re-established. If not fast enough then the ionization decays and the current does not re-establish. It is commonly abbreviated *r.r.r.v.*

Rating. A value assigned by the manufacturer of an item of electrical equipment which expresses the limit of its performance under specified *rated conditions*. For instance, the rated full load of a motor is its maximum output when running at the rated speed on a supply of rated voltage and frequency.

Ratio Adjuster. Alternative name for a *tap changer*. See TAP CHANGING.

Ratiometer. An instrument, the operation of which depends on the ratio between the currents in two separate circuits or branches of one circuit.

Rationalized M.K.S. System. See M.K.S.μ SYSTEM.

Reactance. The voltage/current ratio in r.m.s. terms of a pure energy-storage circuit parameter (inductance L or capacitance C or any combination thereof) under strictly sinusoidal conditions and at a given angular frequency $\omega = 2\pi f$. For a pure inductor the reactance is

$$X_L = V/I = \omega L = 2\pi f L$$

Reactance Drop

For a capacitor it is
$$X_C = V/I = 1/\omega C = 1/2\pi f C$$
In each case V and I represent respectively the r.m.s. voltage and current.

Reactance Drop. The decrease in output voltage of a circuit due to the internal REACTANCE VOLTAGE within the circuit.

Reactance Voltage, E.M.F. The voltage or back-e.m.f. due to a current flowing through the reactance of a circuit. It is the product of the current and the reactance, and opposes changes in the current.

Reaction A.C. Generator. An ALTERNATOR having salient poles without field windings. The exciting current is obtained from an independent a.c. source at the required frequency.

Reaction Turbine. (1) A STEAM TURBINE in which the blading is modified to give pressure compounding with the steam expanding in both fixed and moving blades so that its velocity also increases as it passes through the latter (see Fig. R-2).

(2) A WATER TURBINE sometimes known as the FRANCIS TURBINE (q.v.).

Reactive Component. The component of an alternating current or voltage (considered as vector quantities) which is in quadrature with the current or voltage (see also REACTIVE VOLT-AMPERES).

Reactive Factor. The ratio of REACTIVE VOLT-AMPERES to total volt-amperes.

Reactive Load. A load in which the current and voltage are out of phase at the terminals.

Reactive Volt-Amperes. Product of the current and REACTIVE COMPONENT of the voltage, or of the voltage and reactive component of the current, in each case under sine conditions (symbol: Q; abbreviation: VAr or var).

Reactor. A circuit device whose chief characteristic is the possession of REACTANCE. In an INDUCTOR the reactance is inductive while in a CAPACITOR it is capacitive. The term "reactor" is sometimes used as if it meant only an inductor. See also NUCLEAR REACTOR.

Reciprocal Ohm. See MHO.

Reciprocity Theorem. A theorem employed in NETWORK ANALYSIS. If a given electromotive force in branch A of a network produces a current in branch B, then the same e.m.f.

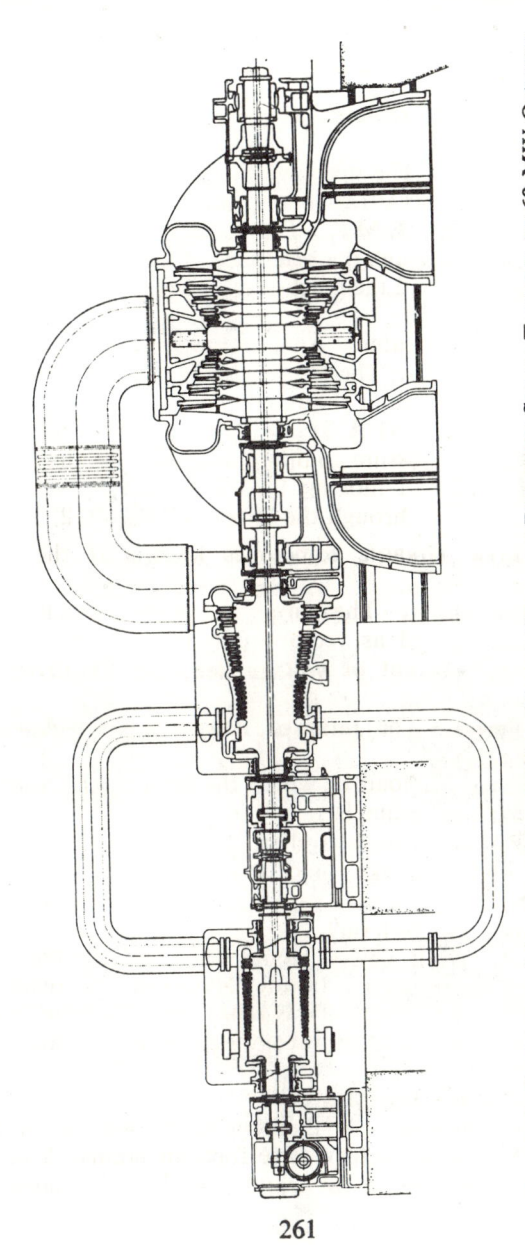

Fig. R-2. Sectional Arrangement through Three-Cylinder Reaction Steam Turbine of 60 MW Output.
(*C. A. Parsons & Co. Ltd.*)

introduced into branch B will produce the same current in branch A. The ratio of the e.m.f. and the current in such a case is called the *transfer impedance*.

Reclosing (Automatic). See AUTOMATIC RECLOSURE.

Recording Instrument. Measuring instrument which represents the reading as a permanent record, such as on a chart showing the value at any time.

Recovery Voltage. The sustained mains-frequency voltage of a power system that appears across the terminals of a circuit-breaker after interruption of the circuit. The term is sometimes used (particularly in the U.S.A.) to cover also RESTRIKING VOLTAGE.

Rectifier. Device for converting an alternating current or oscillating current into a unidirectional or approximately direct current. This may be achieved by the inversion or suppression of alternate half waves. See CONTACT RECTIFIER, ELECTROLYTIC RECTIFIER, MERCURY-ARC RECTIFIER, METAL RECTIFIER, THERMIONIC RECTIFIER.

Rectifier Instrument. A MOVING-COIL INSTRUMENT that has been made suitable for the measurement of alternating current by the addition of a metal rectifier.

Recurrent-Surge Oscillograph. A cathode-ray oscillograph having its time-base synchronized with a regularly repeated transient voltage so that the trace becomes visible. In this way the pattern of an otherwise single transient can be shown by repeatedly initiating it, so long as its duration is substantially shorter than the period of the time-base. The device is used in the impulse testing of power transformers.

Reference Electrode. Electrode used in electrochemistry to measure the potential of another single electrode. The overall potential of this reference system represents the difference between the potentials of the two electrodes, so that it is necessary for the reference electrode to be assigned some arbitrary potential adjusted to a convenient scale, that of the hydrogen electrode being zero.

Reflecting Galvanometer. See MIRROR GALVANOMETER.

Refrigerator. Device for obtaining a reduction in temperature. Three ways of achieving this are: (*a*) *Motor-driven compressor*. This compresses a gas known as the refrigerant. The gas is fed into a condenser where it is cooled and liquefied. The compressed liquid is led into an evaporator through a pressure-reducing device, such as an expansion valve or a capillary tube. In the evaporator the liquid evaporates and in so doing draws its latent heat of evaporation from its surroundings. Thus a cooling effect is produced. (*b*) *Heater-absorber*. In this unit, ammonia is caused to evaporate in the evaporator by being connected to an absorber which absorbs the ammonia into water and causes the necessary reduction in pressure. The water carrying the ammonia is passed into a boiler where it is heated. The heating drives off the ammonia which is then cooled in a condenser and subsequently passed back to the evaporator. (*c*) *Thermoelectric device*. This takes advantage of the PELTIER EFFECT. If a thermocouple is formed from two suitable materials and current is passed through the couple, one end of the loop becomes cold and the other becomes hot.

Regenerative Braking. A method of braking an electric motor utilizing the inherent property of most types of electric motor to generate when run faster than their no-load speed. If the supply is able to absorb power this constitutes a loss of energy, which results in a braking action. The braking torque increases with speed, providing an ideal braking characteristic for motors that have to deal with overhauling loads.

Regulation. A particular change of an output quantity with load. For the governor of a prime-mover the regulation is the change in speed between no-load and full-load prime-mover output, normally expressed as a percentage of full-load speed. For electrical devices such as generators, stabilizers and power packs, it is the change of output voltage with load current between no load and full load. Similar considerations apply to transformers (*voltage regulation*) and motors (*speed regulation*).

Regulator Cell. One of several cells at the end of a battery of accumulators which can be switched into or out of circuit to maintain the total output voltage constant. It is also called an *end cell*.

Reheat. Improvement of steam-turbine efficiency by extracting the steam after it has passed through a portion of the turbine and taking it back to the boiler for reheating to a temperature approaching the initial value. It is then returned to the set to continue the expansion.

Rejector Circuit. An electrical network having a frequency characteristic with an anti-resonance point.

Relative Permeability. The ratio, in any material medium, of magnetic flux density to magnetic field intensity, compared with the same ratio for free space. It is sometimes referred to simply as *permeability*.

Relative Permittivity. The ratio, in an insulating material, of the electric flux density to the electric field intensity, compared with the same ratio for free space. It is frequently referred to as *dielectric constant*.

Relaxation Oscillator. An electronic oscillator whose frequency is determined by the charging and discharging of a capacitor through a resistor, or by an analogous process with an inductor. The most important example is the *multivibrator*.

Relay. An electromechanical device in which the magnetic or thermal effect of applied electric power causes the operation of contacts which are arranged to control the flow of current in one or more separate circuits. In practice, the term is associated with control devices which perform a measuring function, and with small contactors of precision lightweight construction.

Reluctance. The ratio of magnetomotive force to flux for a magnetic circuit, or part thereof (symbol: S; unit, AT/Wb).

Reluctance Motor. A synchronous machine with a wound stator and a cage rotor constructed to provide two axes of widely differing magnetic reluctance. The resulting two fluxes produce a rotating field which enables the rotor to lock on to the supply frequency.

Reluctivity. The reciprocal of PERMEABILITY. It is a measure of the ability of magnetic material to conduct magnetic flux.

Remanence. Alternative name for RETENTIVITY (q.v.). See also B/H CURVE.

Remanent Flux Density. Flux density remaining in a material after an initial magnetizing force has been reduced to zero. Its magnitude is dependent on that of the magnetizing force. If the latter reaches the saturation value of the material, the remanent flux density becomes the RETENTIVITY or *remanence*.

Repulsion-Induction Motor. Motor possessing the starting characteristics of a REPULSION MOTOR and the running characteristics of an INDUCTION MOTOR. The stator is identical with that of a repulsion motor. The rotor has a drum-type lap winding with commutator similar to the rotor winding of a repulsion motor, but in addition it has a device for short-circuiting the commutator segments. This device is operated automatically by a centrifugal mechanism when the speed reaches about 75% of

Repulsion Motor

normal. Alternatively, the motor may have a squirrel-cage winding at the bottom of the rotor slots; this winding takes over at speed and gives induction-motor characteristics.

Repulsion Motor. A single-phase motor in which the field and armature fluxes repel each other to produce a torque in the rotor. If the connections for a single-phase compensated-series motor are changed to those shown in Fig. R-3, the compensating

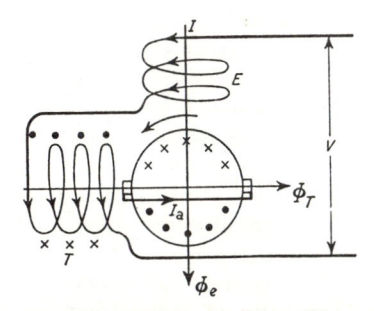

FIG. R-3. REPULSION MOTOR WITH TWO WINDINGS.

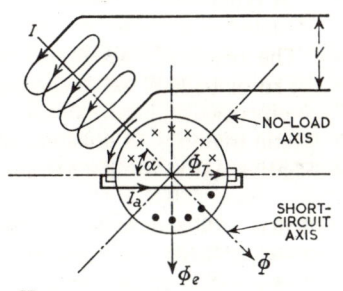

FIG. R-4. REPULSION MOTOR WITH A SINGLE WINDING.

winding acts as the primary winding of a transformer and the armature winding as the short-circuited secondary. The magnetomotive forces in the two windings will be nearly equal and opposite to one another. A repulsion motor has similar characteristics to the series motor as regards variation of speed and torque with current. The torque is produced by the interaction between the exciting flux ϕ_e due to the exciting winding E and the armature current I_a, but in this case the armature current is not taken

Residual Magnetism

directly from the supply but is produced by induction or transformer action from the winding T, known as the transformer winding. The machine with two field windings may be replaced by one having only one winding with its axis at an angle α to the brush axis as in Fig. R–4. This machine is identical in principle to that in Fig. R–3, the m.m.f. of the single winding having a component at right-angles to the brush axis producing the exciting flux, and a component along the brush axis producing by transformer action the current in the armature winding.

Residual Magnetism. The magnetism remaining in a body after the magnetizing force has been removed.

Resin. See EPOXY RESIN, PHENOLIC RESIN, SILICONES.

Resistance. The opposition offered by a conducting material (in the widest sense) to the passage of a conduction current. Unit: ohm (Ω). It is defined under steady conditions by the voltage/current ratio of a homogeneous material or any other device capable of conduction.

Resistance Alloy. An alloy material produced usually to secure a high-resistivity conductor, occasionally for some additional property such as a low RESISTANCE–TEMPERATURE COEFFICIENT.

Resistance Area. Of an EARTH ELECTRODE, the surrounding area within which the voltage gradient is appreciable, i.e. within which practically the whole of the potential difference between the general mass of earth and the electrode occurs when the latter is carrying current.

Resistance Drop. The decrease in output voltage of a circuit due to the internal RESISTANCE VOLTAGE within the circuit.

Resistance Furnace. An electric furnace in which the heat is produced by HEATING RESISTORS. This is probably the most common type of electric furnace to be found in industry. It is used for metal heat treatment of all kinds, and for such other applications as glass annealing and toughening, thermal treatment of chemicals, pottery firing and glazing, and as cremation furnaces. Resistance furnaces are of two main types, *batch* and *continuous*.

The batch furnace can be either horizontal or vertical, depending on the application. A general-purpose hardening furnace is a simple rectangular structure with elements supported on the inside of the walls. Almost any process which can be carried out in a batch furnace can be handled in a satisfactory manner through a continuous furnace, in which the work is progressed through

Resistance-Start Split-Phase Motor. The simplest and cheapest type of SPLIT-PHASE MOTOR in fractional-horse-power ratings. The rotor is of the squirrel-cage type. The resistance is incorporated into the auxiliary or starting winding itself, so that no additional equipment, other than the starting switch, is necessary. As the auxiliary winding is in circuit for only a few seconds, there is little danger of overheating. The winding is cut out of circuit either automatically by a centrifugal switch, or by an auxiliary switch forming part of the starting switch. See also PRIMARY IMPEDANCE STARTING.

Resistance–Temperature Coefficient. The rate of change of resistance of a conductor material (or insulation resistance of an insulating material) with temperature. At a temperature $\theta°$, the resistance R_θ compared with that at 0°C has an expression empirically of the form

$$R_\theta/R_0 = 1 + \alpha_0\theta + \beta_0\theta^2$$

Resistance Thermometer. A thermometer that operates on the principle of increase of resistance with temperature. The element consists of a suitable length of resistance wire wound on a former. The change in resistance with temperature is measured by a WHEATSTONE BRIDGE.

Resistance Voltage. The voltage due to a current flowing through the resistance of a circuit. It is the product of the current and the resistance.

Resistance Welding. A process by which metal parts are welded together, the junction faces being heated by the passage through them of a heavy current followed by their consolidation under high mechanical pressure. The two main types of resistance welding are LAP WELDING and BUTT WELDING.

Resistivity. A term denoting VOLUME RESISTIVITY (q.v.). It is also known as *specific resistance*. See also MASS RESISTIVITY, SURFACE RESISTIVITY.

Resistor. A circuit component, the predominant property of which is RESISTANCE.

Resolver. A form of SYNCHRO having a rotor with two quadrature phases. It is used in radar and for computing. Resolvers generate e.m.f.s in their rotor windings proportional to the angle of the rotor.

Resonance

Resonance. A phenomenon that has its origin in the energy interchanges between parts of a system; the same phenomenon, in fact, as will cause oscillation except that the drive or stimulus is sustained so that the oscillation is a forced sinusoidal one.

Resonant Frequency. The frequency at which, in a particular circuit, the inductive reactance is numerically equal to the capacitive reactance. Its value is $1/(2\pi\sqrt{LC})$ cycles per second where L is the inductance in henrys and C the capacitance in farads. In a series circuit the impedance is a minimum and the current a maximum; in a parallel circuit the impedance is maximum and the current minimum.

Restriking Voltage. A high-frequency transient voltage which appears across the contacts of an interrupting device at a current zero. The important characteristics are the amplitude and the RATE OF RISE OF RESTRIKING VOLTAGE.

Retentivity. Flux density remaining in a material after a magnetizing force, sufficient to saturate the material magnetically, has been reduced to zero (see B/H CURVE and Fig. B–1). It is also known as *remanence*.

Return Feeder. Alternative name for NEGATIVE FEEDER (q.v.).

Reverse Compound-Wound Motor. Alternative term for DIFFERENTIALLY COMPOUND-WOUND MOTOR (q.v.). See also COMPOUND-WOUND MOTOR.

Reverse-Power Protective System. An arrangement of apparatus responsive to a disturbance in a portion of equipment. It isolates the appropriate section if the direction of flow of the electrical energy is reversed.

Rheostat. A RESISTOR that is equipped with means for varying the amount of resistance in circuit.

Rheostatic Braking. A method of braking an electric motor by negative loading, causing the motor to feed power as a generator into a resistance.

Rheostatic Regulator. An automatically operated rheostat that usually controls the main-exciter-field current to maintain constant a.c. generator voltage; today, this type of regulator is generally used for the control of small and medium-sized generators. Simple motor-operation of the rheostat is not normally used, since it is too slow in operation, but the rheostat is usually either in the form of a carbon-stack arrangement for small machines, or a rolling-sector type for the larger machine (see CARBON-PILE REGULATOR).

Ring Main. System of domestic distribution in which, by means of a closed ring circuit, an unlimited number of 13 A socket-outlets may be connected to one 30 A fuse-way serving up to 100m^2 of floor area. In addition to socket-outlets, apparatus such as fixed heaters and water heaters can be taken from the ring by means of *spur boxes*. Cartridge fuses of 2 A, 5 A, 10 A and 13 A are fitted in the plugs, or, for fixed appliances, in the spur boxes. Cable, usually 2·5 mm^2, starts from the 30 A fuse-way, loops into every socket in turn and comes back to the fuse.

Ring System. A distribution system in which the distribution substations are connected in a continuous closed circuit from the main substation (cf. RADIAL SYSTEM). Ring systems offer the possibility of firm supplies at all times, but the cost of the complicated protective gear for the operation of closed rings makes such schemes justifiable only in cases where no loss of supply can be tolerated. Open-ring systems offer restoration of supply after switching operations, and ring systems with intermediate feeders, operated as separate radial feeders, can give high reliability and high utilization.

Ring Winding. Winding consisting of coils wound round an annular magnetic core, each coil being looped through the ring. It may alternatively be known as a *toroidal winding* or a *gramme winding*.

Ripple Control. Method of control, e.g. of street lighting, from a central point, with a control system utilizing the supply network as the signalling channel. The signal current is distinct in frequency from the supply current and is filtered out in a relay fitted instead of a manual switch or time switch at each lamp (or lamp-group) position. The signal current is injected into the network at the central point. Control can be exercised either directly from this point, or from some other position connected to the control point by a twin pilot cable.

Rising Main. Main installation in a multi-storey building, installed vertically with individual feeding points on each floor.

Rochelle Salt. Piezoelectric crystal widely used for microphones and gramophone pick-ups. It has a very high efficiency of conversion of mechanical to electrical energy and can be grown on a commercial scale.

Rod Electrode. The most common form of EARTH ELECTRODE for general use. Copper rods of 12–18 mm diameter, cast-iron pipes 10 cm or more in diameter and not less than 12 mm thickness,

Rod Gap

or galvanized steel water pipe not less than 40 mm diameter, are normal.

Rod Gap. A form of SPARK GAP. A simple rod gap is commonly used for the protection of power systems against lightning and other surge voltages. It comprises a pair of square-section metal rods of ½ in. side. A simple gap of this nature is cheap, but has the disadvantage that the normal system voltage subsequent to the surge will usually be sufficient to maintain the arc across the gap, necessitating the circuit being momentarily switched off. See also HORN GAP.

Roller-Spot Welding. A variety of LAP WELDING similar to seam welding but with the pulses of current so spaced that the spots produced are each separate instead of overlapping.

Rolling-Butt Contacts. Contacts in which the circuit is interrupted and the arc formed at a part of the contact surface different from that used for conduction. Contact is always established along a line. High-melting-point hard inserts can be provided at the tips of the contacts, and silver or silver-alloy inserts can be fitted to the contact base where conduction occurs. The contacts approach each other, the contact tips touch first and the moving contact then carries out a rolling action over the fixed contact surface until it comes to rest with the contact bases firmly pressed against each other. The rolling and sliding action ensures a breakdown of any surface film.

Röntgen. A unit of measurement employing the ionization effect. It is defined as the quantity of X- or gamma-radiation such that the associated corpuscular emission per 1·293 mg of air produces (in air) ions carrying 1 e.s.u. of charge.

Roof Conductor. Of a lightning protective system, a CONDUCTOR connecting several AIR TERMINATIONS, thus extending the zone of protection.

Root-Mean-Square Value. The square root of the average square of a variable, normally a time-dependent variable, over a specified interval. In particular, it is the *effective value* of an alternating current or voltage; that is, the equivalent effect of the alternating current compared with a direct current in terms of a common property, such as heating. The general definition of the r.m.s. value G of a function $g(t)$ over a given interval from $t = 0$ to $t = T$ is

$$G = \sqrt{\left[\frac{1}{T}\int_0^T [g(t)]^2\, dt\right]}$$

For a *sine* wave, $g(t) = g_m \sin \omega t$. Taken over an integral number of half-cycles, $G = g_m/\sqrt{2} = 0{\cdot}707\, g_m$, that is, the r.m.s. value equals $1/\sqrt{2}$ or $0{\cdot}707$ times the peak value.

Rosenberg Generator. A simple form of METADYNE GENERATOR. It has no interpoles, and the control windings surround poles which each carry the flux of the two adjacent polar projections.

Rosenberg Starting. Starting method for synchronous motors. A small auxiliary motor of the squirrel-cage, slip-ring or synchronous-induction type has its stator windings connected in series with those of the main synchronous motor (see Fig. R–5).

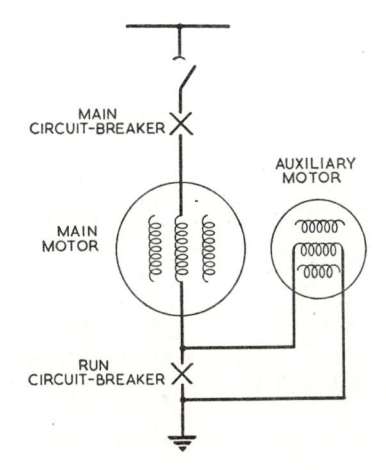

FIG. R–5. LINE DIAGRAM OF ROSENBERG STARTING SCHEME. The auxiliary motor is connected in series to reduce the current while starting.

On closing the main circuit-breaker, with the run circuit-breaker open, full voltage is applied to the two motors in series, the resultant current being reduced to a low value by the relatively high impedance of the auxiliary-motor windings. The auxiliary motor provides most of the torque required to reach slip speed, after which the main motor is synchronized, the excitation adjusted and the auxiliary motor shorted out by the run circuit-breaker.

Ross Courtney Eye. A cable eye or tag. The eye lies open for insertion of the wire and is then bent over a washer to form a tag.

Rotary Converter. See CONVERTER, MOTOR GENERATOR, MOTOR CONVERTER, SYNCHRONOUS CONVERTER.

Rotary Substation. A substation, now obsolescent, containing rotary equipment. Rotary substations were mostly those giving

Rotary Transformer

d.c. supplies, after transformation and conversion from high-voltage a.c. mains, to public supply districts or to traction networks through rotating machinery such as SYNCHRONOUS CONVERTERS or MOTOR CONVERTERS.

Rotary Transformer. A composite machine having a single magnet frame but two separate armature windings, one acting as a generator and the other as a motor, and independent commutators. It is also known as a *dynamotor*.

Rotating Amplifier. A rotating d.c. generator whose function is to convert mechanical to electrical energy in such a way that the electrical power output can be accurately and rapidly controlled by a small electrical signal applied to the control field of the machine. Rotating or rotary amplifiers find their chief application in industrial closed-loop control systems.

The main requirements of a d.c. machine for use as a rotating amplifier are (*a*) a linear relationship between electrical output and control-field input, (*b*) a high ratio of output power to control-field power, and (*c*) rapid response to changes of control-field excitation, i.e. a short control-field time constant. Power amplification and speed of response are closely related in a single-stage d.c. machine. For example, if the field time constant is reduced by increasing the resistance of the field circuit, the power amplification is decreased in exactly the same ratio. Thus, the ratio of power amplification to time constant remains unchanged, and gives a useful figure of merit for the performance of the machine as a rotating amplifier. It is known as the *dynamic amplification factor*. See also AMPLIDYNE, MAGNAVOLT, MAGNICON, METADYNE GENERATOR, ROTOTROL.

Rotating Field. A vector field, the direction in space of which changes with time in a rotary manner. A system of M conductors having geometrical symmetry and to which symmetrical M-phase voltages are applied to charge them or through which symmetrical M-phase currents flow, produces a rotating field, electric or magnetic respectively (see Fig. R–6).

Rotational E.M.F. An electromotive force induced in a short-circuited coil of a rotating machine as a consequence of its cutting an air-gap flux in the commutating zone.

Rotor. The rotating component of a machine. The term is usually applied only to a.c. machines.

Rototrol. Trade name for a type of single-stage ROTATING AMPLIFIER relying on the use of positive feedback.

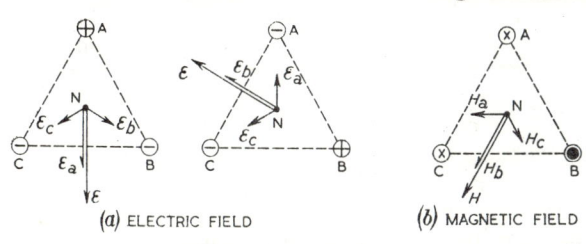

FIG. R-6. ROTATING FIELDS OF SYMMETRICAL THREE-PHASE SYSTEMS.

Rough Service Lamp. Lamp designed for arduous duties. The filament is stouter than in a general-purpose lamp, and it is under-run and provided with more supports.

Rousseau Diagram. Means of determination of the luminous output of a light source from its polar curve. The average height of the Rousseau diagram gives the mean spherical candle power.

Routine Test. A test intended to show that a machine has been assembled correctly, passed the appropriate high-voltage tests and is in sound working order.

Rubber. Insulating material available in a natural or synthetic state. As cable insulation it is used in various forms (see TOUGH-RUBBER SHEATH, VULCANIZED INDIA RUBBER). As pure rubber it has a relative permittivity of 2·6 and a volume resistivity of 10^{15} Ω-cm; a vulcanized rubber has a relative permittivity of 4 and resistivity of 10^{17} Ω-cm.

Ruhmkorff Coil. Alternative name for an INDUCTION COIL (q.v.), from the original designer. It is also known as a *spark coil*.

Russell Angle. Means of determination of the luminous output of a light source from its polar curve. The mean of the candle power at the various angles is multiplied by 4π to obtain the output in lumens.

S

S.I. Units. See SYSTÈME INTERNATIONAL.

SL Cable. Three-core cable as shown in Fig. S–1, with each core sheathed with a *single-lead* cover.

Sag. Of an overhead line, the greatest vertical displacement of the line from the straight line between its points of suspension.

Sag–Tension Relation. Of an overhead line, the dependence of the SAG on the tension of the line. A suspended conductor takes

Salient-Field Winding

FIG. S–1. SL CABLE.

the form of a CATENARY. For all practical applications, and to simplify calculations, the form may be regarded as a parabola. The relationship between sag, span and tension for a specific condition is then

$$S = WL^2/8T$$

where S = sag in feet, W = resultant load (lb/ft), L = span in feet, and T = horizontal tension (lb). If W is in newtons per metre, L in metres and T in newtons, then S will be in metres.

Salient-Field Winding. A field coil around a SALIENT POLE as shown in Fig. S–2, the coil comprising a number of turns

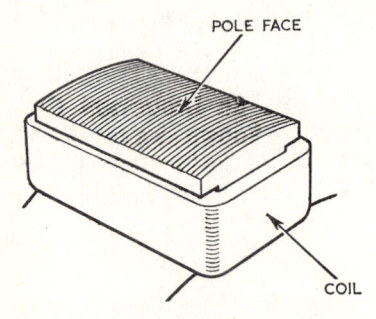

FIG. S–2. SALIENT POLE AND SALIENT-FIELD WINDING.

(between one and several thousand) of wire or strip. Such windings are used for the stationary poles of all d.c. machines, for the rotating poles of low- and medium-speed synchronous machines, and for a few other specialized types.

Salient Pole. A pole piece that projects beyond the magnet yoke towards the armature.

Schenkel Doubler

Salient-Pole Alternator. An a.c. generator whose rotor field system incorporates salient poles projecting outwards from a hub. It is suited to a multipolar arrangement and therefore low speeds, and may be driven by a water turbine or an internal-combustion engine.

Saturable Reactor. The basic amplifying element of a MAGNETIC AMPLIFIER, consisting of iron-cored coils. It is also known as a *transductor*.

Saturable-Reactor Magnetometer. MAGNETOMETER in which a Permalloy or Mumetal wire is subjected to high-frequency magnetization (e.g. 5 kc/s). Even-order harmonic fluxes are produced when a polarizing field is present.

Saturation. A departure from linearity between effect and cause. In ferromagnetic materials successive increments of magnetomotive force produce less and less magnetic flux in the material once a certain flux density has been reached. A comparable effect is the temperature-limited anode current in a diode.

Sawtooth Waveform. Waveform in which the amplitude increases uniformly with time for a period then falls rapidly to zero in a comparatively short time.

Scalar Field. See FIELD.

Scarf Joint. A joint employed where the diameter of the joint must not exceed the diameter of the wire, as on overhead trolley wires. The wires to be joined are cut obliquely and may be grooved to accommodate a locking pin.

Schenkel Doubler. A VOLTAGE DOUBLER. In Fig. S–3, the

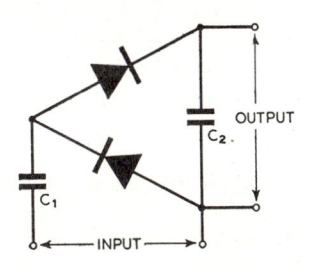

FIG. S–3. SCHENKEL VOLTAGE DOUBLER.

capacitor C_1 is charged to the peak value of the a.c. input voltage. C_2 is connected via two rectifiers so that it develops a potential difference across it of the full input voltage swing, i.e. twice the peak input voltage. Further stages may be added so that the voltage is trebled, quadrupled, etc.

Scherbius Advancer

Scherbius Advancer. A PHASE ADVANCER that resembles the LEBLANC ADVANCER with the addition of a stator winding in parallel with the armature winding (cf. WALKER ADVANCER). The voltage across the stator winding is proportional to the slip of the main induction motor, and its reactance is also proportional to slip. It therefore takes a nearly constant current, produces a constant flux, and generates a constant injection e.m.f.

Scherbius System. A system of motor speed control in which an induction motor has its rotor connected to an a.c. commutator machine operating at slip frequency (*Scherbius machine*) and returning power to the line through an induction generator.

Schering Bridge. A standard equipment for the measurement of capacitance and power factor of insulating materials. It takes two substantially different forms: (*a*) a high-voltage power-frequency bridge, and (*b*) a balanced audio-frequency bridge. The h.v. bridge is shown in Fig. S–4. The unknown capacitor C_x to be

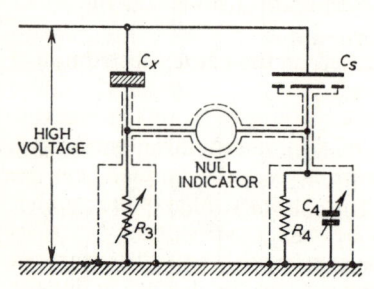

FIG. S–4. POWER-FREQUENCY HIGH-VOLTAGE SCHERING BRIDGE.

measured and the standard capacitor C_s form the high-voltage arms; a resistor R_3 and a parallel resistance-capacitance arm $R_4 C_4$, form the low-voltage arms. The balance conditions are:

$$C_x = C_s(R_4/R_3) \quad \text{and} \quad \tan \delta = \omega C_4 R_4$$

where tan δ is the tangent of the loss angle and ω is the angular frequency.

When dielectric loss measurements are required at audio frequencies, a balanced form of bridge circuit is usual with neither side of the supply earthed and with both supply leads screened. The screens are connected to earth, and one side of the supply is earthed through a fifth arm—the WAGNER EARTH. With the stray capacitances to earth forming the sixth arm, a double bridge network is formed, and the null indicator is switched between one junction and each of the other two junctions in turn. Under the final balance conditions all three junctions are at earth potential without the two junctions on the main network being actually connected to earth.

Schrage Motor. A three-phase a.c. commutator machine having a shunt characteristic and used for variable-speed industrial drives. The primary winding (which is connected to the supply system) is on the rotor and the secondary winding on the stator. The rotor also carries a low-voltage commutator winding, the conductors of which are located in the same slots as, and above, those of the primary winding. The brush gear comprises two movable rockers, each of which is fitted with three brush spindles per pair of poles. The brushes attached to each rocker move over separate portions of the commutator surface to enable these brushes to be placed "in line" or to be moved in either direction, so that more or less commutator segments are included between a brush on one rocker and the corresponding brush on the other rocker.

Since the primary winding is located on the rotor, electromotive forces or slip frequency are induced in the secondary (stator) winding. The e.m.f.s at the brushes are also of this frequency. The e.m.f. induced in each coil of the commutator winding is constant at all speeds, and therefore the e.m.f.s injected, via the brushes, into the secondary winding are proportional to the number of commutator segments included between corresponding brushes on the two rockers, i.e. between the brushes connected to a particular phase of the secondary. Thus when these brushes are in line the secondary winding has no e.m.f. injected into it and the motor will run with its natural slip. When the brushes are moved so that the injected e.m.f. opposes the current, the speed is reduced (positive slip), and when the movement is reversed so that the injected e.m.f. is in the same direction as the current, the speed is increased (negative slip).

Scintillation Counter. Ionization counter comprising a phosphor which receives incident radiation and produces light of a very low order, the light being detected and amplified by a photomultiplier followed by a main amplifier and a scaler or ratemeter. The high sensitivity of the photomultiplier, with its gain of the order of 10^6, renders this instrument of increasing use although its initial cost is high. The sensitivity is good for both alpha- and gamma-rays.

Scott-Bentley Discriminator. A single-step crane LOAD DISCRIMINATOR.

Scott Connection. A three-phase/two-phase connection of transformers. Two single-phase units of different ratings are required, but for interchangeability and the provision of spares it is more usual to employ two identical units, each of which can be used in either position. The arrangement is shown in Fig. S–5.

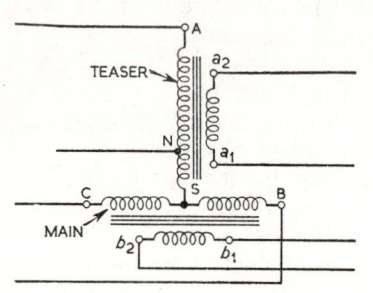

FIG. S–5. SCOTT TRANSFORMER CONNECTION.

Screen-Protected Machine. Machine in which all openings in the casing are covered with a wire mesh so that protection against accidental contact is obtained without serious impairment of ventilation. The mesh holes should have an area of not less than 64 mm^2 ($\frac{1}{10}$ in^2) to avoid clogging.

Screened Cable. A cable which has the whole of the dielectric applied to each core before the cores are laid up. A metal screen is applied over each core insulation, the screens being in contact with each other and the lead sheath. See also HOCHSTÄDTER CABLE.

Screening. Reduction or prevention of penetration by an electric or magnetic field into a given region. *Electrostatic* screening may be achieved by means of a grid or mesh construction (a *Faraday cage*). *Magnetic* screening can be obtained by a sheet of magnetically permeable material, e.g. Mumetal.

Screening Reactor. Alternative name for LINE CHOKE (q.v.).

Sealing End. A closed box, attached to the end of a cable where it connects to an external conductor, to protect the cable insulation from air or moisture. It is also termed a *sealing box* or *sealing chamber*.

Seam Welding. A variety of LAP WELDING in which the work passes between two roller electrodes while the current flows continuously, or intermittently, producing a line of overlapping spots.

Search Coil. A small inductor used for measuring magnetic field flux. It is also known as an *exploring coil*.

Search-Coil Test. Direct method of locating a fault in a cable. In one form (Fig. S–6 (*a*)) an interrupted current is fed into the

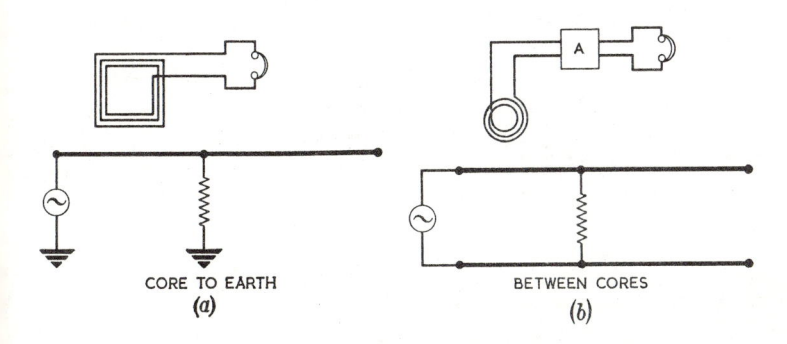

FIG. S–6. SEARCH-COIL TESTS.

faulty conductor, through the fault and back via earth. No signal is heard beyond the fault. This test is often successful on non-metal-sheathed cables but is of no value on metal-sheathed cables unless the conductors, lead and armour are severed at the fault. The second form (*b*) is of a greater value on modern cable systems. A test current is passed along one core, through the fault and back via a second core. The resultant signal, although small, may be picked up between generator and fault with a coil connected to telephones and placed on the cable.

Second. Scientific unit of time (abbreviation: s or sec). It is approximately 1/86,400 of a mean solar day. The accuracy of the second so defined is limited by irregularities in the rotation of the Earth, as indicated by quartz-crystal-controlled clocks or certain atomic or molecular radiations used as standards of frequency (and consequently also of time). The standard second is now based on a natural frequency of the caesium atom.

Secondary Cell. An alternative term for an ACCUMULATOR (q.v.).

Secondary Distribution. Low-voltage distribution from secondary substations to consumers' premises (cf. PRIMARY DISTRIBUTION).

Secondary Electrode

Secondary Electrode. Alternative name for a Bi-polar Electrode (q.v.).

Secondary Emission. The emission of (secondary) electrons from the surface of a solid when it is bombarded by (primary) electrons or ions of sufficient kinetic energy.

Secondary Winding. The winding of a transformer from which the output is taken.

Section Gap. In an overhead conductor for electric traction, an electrical and mechanical gap arranged to provide a continuous path for the current collector. This is usually done by overlapping the adjacent ends in the horizontal plane. It is also known as an *air gap* or *overlap span*.

Section Insulator. In the overhead contact wire of a traction system, a device for dividing the wire electrically into sections while maintaining mechanical continuity.

Sectionalized Busbar. Busbar split into sectons by means of isolators or circuit-breaker and isolators. A single-busbar arrangement is shown as a single-line diagram in Fig. S–7. In-

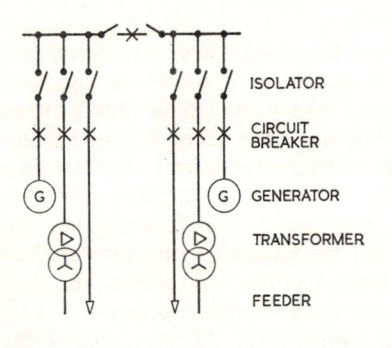

Fig. S–7. Sectionalized Single Busbar.

This may provide better continuity of supplies, and require simpler switchgear.

coming and outgoing circuits may then be shared between two sections so that supplies may be maintained when a section of busbars is shut down. Under certain operating conditions the system may be normally run split, thereby permitting the use of switchgear of lower breaking-capacity.

Seebeck Effect. A thermoelectric effect. If a closed electric circuit is constructed of two (or more) dissimilar metallic conductors, and the junctions between different metals are maintained at different temperatures, an electric current will flow in the circuit. A thermal e.m.f., which depends upon both the metals used and their junction temperatures, is the cause, and the effect is made use of in thermocouples for heat conversion and temperature indication.

Selectivity. Ability of a circuit to respond more readily to signals of a particular frequency to which it is tuned than to signals of other frequencies.

Selenium Rectifier. The most commonly employed METAL RECTIFIER. The cells consist of a steel or aluminium plate, which is coated with a thin layer of selenium to which suitable additions have been made to improve the rectification. The coated plates undergo heat treatment and are then covered with a thin layer of an alloy which usually contains cadmium, since this metal has been found to give the best rectification ratio. After the cells have been completed they are electrically formed by passing a current through them in the reverse direction.

Self-Excitation. The development by an electromagnetic machine of a working flux, without any external source of magnetizing current.

Self-Inductance. The ratio of the magnetic flux-linkage of an electric circuit and the current flowing in that circuit. A circuit has unit inductance of one henry if (*a*) the energy stored magnetically when the current is 1 A amounts to $\frac{1}{2}$ J, (*b*) a current changing at the rate of 1 A/sec induces an e.m.f. of 1 V, or (*c*) the flux linkage totals 1 Wb-turn for 1 A. The first definition is basic; the second provides a means for inductance measurement; and the third can be used for design.

Self-Levelling Device. An automatic device employed in connection with an electric lift car to reduce the speed of the car over a particular zone and to stop it at the level of the building floor. It is sometimes known as a *car-levelling device*. It acts independently of the load in the lift car.

Self-Starting Synchronous Motor. A SYNCHRONOUS MOTOR that has a winding in the pole faces to enable it to be started as a squirrel-cage motor. It then runs in synchronism using d.c. excitation.

Self Surge Impedance. More commonly known simply as SURGE IMPEDANCE (q.v.).

Self-Ventilation. A form of machine ventilation in which the ventilation is achieved without the assistance of any blowing means external to the machine itself. (cf. FORCED-DRAUGHT VENTILATION, INDUCED-DRAUGHT VENTILATION).

Selsyn

Selsyn. A type of electrical machine, similar in many respects to a slip-ring induction motor but of special characteristics to enable two or more units to rotate in mutual synchronism by means of an electrical tie, analogous to an infinitely flexible shaft transmission of unlimited length (see SYNCHRO).

Semiconductor. A material having electrical properties intermediate between those of good electrical conductors and those of insulators. It is a solid that is an insulator at absolute zero but conducts electricity by the passage of electrons at ordinary temperatures. Silicon and germanium are semiconductors of major practical importance. Two important types are the THERMISTOR, whose conductivity varies greatly with change of temperature, the resistance/temperature coefficient being negative, and the unilaterally conductive type which conducts more readily in one direction than in the reverse, and so has a rectifying action. See also the *Glossary of Semiconductor Terms* in the Appendix.

Semiconductor Rectifier. Device utilizing the rectifying properties of a semiconductor junction (see METAL RECTIFIER, GERMANIUM RECTIFIER, SELENIUM RECTIFIER, SILICON CONTROLLED RECTIFIER).

Semi-enclosed Fuse. A fuse in which the element is neither in free air (apart from any external containing case not forming part of the fuse) nor totally enclosed.

Separator. Structure of insulating material employed in an accumulator to separate plates of opposite polarity. It is normally formed of vertical rods or a diaphragm.

Sequence Switch. Another term for NOTCHING CONTROLLER (q.v.).

Series Capacitor. A POWER-SYSTEM CAPACITOR installed at the isolating transformer to improve voltage regulation and reduce the kVA demand at source. Series capacitors are specially helpful in reducing voltage dip due to direct starting of large motors, or light flicker caused by reciprocating loads.

Series-Characteristic Motor. A motor, the speed of which decreases with increasing load, e.g. a series-wound or heavily compound-wound motor. It is also known as an *inverse-speed motor*.

Series Motor. An a.c. COMMUTATOR MOTOR operating with series characteristic on single-phase supply. The smallest sizes are the fractional-horsepower *universal* motors, and are simply d.c. SERIES-WOUND MOTORS adapted to alternating current by having a completely laminated field. Large single-phase series motors

Series-Wound Motor

have been developed to supply the demand for long-distance electric traction.

Series/Parallel Connection. Method of connection of electrical apparatus, in which (*a*) the items are connected alternatively in series or in parallel, or (*b*) some are connected in series and some in parallel.

Series/Parallel Control. Of electric vehicles, a means of control involving the connection of d.c. motors, first in series and then in parallel.

Series/Parallel Starter. A switching starter for two-phase INDUCTION MOTORS. It is connected so that the windings of each phase are in series when in the starting position, and in two parallel circuits in the running position.

Series Transformer. Alternative term for CURRENT TRANSFORMER (q.v.).

Series Trip. Device for releasing the restraining mechanism of a circuit-breaker, incorporating a tripping coil which is energized by the main current passing through the circuit (cf. SHUNT TRIP).

Series-Wound Motor. A D.C. MOTOR in which the field and armature windings are connected in series with respect to the supply (Fig. S–8). The magnetomotive force due to the field

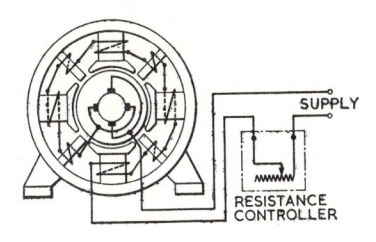

FIG. S–8. SERIES-WOUND MOTOR.

Typical connections for windings and control gear are shown above, and load characteristics on right.

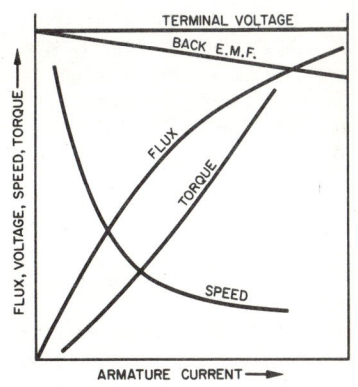

current increases in proportion to the load current. Hence, the main field strength will vary with the load current. But due allowance has to be made for the effect of armature reaction. As

Serving

the load current increases the resistance voltage drop in the armature and field will increase, and for a constant motor terminal voltage the armature e.m.f. will be somewhat reduced, and the speed must decrease as shown in the figure.

Serving. Of a cable, a layer of fibrous material, such as jute, tape or yarn, which is impregnated with a waterproof compound (e.g. bitumen), and applied to protect the metal sheath or armouring.

Servomechanism. A closed-loop amplifying system with negative feedback, wherein the amplifier supplying the output is activated by an error signal which is derived from the difference between input and output. In the more common and restricted sense, the terms *servomechanism* or *servo-system* are usually applied to that apparatus which maintains close angular alignment between a master (controlling) shaft and a more or less remote shaft, the latter being driven (controlled) by an electric—or possibly hydraulic or pneumatic—system.

Any servomechanism can be divided into three main sections. Being an error-actuated device the error must first be sensed by an *error detector*. In general, the resulting error signal will require amplification of both amount and power and the addition of stabilizing terms. These functions are performed by the *error amplifier*. Finally, a *power unit* is required to give a mechanical power output.

Servomotor. A small motor used in a servomechanism. Most are two-phase induction motors having a power output of $\frac{1}{2}$–100 W. D.C. servomotors vary in size from 0·05 hp upwards, and are more generally used in larger power servomechanisms.

Set-Up-Scale Instrument. A measuring instrument which is not deflected until the deflecting force exceeds a preset minimum value. It is also known as a *suppressed-zero instrument*.

Shackle Insulator. An insulator, in an overhead-line system, supported by a pin which passes through it and is attached at each end. It is mainly used for low-voltage applications, but is manufactured for voltages up to 3·3 kV. It is also known as a *bobbin insulator*.

Shaded Pole. An arrangement of a single-phase magnet to produce a "shifting field" across its pole faces. The laminated magnet core carries an a.c. exciting coil. On one pole face (occasionally both) is a slot accommodating a short-circuited coil which embraces part of the pole face. The gap flux will be different in magnitude and phase, that in the unshaded region leading the flux passing through the shading ring. See Fig. S–9.

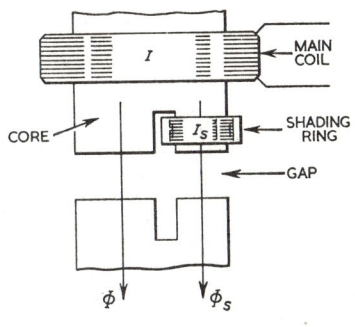

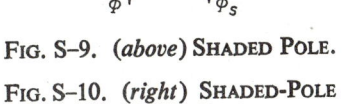

Short-Circuit Characteristic

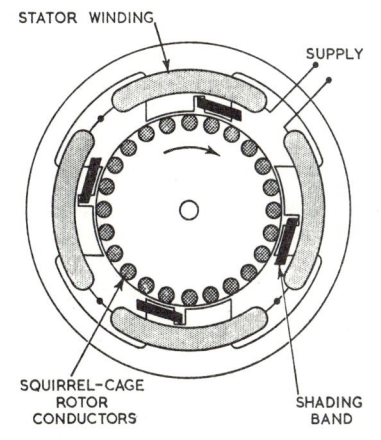

FIG. S–9. (*above*) SHADED POLE.

FIG. S–10. (*right*) SHADED-POLE MOTOR.

Shaded-Pole Motor. A single-phase INDUCTION MOTOR in which the phase-splitting of the stator flux is provided by a permanently short-circuited auxiliary winding displaced in position from the main stator winding. The motor may have salient poles as in Fig. S–10, in which case the auxiliary shading winding is a bare copper band that encircles part of each pole face. Other types have a laminated slotted (*consequent-pole*) stator like a normal induction motor.

Shaft Cable. Cable for vertical installation, as in a mine shaft. It is a NON-BLEEDING CABLE to prevent excess pressure of impregnating medium at the lower end.

Sheet Steel. A major electrical construction material (see MAGNETIC SHEET-STEEL).

Shell-Type Transformer. A transformer in which the core laminations surround and generally enclose the windings (cf. CORE-TYPE TRANSFORMER).

Shock Discharge Test. A method of localizing a fault in a cable by means of the noise produced by a capacitor discharge through the fault.

Short-Circuit, Short. A connection between two points of a circuit, particularly across a source of electrical energy, by a conducting path of low resistance.

Short-Circuit Characteristic. See OPEN- AND SHORT-CIRCUIT CHARACTERISTICS.

Short-Circuit Ratio. For an alternator running at rated frequency, the ratio of the field excitation for rated voltage on open-circuit to the field excitation for rated armature current on short-circuit. As it takes magnetic saturation into account its numerical value is a little greater than the reciprocal of the per-unit unsaturated synchronous reactance. It is an approximate measure of the relative influence on design of the specific magnetic and electric loadings, and determines the degree of stability of the machine. It is abbreviated *s.c.r.*

Short-Circuit Testing. (1) The testing of switchgear to prove its short-circuit rating. This high-power testing, which is an integral part of switchgear manufacture, is carried out in Britain by seven Short-Circuit Testing Stations.

(2) The testing of an electrical machine with its output terminals short-circuited and full-load current flowing.

Short-Circuit Transition. A method of changing d.c. motors from series to parallel connection, alternatively known as SHUNT TRANSITION (q.v.).

Short-Circuited Rotor. Alternative term for SQUIRREL-CAGE ROTOR (q.v.).

Short-Pitch Coil. A DISTRIBUTED WINDING in which the coil span is less than the pole-pitch. It is also known as a *chorded coil*.

Short-Time Current. The current which a circuit-breaker or switch is capable of carrying for a specified short interval of time.

Short-Wave Diathermy. Process used in the treatment of neuromuscular complaints as a means of relieving pain by deep heating. The apparatus consists of a radio-frequency oscillator operating at about 30 MHz and generating up to about 350 W. The part of the patient to be treated is placed between two insulated metal discs which form a capacitor connected to the output circuit of the equipment. Dielectric losses occur which are manifested in the form of heat in the muscle and tissue.

Shot Noise. NOISE occurring in the anode circuit of a thermionic valve because the current is not continuous but consists of a large number of random pulses occurring as each electron reaches the anode.

Shunt. The branch connected in parallel with another branch of a network: in particular, one so connected across an indicating instrument for the purpose of altering its sensitivity.

Shunt Capacitor. A POWER-SYSTEM CAPACITOR installed on the l.v. side of distribution transformers or directly connected

Shunt-Wound Motor

across single-phase motors (with the usual switching and other precautionary measures). Shunt capacitors provide an increase in the active power capacity of the line, a decrease in I^2R losses, and an improvement of voltage regulation.

Shunt-Characteristic Motor. A motor, the speed of which remains more or less constant as the load increases.

Shunt Transition. A method of changing d.c. motors from series to parallel connection, also known as *short-circuit transition*. One motor or set of motors is short-circuited, then open-circuited and connected in parallel with the other motors.

Shunt Trip. Device incorporating a tripping coil which is energized by a low-voltage supply (cf. SERIES TRIP). It is employed to release the restraining mechanism of a circuit-breaker, and is controlled by a relay or pushbutton which may be quite independent of, and remote from, the circuit-breaker.

Shunt-Wound Motor. A D.C. MOTOR in which the field and armature windings are connected in parallel with respect to the supply (Fig. S–11). As the armature current, and hence the torque,

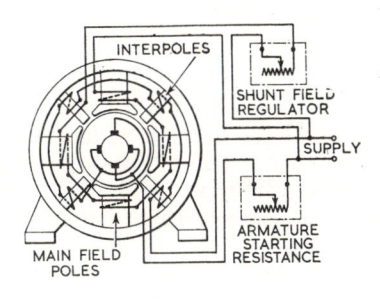

FIG. S–11. SHUNT-WOUND MOTOR.

Typical connections for windings and control gear are shown above, and load characteristics on right.

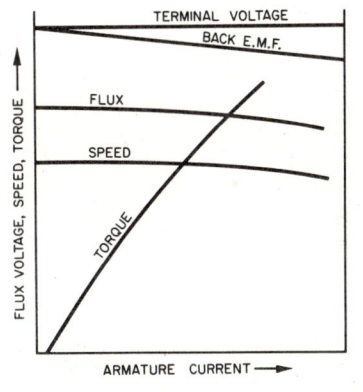

increases, the ohmic drop I_aR_a increases. Also, the total air-gap flux may be slightly reduced by the effect of armature reaction. Therefore the armature e.m.f. decreases, and for stable running it is necessary for the speed to reduce slightly. However, the speed characteristic is sufficiently constant to make the motor suitable

Sideband

for driving equipment that requires even speed and imposes no wide load fluctuations.

Sideband. Band of frequencies extending above and below a carrier frequency, of total width equal to twice the highest modulating frequency.

Sidetone. The reproduction in the receiver of a telephone set of sounds, such as the voice of the speaker, picked up by the microphone of the same set. It is reduced by the use of HYBRID COILS.

Siemens. Suggested name for the unit of conductance or admittance, more commonly known as a MHO.

Siemens Dynamometer. A dynamometer designed for the measurement of current or power. The electromagnetic forces are balanced against the torsion of a spiral spring.

Signal Control. Of a lift-car, control method in which the car is started from inside the car, but stopped by signals from any button, either on a landing or in the car.

Signal/Noise Ratio. Ratio of the strength of a wanted signal to that of the noise interference present; it is usually expressed in DECIBELS.

Silent Discharge. A noiseless high-voltage discharge, involving considerable energy.

Silica Gel. Desiccating agent, commonly used in a transformer BREATHER. It is clean, does not clog or cake, is easy to reactivate by baking and, as it changes colour when saturated, provides its own indication of when maintenance is required.

Silicon. A tetravalent semiconductor element which is widely used in the manufacture of semiconductor devices, particularly diodes, transistors, f.e.t.s, thyristors and integrated circuits. Silicon is physically and chemically similar to GERMANIUM, another commonly used semiconductor material, but it has superior high-temperature characteristics and a higher avalanche breakdown voltage. It can thus withstand greater peak inverse voltages, which is a prime requirement when operating at higher power levels.

Silicon Controlled Rectifier. A four-layer p–n–p–n device having a reverse characteristic similar to that of a normal rectifier and a forward characteristic such that, if no signal is applied to a trigger terminal, it will block positive "anode" to "cathode" voltages. If a signal is applied to the trigger terminal, the device is made to conduct and will go on doing so until the anode current is reduced below a critical level, known as the *holding current*. (See Fig. S–12.) It is now more commonly referred to as a THYRISTOR (q.v.).

Simple Catenary Suspension

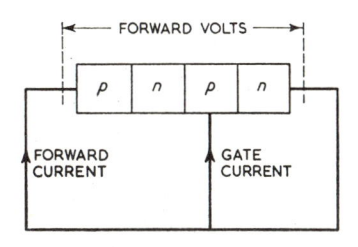

Fig. S-12. Principle of a Silicon Controlled Rectifier.

Silicon Rectifier. A rectifier employing the semiconducting properties of silicon. It is similar in construction and operation to a GERMANIUM RECTIFIER.

Silicones. Semi-inorganic thermosetting plastics manufactured as fluids, greases, resins and rubbers. All have extremely low-loss electrical insulating properties and, in combination with mica, glass fibre or asbestos, form a full range of insulating components that can be considered for continuous application within the temperature range of $-50°$ to $+200°C$, and for intermittent operation up to 300°C. These materials meet the requirements for H-CLASS INSULATION. They are as water-repellent as paraffin waxes and far more durable. They are chemically inert and resistant to a wide variety of industrial chemicals.

Silistor. A temperature-sensitive silicon resistor. Unlike a normal THERMISTOR, it has a positive resistance-temperature coefficient.

Silmanal. Permanent-magnet material, being an alloy of non-ferromagnetic substances but having so large a coercive force that the magnet-strength is unaffected by fields up to about 3,000 AT/cm.

Silver. A white metallic element that will not oxidize in air and has a specific resistivity of $1·62$ $\mu\Omega$ per centimetre cube. It is alloyed and mixed with other metals to overcome a relative softness and to raise its melting point. One useful alloy is STANDARD SILVER.

Simple Catenary Suspension. In electric traction, a form of construction for overhead conductors in which the contact wire is **suspended by means of droppers at intervals of $4\frac{1}{2}$–6 m from a** support wire which is itself suspended as a catenary over the track (Fig. S-13). The contact wire is kept at a uniform height above the track by varying the length of the droppers. See also DOUBLE CATENARY SUSPENSION, COMPOUND CATENARY SUSPENSION and STITCHED CATENARY SUSPENSION.

Simplex Winding

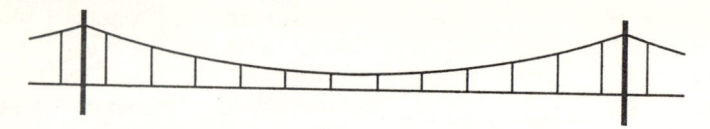

FIG. S–13. SIMPLE CATENARY SUSPENSION.

Simplex Winding. An armature winding in which there is one electrical path only for each pole.

Sine Wave. A waveform having the general equation

$$y = a \sin 2\pi\left(\frac{t}{T} - \frac{x}{\lambda}\right)$$

where y is the displacement at distance x from the origin, t is the time to travel distance x, a is amplitude, T is period, and λ is wavelength. It is the ideal waveshape of alternating current or voltage. It may be alternatively defined as a curve, the amplitude of which at any instant is proportional to the sine of the angular displacement of a point which travels at constant angular velocity on a circular path.

Sine-Wave Impedance. See IMPEDANCE.

Single-Break Circuit-Breaker, Switch. A circuit-breaker or switch in which the circuit is broken at one point only in each pole or phase.

Single-Electrode System. Of an electrolytic cell, one electrode and the portion of the electrolyte in contact with it. It is also known as a *half element* or *half cell*.

Single-Layer Winding. A winding having a single coil side in each slot.

Single-Phasing. The effect in a three-phase system when two of the phases are rendered ineffective; in particular the disconnection of one line of a polyphase induction motor when the machine is running. Deliberate single-phasing to produce negative-sequence braking is sometimes employed for light duty, as in the operation of machine tools.

Single Potential. Alternative term for ELECTRODE POTENTIAL (q.v.).

Single-Wire Overhead Distribution. Single-phase distribution system utilizing only one overhead line conductor and making use of earth as the return path of the electric circuit. Elimination of the second conductor simplifies line and substation construction

Sinor. A term describing the representation of a steady-state sine-varying quantity (such as an alternating current or voltage) as a complex number (see COMPLEXOR).

Skein Winding. A winding which has been reintroduced after becoming obsolete. An example is shown in Fig. S-14.

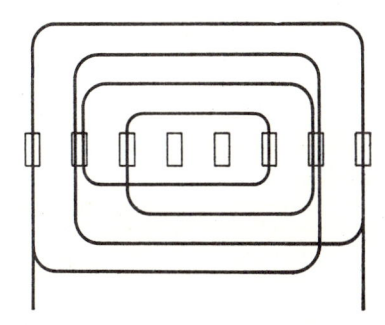

Fig. S-14. A Skein Winding.

Skin Effect. Phenomenon whereby, with alternating currents, particularly if the frequency is high, the current carried by a conductor is not uniformly distributed over the available cross-section, but tends to be concentrated at the conductor surface. It is due to magnetic flux that links part, not all, of the conductor. Those parts of the section with greater linkage have in consequence a greater inductance (and so a greater inductive reactance) than other parts. See Fig. S-15.

Slave Clock. An auxiliary clock used in conjunction with a MASTER CLOCK to take load from the master pendulum.

Sleeve. (1) A short length of unthreaded tubing used to connect together the ends of two lengths of plain conduit.

(2) A casing of lead or copper placed round joints in cables and filled with compound.

Slip. The fractional speed-difference between a driving and a driven member, such as a pair of pulleys connected by a belt, the two parts of a clutch, or the stator and rotor of an induction motor. If the driving speed is n_1 and the speed of the driven member is n,

Slip Regulator

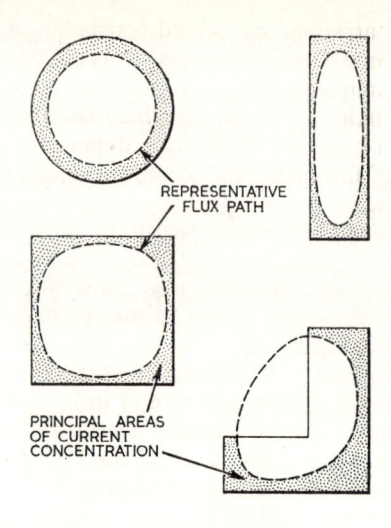

FIG. S–15. SKIN EFFECT IN ISOLATED CONDUCTORS.

then the slip is
$$s = (n_1 - n)/n_1$$

Slip Regulator. A device for varying the slip of an induction motor rotor, generally used to decrease the motor speed when heavy loads occur. Slip is the difference between the speed of an induction motor and that of its rotating field and can be increased by inserting resistance in the rotor circuit.

Slip Resistance. A secondary resistance step incorporated in a motor controller to limit the current taken from the supply at the instant when the peak load is applied to the motor in applications such as press drives, guillotines, etc. As the ohmic value is small, it is usual to have a conventional starter so arranged that the last step of resistance is not cut out when the starting handle is right home. This last step of resistance is continuously rated.

Slip-Ring. A conducting ring connected with a winding and rotating with it, enabling an electrical connection to be maintained with an external circuit by means of a stationary brush resting on the ring.

Slip-Ring Motor. An INDUCTION MOTOR having a wound rotor with the connections from this made to slip-rings. It is used for duties where the motor has to start against a fairly heavy load, and

Small-Oil-Volume Circuit-Breaker

the slip-rings are arranged for added resistance to be inserted in the rotor circuit for starting purposes. It is sometimes referred to as a *wound-rotor motor*.

Slow-Break Switch. A switch in which the speed of operation of the contacts in breaking a circuit is dependent entirely on the speed of action of the operator, i.e. no mechanical aid such as a spring is included.

Slow Butt Welding. Alternative term for UPSET BUTT WELDING (q.v.).

Small Bayonet Cap. A lamp cap consisting of a BAYONET CAP with a cap diameter of about 15 mm ($\frac{5}{8}$ in.). It is frequently used for automobile lamps.

Small Centre Contact. A CENTRE-CONTACT CAP used for lamps and having a diameter of about 15 mm ($\frac{5}{8}$ in.).

Small-Oil-Volume Circuit-Breaker. CIRCUIT-BREAKER in which arc extinction takes place in oil while air is used for insulation between phases, and sometimes to earth (Fig. S–16). The quantity

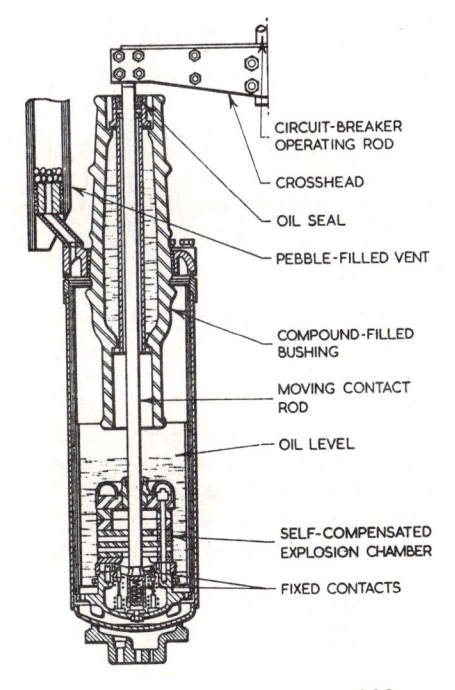

FIG. S–16. SECTIONAL DRAWING OF AN OIL VESSEL OF A 33 kV, 1,200A SMALL-OIL-VOLUME OIL CIRCUIT-BREAKER.

There are a pair of these vessels in each phase of the breaker.

(*Associated Electrical Industries Ltd.*)

of oil is considerably less than in dead-tank oil circuit-breakers and this results in less costly oil-handling plant. The oil is, however, carbonized more rapidly, requiring more frequent attention. S.O.V. breakers have been manufactured for installation in stone cells for alternating voltages of 6·6 to 33 kV.

Smoke Detector. Device for the indication and recording of smoke density as a guide to the prevention of atmosphere pollution. It usually employs photoelectric cells for control purposes.

Smooth-Conductor Cable. Cable in which a smooth surface is provided for the conductor (as opposed to the ridged surface of a stranded or bunched conductor) by means of a smooth metallic layer laid closely over the conductor below the dielectric.

Smoothing. Attenuation of ripple components, such as are present in the output of low-power rectifiers, to obtain a smooth direct-voltage or -current output. It can be achieved by a capacitance-resistance filter circuit.

Snail Clamp. A variant of an ANCHOR CLAMP designed for use with heavy overhead-line conductors. The conductor is taken round a spiral groove, the resultant static friction enabling a comparatively small clamping plate to be used at the centre of the spiral.

Snap Switch. Alternative name for a QUICK MAKE-AND-BREAK SWITCH (q.v.).

Snatch Off. Method of pulsing an automatic voltage regulator by the application of a back-e.m.f. to reverse the force on the voltage sensing device. It is an alternative to the motor-driven cam referred to in PULSE-TYPE REGULATOR.

Socket-Outlet. Device commonly employed in domestic installations to facilitate the connection to the main supply of portable lighting fittings and other appliances. It is fixed and designed to receive a plug that carries protruding metal contacts corresponding to recessed contacts in the socket-outlet.

Sodium-Vapour Lamp. A DISCHARGE LAMP containing sodium vapour, useful light being emitted from or excited by the discharge through this vapour. It is long in form and is therefore made up in a U-shape with a bayonet cap at the dual end; an earth strip is inserted between the stems of the U-shape to assist in starting the lamp. Also provided is a clear-glass vacuum jacket in which the lamp rests and this serves to retain the heat in the lamp, so keeping the sodium metal in vapour form. On starting, the lamp takes on a red hue since the discharge is struck in neon gas of low breakdown potential; the colour then changes to yellow as the sodium takes over the discharge. The yellow output is a highly efficient light, but it limits the usefulness of the lamp to applications such as street lighting and floodlighting. The high voltages required are obtained from step-up transformers and these have a leaky magnetic field so that the regulation serves as a high impedance to limit the current in the discharge.

Soft-Iron Instrument. Indicating instrument preferably known as a MOVING-IRON INSTRUMENT (q.v.).

Soil Warming. Heating of the soil for horticultural purposes. Most installations now consist of transformer-fed low-voltage systems, with a grid of heating wires laid just below the surface of the ground.

Solar Cell. Device for the direct conversion of solar energy to electrical energy by a thermocouple or a photovoltaic cell. Powers between about 20 and 120 watts per square metre (i.e. efficiencies of 2–10%) can be obtained, and such devices have been used for supplying power to telephone repeaters located in isolated situations, and also for supplying space vehicles.

Solder-Pot Relay. A form of thermal relay, frequently fitted in motor control gear.

Solenoid. A coil of insulated wire intended for connection in an electric circuit to produce a concentrated magnetic field. A standard solenoid is designed with a large length/radius ratio so that a known flux density can be produced along its axis for a known current, when it forms the basis of a standard magnetic flux linkage. Applications of solenoids are to devices (such as contactors and relays) in which the magnetic flux is augmented by a partially ferromagnetic circuit, and applied to produce a force of attraction. Solenoids are designed for operating on either direct or alternating current. *D.C. solenoids* generally have a solid iron construction with a solid iron plunger working in a close-fitting brass tube. *A.C. solenoids* generally have laminated cores to reduce eddy-current loss, although some smaller a.c. solenoids have a solid iron construction. Single-phase solenoids invariably have a shading ring.

Solid Carbon. A homogeneous carbon rod for use as an electrode in a carbon arc lamp.

Solid-Core Insulator. See POST INSULATOR.

Solid End

Solid End. Of a cable, an end which has been hermetically sealed, with all the conductors sweated to a lead cap at the end of the sheath.

Solidal Cable. Trade name for a cable having solid, shaped conductors of soft aluminium.

Solidly Earthed. Connected effectively to earth without intervention of any device such as a fuse, switch, circuit-breaker, resistor, reactor or solenoid.

Space Factor. Of bunched cables, the ratio of the sum of the effective overall cross-sectional areas of the cables to the internal cross-sectional area of the conduit, duct, trunking, etc. in which they are installed. The effective overall cross-sectional area of a non-circular cable is taken as that of a circle of diameter equal to the major axis of the cable.

Space Heating. The maintenance of a body of air, particularly the interior atmosphere of a building, at a comfortable temperature. Electric heating may be RADIATION HEATING, either at high temperature or low temperature, or CONVECTION HEATING. See also FLOOR WARMING.

Spacistor. A TRANSISTOR in which charge carriers are injected into the space-charge region of a reverse-biased junction. This makes the operation independent of charge carrier lifetime, and frequencies up to 10,000 MHz have been mentioned.

Span-Length. Of an overhead conductor, the horizontal distance between two adjacent points of support.

Spark. An impulsive discharge between electrodes in a gas, accompanied by the production of light and heat in the spark path. It has been defined as an unstable, irreversible, transient phenomenon sometimes marking the transition from one more or less stable condition of current between electrodes in a gas to another more stable one under imposed conditions. Thus, a spark may mark the transition from a glow discharge to an arc; or it may be a term properly applied to a discharge of an instantaneous nature that discharges the associated capacitance even when the conditions in the external circuit preclude the continuation of the new stable state. See TOWNSEND DISCHARGE.

Spark Coil. Alternative name for INDUCTION COIL (q.v.).

Spark Gap. A pair of electrodes so designed that a spark or an arc can safely pass between them when the voltage across exceeds the *breakdown value*. The chief purposes of such gaps are for measurement of high voltages and for protection of apparatus

Splitter

against damage due to excessive voltages. Common types of spark gap are the SPHERE GAP, NEEDLE GAP, ROD GAP and HORN GAP.

Spark Machining. A form of ultrasonic machining in which minute fragments are cracked off the work by the thermal shock produced by a spark discharge under oil between the work and an electrode having the appropriate profile. The electrode is fed into the workpiece by a servomechanism which maintains a small clearance. Each spark produces a shallow depression.

Sparking Plug. Device used in automobile electrical engineering consisting of a spark gap connected between earth or the chassis of a vehicle and an insulated central electrode supplied from the h.t. DISTRIBUTOR. It is screwed into the cylinder head of a petrol engine to provide ignition.

Specific Inductive Capacity. An alternative term for the relative permittivity or dielectric constant of a dielectric (see RELATIVE PERMITTIVITY).

Specific Loading. In electrical machine design, the mean flux density over the air-gap surface where the electromotive force is induced. The specific electric loading is the number of ampere-conductors (product of number of conductors and the r.m.s. current per conductor) for unit length of armature or stator periphery.

Specific Resistance. Alternative name for *resistivity* or VOLUME RESISTIVITY (q.v.).

Sphere Gap. A SPARK GAP having spherical electrodes of brass, copper or aluminium. It is widely used as a sub-standard voltage-measuring device for voltages between about 2 and 2,500 kV.

Split Fitting. An angled conduit connection fitting which is split longitudinally and held together by screws. It can be placed in position after the wires have been drawn into the conduit.

Split-Phase Motor. A single-phase INDUCTION MOTOR fitted with an auxiliary stator winding displaced in magnetic position from, and connected in parallel with, the main stator winding. The circumferential displacement of the two stator fluxes is obtained by fitting the coils of the auxiliary winding midway between the main stator coils. See also RESISTANCE-START SPLIT-PHASE MOTOR.

Splitter. A consumer's distribution unit containing distribution fuses and main switch. It is suitable where the total connected load is not in excess of about 60 A.

Spot Welding

Spot Welding. A variety of LAP WELDING used generally in place of rivets in joining sheet metal or pressings, the current being concentrated at the weld by means of electrode tips that have limited area.

Spur. An extension from a RING MAIN. When it is inconvenient to include all the sockets on the ring, a limited number may be connected as spurs. Such spurs, containing not more than two sockets, may be looped either from a socket on the main ring or from a junction box.

Square-Loop Characteristic. A hysteresis loop for a material such that an abrupt change in response occurs at a particular flux density. Such a characteristic, for moderate applied field intensities, becomes saturated in one direction or the other, giving a two-state property useful in magnetic amplifiers and computers.

Square Wave. Alternating current or voltage, the waveform of which is approximately square or rectangular.

Squirrel-Cage Motor. An INDUCTION MOTOR having a rotor with a SQUIRREL-CAGE WINDING. It is the most reliable type of induction motor, but has a poorer starting performance in terms of the ratio of starting torque to starting current than a SLIP-RING MOTOR.

Squirrel-Cage Rotor. Rotor of an INDUCTION MOTOR having on it a SQUIRREL-CAGE WINDING. It is also known as a *cage rotor* or *short-circuited rotor*.

Squirrel-Cage Winding. A type of rotor winding commonly employed in an INDUCTION MOTOR. A series of bars is accommodated in the rotor slots, all bars being connected at each end to a common conducting ring. The bars and end-rings then form a construction resembling a cage. This cage may be regarded as a superimposed series of full-pitch turns formed by pairs of bars a pole-pitch apart, joined together into a closed loop by the end-rings. Provided that there is an adequate number of bars, the cage winding accommodates itself to any number of poles on the stator. It is often referred to simply as a *cage winding*.

Stability. The ability of a system to return to a normal condition after being subjected to a disturbance. On a power system under normal conditions the various machines run in synchronism with each other, the relative angular positions of their rotors being determined by the power transfers between them. The effect of a system disturbance is to alter the power flow

Standard

between the machines and thereby cause their rotors to oscillate with respect to each other. This *swinging* of the machines causes power surges and voltage fluctuations in the system. The system is said to be *stable*, if, after such a disturbance, all machines return to a condition of synchronism, and to be *unstable* if one or more of the machines loses synchronism as a result of the disturbance. System stability is frequently divided into STEADY-STATE STABILITY and TRANSIENT STABILITY.

Stabilizer. (1) Device included in a circuit to maintain a constant potential difference between two points (see VOLTAGE STABILIZER).

(2) Equipment to reduce automatically the rolling motion of a ship. It consists of fins extending out below the water line and tilted by an electrohydraulic servomechanism.

Stabilizing Winding. A delta-connected winding employed on star/star-connected transformers, or auto-transformers, to facilitate the flow of zero-phase-sequence currents, reduce third-harmonic voltages, reduce interference due to third-harmonic currents in the lines and earth, and stabilize the neutral point of the fundamental frequency voltages.

Stalloy. Trade name of a common form of silicon steel sheet (see MAGNETIC SHEET-STEEL).

Standard. A unit of reference, usually defined legally. The simplest kind of standard is a physical object having the desired property. Electrical and magnetic standards cannot exist in this form and must be *derived* from fundamental units. Day-to-day calibrations at commercial level can be direct, in the sense that comparisons can be made with *sub-standards* checked periodically with *reference standards*, which in turn are verifiable by absolute measurements.

Standard resistors of manganin can be made constant to a degree of accuracy greater than that to which the absolute ohm can be determined, and will vary by only a few parts in a million over a period of years. The reference standard for voltage is the WESTON CELL used with a potentiometer the resistances of which are accurately known; the same equipment with a standard resistor serves for a current reference. A mutual inductor is employed for both self- and mutual-inductance references. Capacitance can be measured accurately in terms of mutual inductance and resistance by a bridge method.

Standard Cell

Standard Cell. A special form of PRIMARY CELL that will generate a constant electromotive force under varying conditions of use for a long period, provided that only a very small current is drawn (see WESTON CELL).

Standard Ohm. See INTERNATIONAL OHM.

Standard Silver. A hardened form of silver, containing $7\frac{1}{2}\%$ of copper, used as a material for contacts, sometimes in the form of inserts attached to heavier copper contacts.

Standard Solenoid. SOLENOID used with a search coil to form a standard of magnetic flux linkage. It has usually a large length/radius ratio in order that a known flux density can be produced along its axis for a known current.

Standard Wire Gauge. The legal British system of designating the diameter of wires by numbers. It is also known as the *British Standard Wire Gauge (S.W.G.)*. See WIRE GAUGE.

Star Connection. The arrangement of phase windings into a group by connecting to a *star point*, or common terminal, all the corresponding ends of the phase windings. Normally the phase voltages differ only in time phase in a symmetrical manner. If the phase voltages are

$$V_1 = V \angle 0$$
$$V_2 = V \angle -(2\pi/m)$$
$$\cdots \quad \cdots$$
$$V_m = V \angle -(m-1)(2\pi/m)$$

then the voltage AB in Fig. S–17 is $V_1 - V_2$, the voltage AC is $V_1 - V_3$, etc. The current leaving line A is the same as the current in phase 1, etc.

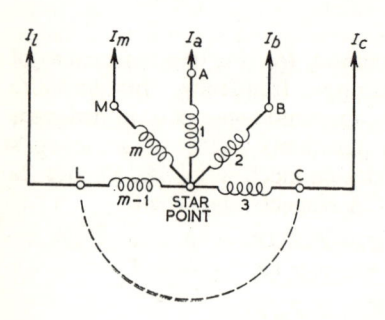

FIG. S–17. STAR CONNECTION.

Star/Mesh Conversion

Star/Delta Starting. A method of reducing starting current of three-phase motors. It is used in conjunction with a motor designed to operate with the primary winding connected in delta, but with six terminals brought out for connection in star during starting. The starting current and torque are one-third the value obtained with DIRECT-ON-LINE STARTING. The method is common in machines of 5–50 hp, but on larger sizes the transient voltages at the instant of changeover from start to run might be harmful. The connections of a typical star/delta starter are shown in Fig. S–18.

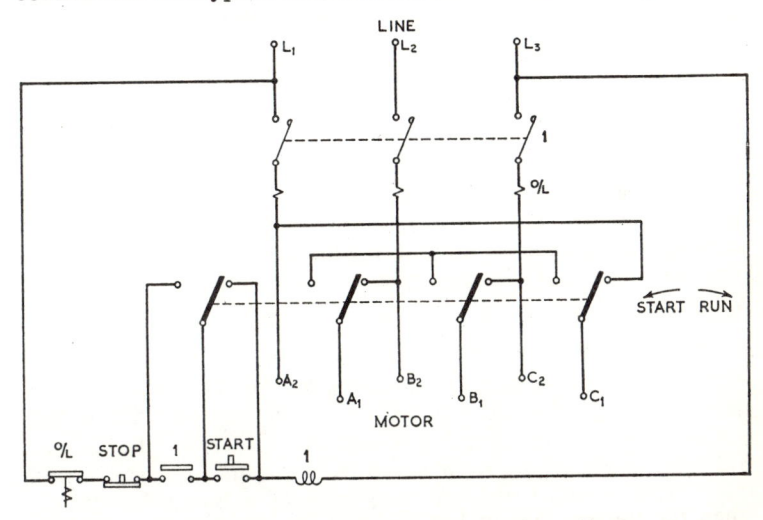

FIG. S–18. STAR/DELTA STARTER INCLUDING LINE CONTACTOR AND HAND-OPERATED START/RUN SWITCH.

The start button must be depressed until the switch is moved into the run position.

Star/Mesh Conversion. A method for the simplification of networks operating at a given single frequency. In *star/delta* (three-branch) conversion, a star-connected network of impedances Z_a, Z_b, Z_c connected between terminals ABC in a network (Fig. S–19) are replaced by a delta connection of Z_1, Z_2 and Z_3 between terminals AB, BC and CA respectively, where

$$Z_1 = Z_a + Z_b + Z_a Z_b / Z_c$$
$$Z_2 = Z_b + Z_c + Z_b Z_c / Z_a$$
$$Z_3 = Z_c + Z_a + Z_c Z_a / Z_b$$

Star Point

A delta connection can be converted to a star if $Z_a = Z_3Z_1/Z$; $Z_b = Z_1Z_2/Z$; and $Z_c = Z_2Z_3/Z$ (where $Z = Z_1+Z_2+Z_3$). The star and delta are then equivalent in the sense that the impedance measured between any pair of terminals is the same for each.

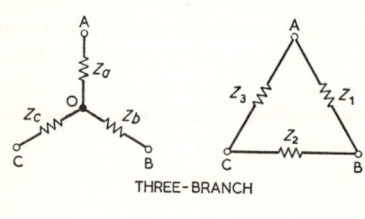

FIG. S–19. STAR/MESH CONVERSION.

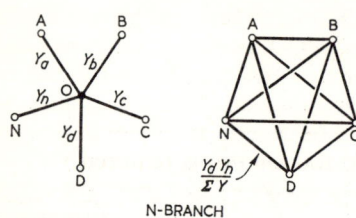

In general star/mesh conversion, an n-branch star network is replaced by a mesh of $\frac{1}{2}n(n-1)$ branches. If Y_a, Y_b ... are the admittances of branches AO, BO ... of the star, then the admittances in the corresponding mesh are Y_aY_b/Y, Y_aY_c/Y ... Y_bY_c/Y ..., between terminals AB, AC ..., BC ... respectively, there being a branch between every pair of terminals (except O). The value of Y is $Y_a + Y_b + ... + Y_n$. In general the number of mesh branches exceeds that of the original star branches.

Star Point. The point at which the branches of a winding in STAR CONNECTION are joined together, and sometimes to earth.

Star Voltage. The voltage between any line of a three-phase or six-phase system and the neutral point of the system. An unsymmetrical system may have more than one value of star voltage. It is also known as *Y-voltage* or *phase-voltage*.

Starter Motor. In automobile electrical engineering, a motor having a small pinion mounted on an extension of its armature shaft which engages with a gear ring on the periphery of the flywheel to rotate the latter. The motor is usually a straightforward four-pole motor, either series-wound or series-parallel according to the duty it is called on to perform.

Static Regulator

Statampere, Statcoulomb, Statvolt, etc. Names given to units in the absolute electrostatic C.G.S. system to distinguish them from the "practical" ampere, coulomb, volt, etc. They are related to the practical units by powers of 10 and of c, the velocity of free-space electromagnetic waves.

Static Balancer. An auto-transformer or reactor with windings so interconnected that the voltage of an a.c. or d.c. system is divided equally between the wires. A common application for static balancers is to convert a supply from two- to three-wire, or from three- to four-wire.

Static Balancing. A means of determining and correcting unbalance in rotating machinery, by revolving it on frictionless bearings such as knife-edge ways. Under these conditions it will revolve until the centre of gravity is perpendicularly under the axis. Correction is then made by removing material from the heavy side or adding material on the opposite side. Static balancing is applicable to apparatus having an axial length small in relation to diameter. In other cases DYNAMIC BALANCING is required.

Static Electrification. A surface contact potential difference that exists across the boundary when two bodies of different materials are in intimate contact. The potential is usually of the order of a fraction of a volt, but if one (or both) of the materials is a non-conductor, the charges will be retained when the surfaces are again separated.

Static Machine. Alternative name for ELECTROSTATIC GENERATOR (q.v.).

Static Regulator. An AUTOMATIC VOLTAGE REGULATOR that possesses no moving parts. It usually employs a MAGNETIC AMPLIFIER. A typical equipment for use in controlling the voltage of a large steam-driven a.c. generator is shown in Fig. S–20. The a.c. generator voltage obtained from a three-phase voltage transformer is rectified (2), and supplied to a voltage-sensitive bridge (3) via a setting rheostat (R_5). When the voltage varies a signal is fed into a magnetic amplifier (4), and the output is fed into the booster control field. A positive control field on the booster, adjusted to 50% of the maximum negative output of the magnetic amplifier, is provided by direct energization from the pilot exciter. Hence, when the output of the magnetic amplifier is 50%, the resultant output of the booster is zero. Peak output from the magnetic amplifier results in peak negative output from the booster. Zero

Static Relay

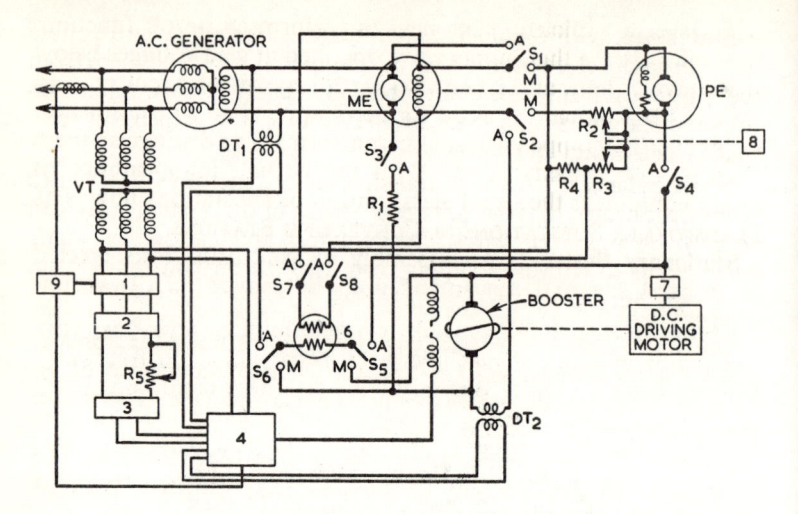

FIG. S-20. SCHEMATIC DIAGRAM OF A STATIC MAGNETIC-AMPLIFIER TYPE OF AUTOMATIC VOLTAGE REGULATOR.

1. Three-phase no-volt relay.
2. Rectifier.
3. Voltage-sensitive network.
4. Magnetic amplifier.
5. Booster.
6. Indicator.
7. Starting gear.
8. Differential follow-up unit (q.v.).
9. Low excitation network.
ME. Main exciter.
PE. Pilot exciter.
AM. Auto/manual switch.
DT_1, DT_2. Damping transformers.

(*Associated Electrical Industries Ltd.*)

output from the magnetic amplifier results in peak positive output from the booster.

Such an arrangement ensures suitable operation, even under conditions of system short-circuit, or the equivalent. The armature of the booster is connected in series with the shunt field of the main exciter, which is so arranged by means of resistance R_1, that at full load the machine is self-excited, and requires no output from the booster. The main exciter feeds the a.c. generator field windings.

Static Relay. A rectifier device that can be used as a relay without moving parts. It depends for its action on the fact that the impedance presented to a small alternating voltage may be controlled by the value and direction of a steady direct voltage on which the alternating voltage is superimposed. Provided that the input level is sufficient, such devices perform an on/off function. Examples are the thermionic valve operated at grid voltages below and above cut-off, and its counterpart in transistor circuitry.

Static Substation. A term formerly applied to substations on a.c. networks equipped with static transformers, to distinguish them from ROTARY SUBSTATIONS. With the virtual disappearance of rotary equipment the usual classification of substations now is as TRANSFORMER SUBSTATIONS or CONVERTING STATIONS.

Stationary Battery. An assembly of ACCUMULATORS erected on a fixed site and not intended to be moved (cf. TRACTION BATTERY).

Statistical Lag. In a spark gap, the time between the application of a breakdown voltage and the chance occurrence of an electron in a position suitable to initiate breakdown. See TOWNSEND DISCHARGE.

Stator. The fixed or stationary component of a machine. The term is usually applied only to the stationary magnetic parts of a.c. machines and the associated windings.

Stator-Fed Shunt Motor. An a.c. COMMUTATOR MOTOR having characteristics similar to a SCHRAGE MOTOR. The primary winding is on the stator as in a normal INDUCTION MOTOR; there are no slip-rings, and speed variation is accomplished by means of an induction regulator connected between the brush gear and the line. A *stator-fed series motor* is similar but has a transformer instead of the induction regulator with speed control by brush-gear movement.

Steady-Arm, Steady-Brace. Fitting used with CATENARY SUSPENSION to keep the contact wire in its correct lateral position.

Steady-State Characteristic. Of an ARC, the relation between the voltage gradient and the current flowing through the arc.

Steady-State Stability. STABILITY following small or gradual disturbances, such as a slow increase in load on a power system.

Steam Turbine. A device for converting the potential energy stored in steam under pressure into mechanical work in a form suitable for driving electrical generators, compressors or other machinery. The conversion is made in two stages: first, kinetic energy is obtained by allowing the steam to expand, and secondly the resultant jets of high-velocity steam impinge on blades which are forced to rotate in the same direction as the steam jets, so that power is transferred to a shaft. See also IMPULSE TURBINE, REACTION TURBINE.

Steel

Steel. An alloy of IRON with less than 2% of carbon and smaller amounts of manganese, silicon, phosphorus, sulphur and oxygen. It is used as a conductor when mechanical strength is important. It has a wide application, in varying forms, as a permanent-magnet material, and for cases for transformers, electromagnets, etc.

Steel–Alkaline Cell. A storage cell or ACCUMULATOR which employs an alkaline electrolyte. There are two distinct types, the NICKEL–IRON CELL and the NICKEL–CADMIUM CELL, both of which employ an electrolyte of dilute potassium hydroxide (caustic potash) and have a positive plate of nickel hydrate. In both types the plates consist of assemblies of steel tubes or pockets perforated minutely all over, enclosing the active materials. The average voltage per cell on discharge at a 5–10 hr rate is approximately 1·20 V.

Steel-Cored Aluminium. Conducting material consisting of layers of aluminium wire encircling a core of galvanized steel strands. It is commonly abbreviated *s.c.a.*

Steel-Tank Rectifier. Type of MERCURY-ARC RECTIFIER in which the arc occurs in a steel tank.

Stefan-Boltzmann Law. If a black body at absolute temperature T_1 is surrounded by another black body at absolute temperature T_0, the amount of radiant energy lost per second per unit area of the former is $p_r = \sigma(T_1^4 - T_0^4)$. Here σ is the Stefan or total-radiation constant, $5 \cdot 6697 \times 10^{-8}$ W/m²°K⁴.

Step Tariff. A tariff similar to a BLOCK-RATE TARIFF. In this case, when each quantity of units is exceeded, the total number of units supplied are charged at the lower rate.

Stitched Catenary Suspension. In electric traction, a form of construction for overhead conductors that has been used as an alternative to COMPOUND CATENARY SUSPENSION. It is similar to SIMPLE CATENARY SUSPENSION but the contact wire near a support structure is held by an additional wire attached to the catenary as shown in Fig. S–21.

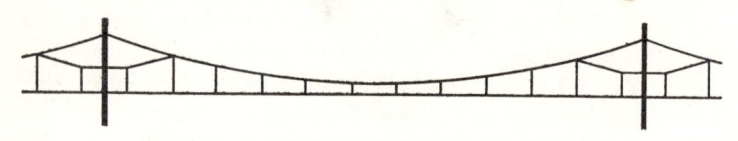

FIG. S–21. STITCHED CATENARY SUSPENSION.

Straight-Through Joint

Stockbridge Damper. The most common vibration damper employed to prevent the vibration of an overhead line. It consists of two weights at the end of a short length of stranded steel cable suspended from the conductor at a point midway between two nodes of a vibration, where the amplitude would be a maximum.

Storage Cell. An alternative term for an ACCUMULATOR (q.v.).

Store. In an electronic computor, a component that retains data for subsequent reference. It may store the data as signals on a magnetic drum or tape, on a nickel wire, in a column of mercury, in a mesh of ferrite cores, or on the screen of a cathode-ray tube. A main store, or *memory*, may hold many million bits of information simultaneously. A cathode-ray storage tube can contain as many as three electron guns, and the possibility of deflecting the beam rapidly to any particular part of the screen allows more rapid operation of the computor than does, for example, a mercury delay line or a magnetic drum. A magnetic-core store may contain about 114 ferrite cores per square inch. Each core may be charged in either direction to provide a binary digit.

Straight-Through Joint. A form of cable joint in which the ends of two abutting cables are joined to form one continuous length (Fig. S–22). Straight joints are usually enclosed in extruded lead sleeves of the highest purity and of sufficient thickness for

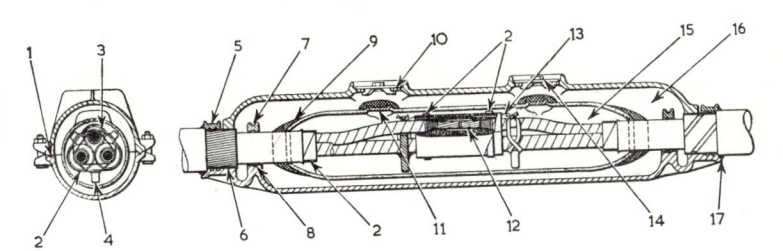

FIG. S–22. STRAIGHT-THROUGH HIGH-VOLTAGE JOINT.

1. Vee-grove joint.
2. Impregnated cotton tape binder.
3. Impregnated paper separator.
4. Porcelain spreader.
5. Cast-iron armour clamp.
6. Plain lead tape.
7. Cast-iron clamp.
8. Corrugated lead tape.
9. Wiped joint.
10. Cast-iron filling plate.
11. Filling cap wiped over.
12. Weak-back jointing ferrule.
13. Separator positioning tape tied to spreader.
14. Compound-filled recess.
15. Compound-filled lead sleeve.
16. Compound-filled cast-iron box.
17. Jute packing.

Strain Gauge

dressing down and to withstand handling without damage, but solid-drawn copper sleeves are more often used for 33 kV cables.

Strain Gauge. A device employed primarily for detecting and measuring small variations in the dimensions of the surface to which it is attached. The pick-up converts mechanical movement into change in an electrical quantity, the variations of which are subsequently amplified and impressed on a recorder.

Strain Insulator. An insulator, in an overhead-line system, which is capable of transmitting the tension of the conductor to the supporting tower. It is also known as a *tension insulator*.

Stranded Conductor. A CONDUCTOR composed of several wires twisted together. The direction of twist of adjacent layers is reversed. Cf. BUNCHED CONDUCTOR.

Stray Load Loss. The additional loss in an electrical machine, wherever occurring, caused by the load current and due to changes in flux distribution and to eddy currents.

Streamer Discharge. A suggested mechanism for the formation of a spark between electrodes in a gas (see also TOWNSEND DISCHARGE). It involves secondary ionization of the gas by photons from the initial avalanche.

Striae. Transverse bands, alternately dark and luminous, observable in an electric discharge through a gas at a particular low pressure.

Striking Voltage. Voltage at which a discharge is initiated when the applied voltage between two electrodes is increased.

String Electrometer. A form of ELECTROMETER in which a metallized quartz fibre is hung between fixed plate conductors, held at a high potential difference (Fig. S–23). A small test voltage

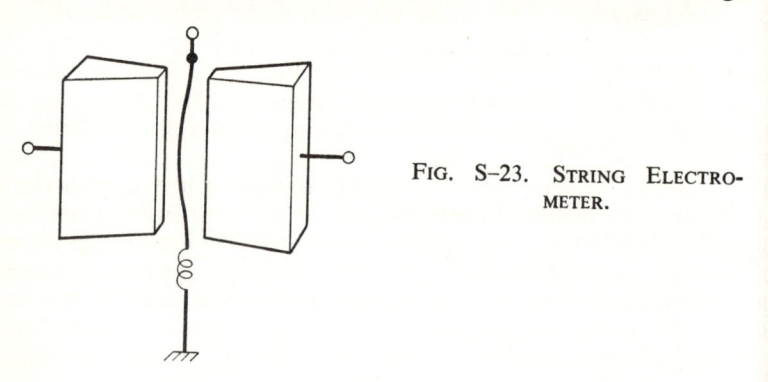

FIG. S–23. STRING ELECTROMETER.

applied to the fibre causes it to move towards one or other fixed plate. The deflection is observed by means of a microscope.

Stringing. (1) Running and tensioning the conductors on the supports of an overhead line.

(2) Assembling suspension insulators into units or *strings*.

Strip Electrode. A form of EARTH ELECTRODE. Untinned copper strip about $25 \times 1\frac{1}{2}$ mm ($1 \times \frac{1}{16}$ in.) is normal, but scrap lengths of bare stranded copper are also suitable.

Stroboscope. A device for viewing a rotating object at regularly recurring intervals, by means of either (*a*) a rotating or vibrating shutter, or (*b*) a suitably designed lamp which flashes periodically. If the period between successive views is exactly the same as the time of one revolution of the revolving object or any multiple of it, and the duration of the view very short, the object will appear to be stationary. If the period is some sub-multiple, say one nth, then the object will be seen stationary in n different positions.

Stubbs–Perry System. Method of load equalization similar to WARD LEONARD/ILGNER CONTROL. It is employed when the generators are driven by steam turbines.

Submerged Arc Welding. A form of consumable-electrode ARC WELDING that uses a bare wire in conjunction with an overburden of granulated flux. The arc is submerged beneath the flux, part of which fuses to form a slag, the unfused portion being recovered and re-used.

Submersible Motor. A machine employed for driving underwater and borehole pumps. The *dry* type requires a chamber intermediate between the pump and the motor. The *wet* type is submersible in the literal sense.

Sub-standard. A unit of reference, whose value is determined and regularly checked by comparison with a STANDARD.

Substation. An assemblage of equipment including any necessary housing, for the conversion, transformation or control of electric power. The term substation was originally employed to denote those points in an electrical network where feeders from the central generating station terminated, and where transformers, converters, switch and fuse gear were installed to serve the needs of local distribution. Nowadays, it embraces all places outside power stations where electricity is handled: these range from the simple pole-mounted transformer of a few kilowatts capacity to the very large transformer and switching installations on the national grid system; and from the converter substations of various sizes serving traction networks to the large and complex substations associated with high-voltage d.c. transmission schemes.

At one time substations were roughly classified as *static* or *rotary*; static substations being those on a.c. networks established at points where a change in voltage was required and which were therefore equipped with static transformers, while rotary substations were mostly those giving d.c. supplies, after transformation and conversion from high-voltage a.c. mains, to public supply districts or to traction networks through rotating machinery such as synchronous or motor converters. The latter forms of conversion equipment have been almost entirely replaced by static equipment in the form of mercury-arc rectifiers. It is customary now to classify substations as TRANSFORMER SUBSTATIONS and CONVERTING STATIONS, the latter term having come to imply the installation of rectifiers only.

Sub-Transient Reactance. Reactance associated with the analysis of a synchronous machine that determines the maximum fault current under symmetrical-fault conditions. It corresponds to the leakage flux that occurs in the initial stage of a short-circuit.

Sulphur Hexafluoride. A heavy, very stable, colourless, odourless, non-toxic, non-inflammable ELECTRONEGATIVE GAS. Its dielectric properties and considerable arc-quenching ability suit it for use in high-voltage circuit-breakers.

Superconductive Device. A device such as the *cryotron*, which depends for its action on the effect of a magnetic field on the critical (or transition) temperature below which certain materials exhibit SUPERCONDUCTIVITY. The critical temperature decreases with increase in magnetic field, so that if a superconductive material is maintained at a temperature just below its critical value, the resistance may be switched from zero to a finite value by the application of the field. A plot of transition temperature as a function of applied magnetic field is roughly parabolic, levelling out as absolute zero of temperature is approached.

Superconductivity. Phenomenon associated with materials at very low temperatures. At absolute zero ($0°K = -273 \cdot 2°C$) the atoms cease to vibrate and free electrons can pass through the lattice with little hindrance. At about $5°K$ the resistance of certain metals becomes precisely zero, so that a current, once started, may persist for a long time (because of its associated magnetic energy storage and complete absence of resistance dissipation) without any applied electric field. This phenomenon has not been so far satisfactorily explained.

Superexcitation. Application of the principle of quick-response excitation to synchronous machines. It leads to rates of voltage build-up of the order of 6 or 7 kV/sec. Exciters designed for superexcitation are of high rated voltage (e.g. 600 V for a 250 V excitation) and have a correspondingly high ceiling voltage, approximately 1,000 V.

Supergrid. Transmission lines incorporated in the British GRID SYSTEM and rated at 275 kV and above.

Superposition. A principle employed in NETWORK ANALYSIS. The effect on, or the reponse of, any linear system due to the simultaneous action of a number of impressed causes or disturbances may be found by considering the effect due to each cause taken separately and summing them. Applied to an electric network, this means that the currents produced in various branches by a number of sources are obtainable by superposing the currents resulting from the generators acting separately. The only restriction on the superposition process for a linear system is that the network configuration must not be disturbed—i.e. generators left out of consideration during the step-by-step application must be represented by their internal impedances or admittances. The principle underlies several network theorems which rely implicitly on the validity of superimposed effects, for example the HELMHOLTZ THEOREMS, the MILLMAN THEOREM and the method of SYMMETRICAL COMPONENTS.

Supersonics. See ULTRASONICS.

Supervisory Control. System of control of the GRID SYSTEM, railways, etc., which allows a controller, located at a convenient point in the network, to control electrical plant at one or more remote points, and receive continuously from these points indications of the condition of the plant. The essential feature of such a system is that the control point is connected to the controlled point by a number of pilot wires that is substantially less than the number of operations performed and indications received.

Supplementary Anode. In electroplating, a small anode that is provided near deeply recessed parts of a cathode surface to facilitate a uniform deposition.

Suppressed-Zero Instrument. A measuring instrument that is not deflected until the deflecting force exceeds a preset minimum value. It is also known as a *set-up-scale instrument*.

Surface-Barrier Transistor. TRANSISTOR produced by an electrolytic process, having a very thin base layer and small junction area. It is operable up to about **70 MHz**.

Surface Resistivity. The resistance between the opposite edges of a square of unit dimensions on the surface of insulating material. The value is independent of the size of the square, provided that the surface has uniform properties: the resistivity is therefore commonly expressed in MΩ "per square".

Surge. A transient overvoltage on an electrical network, usually in the form of a travelling wave on a transmission line. In such a case the surge may be initiated by lightning strokes or switching operations.

Surge Absorber. A protective device connected in series with an overhead line at substation terminals for the purpose of absorbing some of the energy of a SURGE voltage.

Surge-Current Indicator. Alternative name for a MAGNETIC LINK (q.v.).

Surge Diverter. A device connected between a transmission line and earth in order to divert to earth a momentary high-voltage SURGE. It normally comprises one or more gaps in series with non-linear resistors. If the voltage across it exceeds a predetermined value (slightly above the normal voltage to earth of the line) the gaps break down and the high voltage of the surge results in the non-linear resistor assuming a low resistance value, thus enabling the diverter effectively to discharge the surge to earth.

Thyratrons, ignitrons, and cold-cathode electronic devices may be employed in some situations as surge diverters.

Surge Generator. See IMPULSE GENERATOR.

Surge Impedance. The ratio voltage/current in a single SURGE travelling in one direction on a transmission line. For a line of inductance L and capacitance C per unit length, the surge impedance in ohms is $Z_0 = \sqrt{(L/C)}$. This is strictly the *self surge impedance*. The *mutual surge impedance* is the ratio of the voltage in a surge along one line to the current in the associated surge along another line.

Surge-Limiting Electrolytic Capacitor. An electrolytic capacitor designed to restrict the maximum voltage which can be maintained across its terminals.

Surge-Proof Electrolytic Capacitor. An electrolytic capacitor designed to withstand momentary or intermittent surges of voltage in excess of the rated working voltage. It is not necessarily a SURGE-LIMITING ELECTROLYTIC CAPACITOR.

Surgical Diathermy. The use of surgical equipment consisting principally of an r.f. oscillator capable of delivering an output of up to 200 W at a frequency between 1 and 3 MHz. The output is fed to a cutting electrode, a thin needle which, when brought to within about half a millimetre of the skin, produces an intense arc. This arc destroys the cells over a very small area so that when the tip of the needle is moved a fine cutting action results. The heat from the arc causes local coagulation and permits extensive cutting without bleeding.

The coagulation of areas of bleeding or the sealing of individual blood vessels is achieved with a special electrode that is brought into contact with the area to be coagulated. No arcing takes place, but local heating results from the r.f. current flowing into the body at that point. This accelerates the coagulating action.

The technique is also known as *fulguration*.

Susceptance. That part of ADMITTANCE concerned with reactive elements. An admittance $Y = G + jB$ is composed of two branches in parallel, one of conductance G, concerned with dissipation, and the other of susceptance B.

Susceptibility. In a magnetic circuit, the ratio of the intensity of magnetization to the magnetizing force. It is also known as *volume susceptibility*; symbol κ.

Suspension Insulator. An insulator, in an overhead-line system, which hangs freely from the cross-arm to support the line conductor. It normally consists of a number of separate units linked to form a flexible string.

Swan-Neck Insulator. A PIN INSULATOR having the pin shaped to bring the insulator into the same horizontal plane as the support.

Switch. A mechanical device for making or breaking non-automatically the load current in a circuit. It differs from a CIRCUIT-BREAKER in that it is designed to make but not break the fault current in the circuit, and it is usual to protect the circuit which the switch controls by means of fuses or a circuit-breaker.

Switch-Fuse. A combination of a switch and one or more fuses. The fuses are *not* carried on the moving part of the switch as in the FUSE-SWITCH. This definition applies strictly to equipment for operation at medium and low voltages. Manufacturers of high-voltage equipment tend to use their own terminology.

Switch-Type Voltage Regulator

Switch-Type Voltage Regulator. An electromagnetic device having a primary winding in parallel and a secondary winding in series. The voltage impressed on a load is adjusted by varying the number of turns in one or both of the windings.

Switchboard. An assembly of SWITCHGEAR.

Switchgear. Apparatus for controlling the distribution of electrical energy or for controlling or protecting apparatus connected to a supply of electricity. The main components are switches and circuit-breakers with ancillary apparatus such as current and voltage transformers and cable-sealing ends. Switchgear can be classified as *open* or *enclosed*, and each may be designed for *indoor* or *outdoor* application (see Fig. S–24).

Switching Station. A SUBSTATION that includes associated switchgear and busbars.

Switching Transient. TRANSIENT imposed on a circuit by the operation of a switch. The closing of a switch is equivalent to the injection across the switch terminals of a voltage equal and opposite to that existing across the switch when open. Because closure reduces the voltage across the switch to approximately zero, the effect may be simulated by the cancelling effect of the injected voltage at the instant of switch closure. The opening of a switch is equivalent to the injection across the switch of a current equal and opposite to that flowing in it at the instant immediately preceding switch operation.

Symmetrical Breaking Capacity. The r.m.s. value of the a.c. component of current that can be broken at a stated voltage simultaneously by all poles of a circuit-breaker.

Symmetrical Components. A method of dealing analytically with unbalanced conditions in an electrical network or a machine, with a facility approaching that associated with balanced operation. It makes use of the symmetrical phase-sequence components of the currents and voltage. Any unbalanced system can be resolved into three symmetrical systems, of POSITIVE PHASE SEQUENCE, NEGATIVE PHASE SEQUENCE and ZERO PHASE SEQUENCE, which are mutually independent and which can be manipulated on a single-phase basis.

Symmetrically Cyclic Magnetic State. A condition of a magnetic material when it is in a cyclically magnetized state, and the limits of the applied magnetizing forces are equal and of opposite sign, so that the limits of flux density are also equal and of opposite sign. It is abbreviated *s.c.m.*

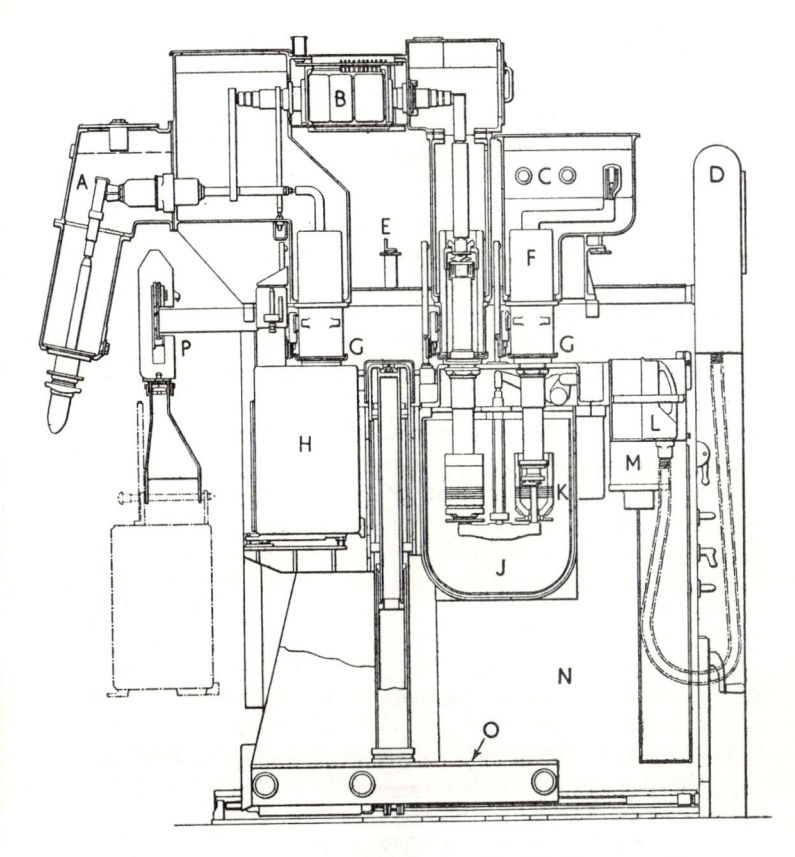

Fig. S–24. General Arrangement of 33 kV Vertical Isolation Switchgear Unit.

A. Cable-box and adaptor chamber.
B. Current transformers.
C. Busbar-chamber.
D. Cabinet for secondary equipment.
E. Earth contacts.
F. Isolating contacts.
G. Locking-off doors.
H. Voltage transformer.
J. Circuit-breaker.
K. Turbulator arc-control device.
L. Secondary plug-box.
M. Solenoid operating-mechanism.
N. Frame-standard.
O. Captive traversing-carriage.
P. Monorail for voltage transformer.

(*A. Reyrolle & Co. Ltd.*)

Synchro

Synchro. A small instrument, up to about 4 in. diameter, built like an induction motor and used for the transmission and reception of both electrical and position data. In general, the term SELSYN applies to devices of substantial power rating, while *synchro* is used for small versions of the same basic principle applied, for example, to servomechanism error pickups and position indicators.

The basic synchro consists of a symmetrical three-phase stator and a single-phase rotor. One end of each stator "phase" winding is brought to a star point (special synchros may have a delta connection), and the other end is brought to a terminal. The rotor winding is brought to two slip-rings and is energized from an a.c. supply of some frequency between 15 and 1,000 Hz according to the design. A voltage is generated in each stator winding of value determined by its axis relative to that of the rotor winding.

Synchrocyclotron. An ORBITAL ACCELERATOR in which the relativistic energy limitation of the SYNCHROTRON has been removed, by modulating the frequency of the oscillator. The frequency is high initially when a bunch of ions starts near the centre of the magnet and decreases appropriately as the ions gain in energy and relativistic mass. The desired frequency modulation of the oscillator may be obtained by attaching a rotating variable capacitor with suitably shaped plates into the oscillator anode circuit. Such a rotating capacitor may weigh several tons and rotate at up to 2,000 rev/min. A single "Dee" is used and the accelerating voltage applied between this and a grounded sheath. The 156 in. (3·96 m) machine at Liverpool University (Fig. S–25) accelerates protons to 410 MeV.

Synchronizer. Alternative term for SYNCHROSCOPE (q.v.).

Synchronizing. The process of connecting two a.c. supplies together in parallel; the selection of the appropriate instant for switching a synchronous a.c. generator on to energized busbars or into parallel with another, normally running, synchronous machine. Before a generator can safely be connected to live busbars it is necessary for the voltage and frequency to be approximately the same as that of the busbars. In the case of a polyphase machine, the phase sequence also must be identical.

Synchronous Clock System. System in which clocks are operated from a.c. mains. The latter must be frequency-controlled, otherwise they are useless for clock operation.

Synchronous Converter

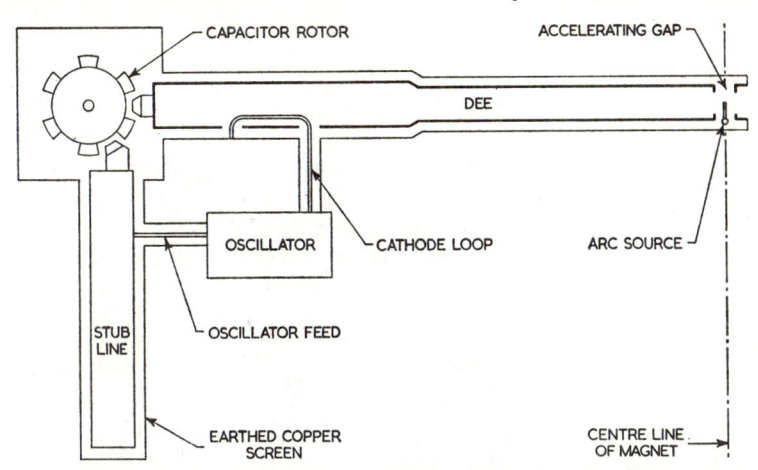

Fig. S–25. R.F. System of the Liverpool Synchrocyclotron.

Synchronous Condenser. A machine which is connected to a supply system to provide reactive MVA, normally leading but sometimes also lagging, without at the same time furnishing mechanical power to external plant. The output of the unit is adjusted, both in sign and magnitude, by alteration of the exciting current, invariably under the control of an automatic regulator. The primary use is in the correction of power factor and in the improvement of transmission-line stability.

Synchronous Converter. A form of CONVERTER in which the armature has connections to slip-rings for the a.c. input (two for single-phase and three or six for three-phase) and to a commutator for the d.c. output. Since the circuits are electrically connected, the voltage ratio is fixed and a transformer is usually necessary at the input end. Excitation can be provided from the d.c. output or from an exciter. Fig. S–26 shows the normal connections of a six-ring synchronous converter. For this arrangement, the r.m.s. phase voltage V_a and the direct output voltage V_d are related by

$$V_a = (V_d/\sqrt{2}) \sin \tfrac{1}{6}\pi = 0\cdot 35\ V_d$$

while the transformer secondary phase current I_a is

$$I_a = 2\sqrt{2} I_d/6 = 0\cdot 47 I_d$$

for an output current I_d, for unit power factor and efficiency.

Synchronous Generator

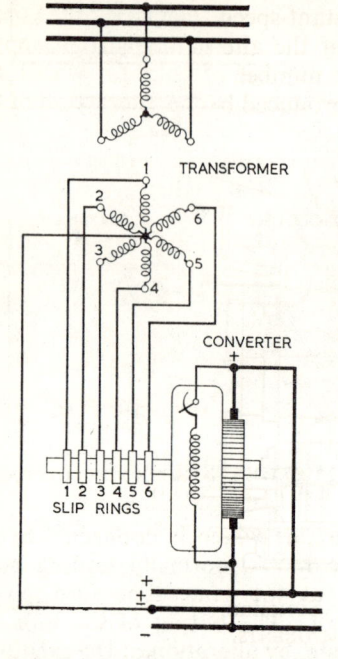

Fig. S–26. Diametrally Connected Six-Ring Synchronous Converter.

The a.c. supply is fed to the slip-rings; the d.c. output is taken from the commutator brushes. Commutation is normally simpler than in a d.c. machine, because the d.c. and a.c. armature reactions almost neutralize each other.

Synchronous Generator. An a.c. generator or ALTERNATOR that is driven at a constant speed corresponding to the output frequency required.

Synchronous Impedance. The ratio of the short-circuit field current to the open-circuit current of a synchronous machine under specified conditions. It is usually expressed as a percentage.

Synchronous-Induction Motor. A compromise between the SLIP-RING MOTOR and the SYNCHRONOUS MOTOR. The good starting performance of the induction motor is combined with the high running efficiency and variable power factor of the synchronous motor. It is started as a slip-ring motor, torque being provided by the interaction of a rotating magnetic field produced by the primary windings and the polyphase currents induced by this field in the rotor windings. When full speed as an induction motor has been reached, d.c. excitation is applied to the secondary and the rotor pulls into synchronous speed. The required power factor of the load is then obtained by adjustment of the field-excitation current.

Synchronous Motor. A constant-speed machine, the speed depending upon the frequency of the alternating-current supply to which it is connected, and the number of poles for which it is designed (Fig. S-27). Torque is produced by the interaction of the

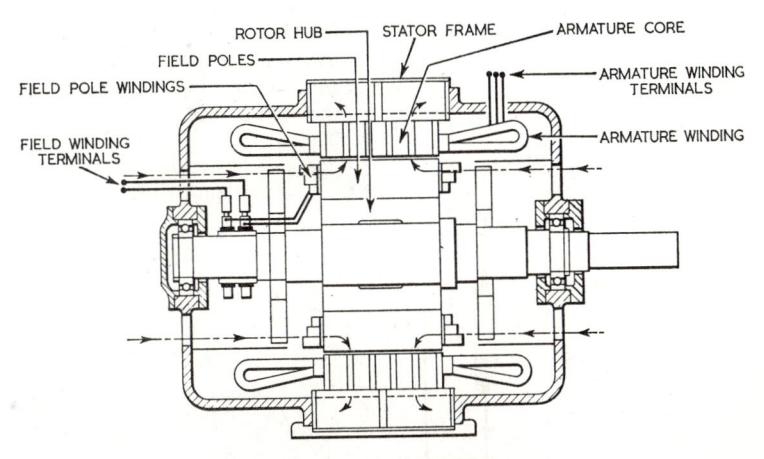

FIG. S-27. SECTIONAL VIEW OF SYNCHRONOUS MOTOR SHOWING MAIN COMPONENTS.

flux from the field poles of fixed polarity, and the field of the armature revolving at synchronous speed. Usually the primary, referred to as the armature, is mounted in the stator frame, and the field poles are located on the rotor shaft or rotor hub.

Synchronous Reactance. The vector difference between the SYNCHRONOUS IMPEDANCE of a synchronous machine and its effective armature resistance.

Synchroscope. A dial-and-pointer instrument used for synchronizing an incoming machine. The pointer shaft carries a rotor acted upon by the magnetic field of two coils, energized respectively by the machine and the busbar voltages. One form is shown in Fig. S-28. The fixed coils are energized from the busbar voltage. The coils on the rotor are fed from the incoming machine, one through a resistor and the other through an inductor, and they magnetize separate cores with polar vanes, the latter forming two groups having magnetic fields with a 90° space phase and (approximately) a 90° time phase. As a result, the rotor system

Synchrotron

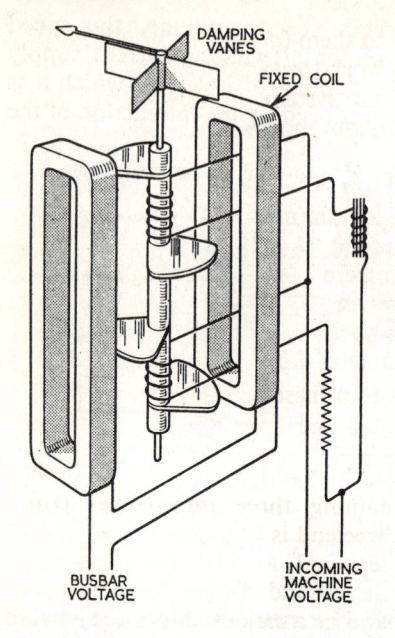

FIG. S-28. SYNCHROSCOPE.

The pointer is stationary and at zero when the frequencies are the same and in time phase. The direction of pointer rotation depends on the sign of the frequency difference.

rotates at a speed corresponding to the frequency difference between the busbars and the incoming machine.

Synchrotron. An ORBITAL ACCELERATOR that was developed from the BETATRON. It uses an annular magnetic guide field that increases as the particles gain energy, so that they maintain a constant orbit radius.

Synthesis. See NETWORK SYNTHESIS.

Synthetic-Resin-Bonded Paper. Laminated sheet insulating material consisting of papers bonded together under heat and pressure with synthetic resins. It is available in thicknesses from 0·2 to 50 mm (0·008 to 2 in.) in several grades, and is used for panels, terminal boards, coil flanges, etc. It is abbreviated *s.r.b.p.*

System Function. A network function that describes the kind of response that will result from a given stimulus (e.g. applied voltage). It takes a form similar to the complex impedance or admittance of a network in the analysis of networks under transient conditions. It is expressed as a function of the complex variable $s = x - j\omega$, and is normally a polynomial in s. The system function may express, for example, the current input at the terminals of a network when a voltage is applied to them (*driving-point function*) or the response elsewhere in the network (*transfer function*).

Système International. A coherent system of units agreed in 1960, and now being adopted throughout most of the world. There are seven primary S.I. units:

length	metre	m
mass	kilogram(me)	kg
time	second	s
electric current	ampere	A
temperature	kelvin	K
luminous intensity	candela	cd
amount of substance	mole	mol

All physical quantities are derived from these primary units, and a list is given in the Appendix.

T

T-Network. A network containing three impedances connected together at one end. One free end is connected to an input terminal, another to an output terminal, and the third free end to both the second input and the second output terminals, as shown in Fig. T–1. It is also known as a *star* or *Y-network*.

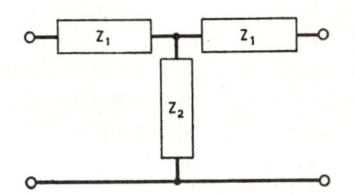

FIG. T–1. SYMMETRICAL UNBALANCED T-NETWORK.

Tachometer Generator. An electromechanical device employed in servomechanisms. It resembles a small motor and has an output voltage proportional to its shaft speed. Units that produce a direct-voltage output usually have a permanent-magnet field excitation. A.C. units are excited by the alternating supply, and produce an output voltage of supply frequency, the polarity of which is dependent on the direction of shaft rotation. It is also known as a *rate generator*.

Tandem-Compound Turbo-alternator. A TURBO-ALTERNATOR in which the turbine has two or more cylinders in line driving a common shaft (cf. CROSS-COMPOUND TURBO-ALTERNATOR).

Tank Capacitor. An INDUSTRIAL CAPACITOR in which the

Tank Circuit

dielectric and electrodes are laid up in a roll, these elements then being joined in series-parallel to suit the rated operating voltage and output. The tank allows adequate space for free thermal circulation of the oil, and its size is determined by the heat to be dissipated from the capacitor losses, which are normally 0·2–0·4% of the capacitor rating. Tank capacitors are rated up to 500 kVAr or above.

Tank Circuit. In an electronic oscillator, the tuned LC circuit in which the stored energy circulates at a frequency determined by the circuit parameters. The energy is dissipated in tank-circuit losses, and may also be drawn upon for an output load (e.g. for high-frequency heating or for aerial radiation).

Tap Changing. A method of controlling the voltage ratio of a power transformer, by tapping the windings in such a way as to alter the number of primary or secondary turns. Tap-changing may be done when the transformer is out of circuit; or, particularly with short-period and daily load cycles requiring voltage adjustment, when on load. The latter condition is more complicated and needs special equipment (see ON-LOAD TAP CHANGING).

Tariff. The rate at which a charge is made by an electricity supply authority for the supply of electrical energy. It may be in the form of a FLAT-RATE TARIFF or a TWO-PART TARIFF. Modifications of the two-part tariff are the BLOCK-RATE TARIFF and STEP TARIFF. Other available tariffs include a MAXIMUM-DEMAND TARIFF, MULTIPLE TARIFF and OFF-PEAK TARIFF.

Teaser Transformer. One of the units in a SCOTT CONNECTION of transformers for three-phase/two-phase transformation.

Tee Joint. A form of BRANCH JOINT in which the branch cable is taken away at right-angles to the main cable. It is enclosed in a pressed lead or copper box of suitable shape.

Telemetering. The remote indication of electrical quantities.

Temperature Classification. A classification of insulating materials in accordance with their operating temperature. The classes are: Y (90°C), A (105°), E (120°), B (130°), F (155°), H (180°), C (>180°). See also entries under individual classes.

Tension Insulator. An insulator, in an overhead-line system, which is capable of transmitting the tension of the conductor to the supporting tower. It is also known as a *strain insulator*.

Terminal Tower. A LATTICE TOWER situated at the end of an overhead line, and designed to withstand the longitudinal load of the phase conductors. It is also known as a *dead-end tower*.

Tertiary Winding. An auxiliary winding, additional to the normal primary and secondary. Power transformers may be constructed with tertiary windings for any of the following reasons: (*a*) for an additional load that for some reason must be kept conductively separate from that of the secondary, (*b*) to supply phase-compensating devices, such as capacitors, operated at some voltage different from that of the secondary or primary or with some different connection, (*c*) as a voltage-indicating coil in a testing transformer, (*d*) to load a large split-winding generator, (*e*) to interconnect three supply systems of the same frequency but different operating voltages, or (*f*) in star/star-connected transformers, to allow sufficient earth-fault current to flow for the operation of protective gear, to suppress harmonic voltages, and to limit unbalance when the main load is asymmetrical; in each such case the tertiary winding is delta-connected.

Tesla. The M.K.S. (and S.I.) unit of magnetic flux density (abbreviation: T). It is a density of one weber of magnetic flux per square metre.

Tesla Coil. Induction coil used to develop a high-voltage discharge at a very high frequency. A high-voltage transformer (Fig. T–2) promotes a discharge across the gap G_1 and charges

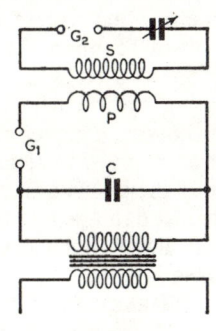

FIG. T–2. TESLA COIL.

capacitor C. Low-frequency high-amperage currents oscillate in winding P, inducing high-frequency high-voltage oscillations in S. These may produce discharges across the gap G_2. It is also known as a *Tesla transformer*.

Test Set. A multi-range instrument for measuring two or more quantities, e.g. voltage, current and power. Both single- and double-instrument test sets are in use: the former is cheaper but the latter enables two quantities to be read simultaneously. For d.c. test sets, moving-coil instruments are always incorporated. For a.c., moving-iron instruments are usual, but when power is to be measured in addition then an electrodynamic or induction wattmeter must be added; sometimes all three instruments are of the same type.

Test Shield. Of a cable, a metal sheath situated under the lead and insulated from it (cf. EARTH SHIELD).

Testing Joint. Of a lightning protective system, joint in the system designed to facilitate the measurement of resistance.

Tetrode. An electronic device with four electrodes, normally cathode, anode, control grid and screen.

Thermal Breakdown. A form of DIELECTRIC BREAKDOWN caused by self-heating of a material from its internal losses creating runaway conditions, or by conducting contaminants overheating to produce carbon and forming conducting tracks through the material.

Thermal Noise. See JOHNSON NOISE.

Thermal Ohm. A measure of thermal resistance. It is the thermal resistance of a body which has a temperature difference of 1 °C between opposite faces when heat flows at the rate of 1 W.

Thermal Overload Relay. An OVERLOAD RELAY that makes use of the heating effect of current for operation. Many incorporate a bimetal strip the deflection of which, when excess current is passed, permits an associated mechanism to open the normally closed trip contact. Cf. MAGNETIC OVERLOAD RELAY.

Thermal Power Station. A generating station where electrical energy is derived from thermal prime-movers, e.g. steam turbines. The heat may be obtained from fuel-fired plant or nuclear reactors. Thermal power stations produce most of the world's electricity.

Thermal Protection. Protection against damage by overheating. This may be achieved with an OVERLOAD RELAY such as a THERMAL OVERLOAD RELAY, or a simple THERMAL RELAY.

Thermal Reactor. A type of NUCLEAR REACTOR. If the neutrons resulting from fission of a fissile material are permitted to make successive collisions with non-fissile nuclei they will eventually be slowed down to a velocity of about 22 m/sec at which they will have an energy of 0·025 eV, these values corresponding to those for molecules in a gas at normal temperature and pressure. Neutrons slowed down in this way, known as "slow" or "thermal" neutrons, are much more likely to cause fission than are fast neutrons, and a chain reaction can be made to occur with them in natural or only slightly enriched uranium (i.e. uranium with up to a few per cent of uranium-235 instead of only 0·7% as in natural uranium). A reactor incorporating such a slowing-down process is known as a thermal reactor. Because of the material included in the core to retard the neutrons and also because of the non-fissile uranium-238 present, the size of the core for a given heat output will be 100–1,000 times as large as that of a FAST REACTOR, and will give 15–150 kW/ft^3 (500–5,000 kW/m^3).

Thermal Relay. A relay that disconnects apparatus when the temperature exceeds the safe value. Such a device may operate directly from an electrical signal produced by a hot-bulb or resistance thermometer placed at a suitable point in the apparatus to be protected. See also THERMAL OVERLOAD RELAY.

Thermal Storage. Storage of heat in suitable media such as concrete or water, applied particularly to heating on an OFF-PEAK TARIFF.

Thermionic Emission. The emission of electrons from a substance when heated. Emission takes place from a hot metal or semiconductor when its electrons are given sufficient thermal (kinetic) energy to surmount the surface potential barrier.

Thermionic Rectifier. A thermionic valve in which the unidirectional electron flow is utilized for rectification. A vacuum diode can be used for obtaining a d.c. output up to a few hundred milliamperes at about 500 V from an a.c. supply. Frequently two anodes are enclosed in one envelope to provide full-wave rectification. The vacuum type can also be used at considerably higher voltages when only a small current is required, for example for a cathode-ray tube or for the production of direct voltages up to about 50 kV for high-voltage testing. For higher powers the gasfilled diode has several advantages over the vacuum type, and is generally preferred. The gas most frequently used is mercury vapour.

Thermionic Relay. Thermionic valve, usually a THYRATRON, operated so that the application of a comparatively small impulse to the control electrode permits the passage of a large current.

Thermistor. A temperature-sensitive resistor, comprising a synthetically prepared SEMICONDUCTOR with a large negative resistance/temperature coefficient. The resistance decreases by 3–6% for 1°C rise in temperature at room temperature. Thus a thermistor can be used in circumstances which require a resistance that varies with power dissipation. It is also used for the measurement of small temperature changes.

Thermocouple

Thermocouple. A device which utilizes the SEEBECK EFFECT to measure temperature. Of the available dissimilar metals, antimony/bismuth give the highest output, but iron or copper/Constantan are more durable and stable. Other useful combinations are platinum/platinum–10% rhodium, Chromel/Alumel and Chromel/Constantan. Couples including platinum can be used up to temperatures of 1,500°C, copper/Constantan over the range −100 to +350°C, and the others up to about 1,000°C. The e.m.f. can be read on a potentiometer or on a sensitive galvanometer-type indicating instrument.

Thermocouple Generator. A form of THERMOELECTRIC CONVERTER. The normal thermocouple is inefficient as a converter because of the heat-conduction loss between the hot and cold junctions. If one conductor is replaced by a plasma, or semiconductor materials are used, a good electrical conductivity can be combined with a poor thermal one, so increasing the thermal-electrical conversion efficiency.

Thermocouple Relay. A combination consisting of a thermocouple connected to operate a sensitive relay when some pre-arranged temperature is reached. The thermocouple may be associated with a heater carrying the control current so that operation is obtained as for the THERMAL RELAY.

Thermoelectric Converter. A device for the direct conversion of heat into electrical energy. Various approaches include the FUEL CELL, THERMO-ELECTRON CONVERTER, THERMOCOUPLE GENERATOR and MAGNETOHYDRODYNAMIC GENERATOR.

Thermoelectric Effect. See SEEBECK EFFECT.

Thermo-electron Converter. A form of THERMOELECTRIC CONVERTER. When electrons are emitted from a hot cathode, a calculable fraction of their number will possess a velocity high enough to reach a nearby anode at a small negative potential, in spite of the opposing potential barrier. The electrons give up their kinetic energy at the anode surface, and the potential energy that results can be used to drive a current through a load circuit connected between anode and cathode.

Thermometal. Alternative name for BIMETAL (q.v.).

Thermonuclear Reaction. Reaction involving NUCLEAR FUSION. It is induced by heat.

Thermopile. A device for converting heat directly into electrical energy. It generally consists of a battery of series-connected THERMOCOUPLES.

Thermostat

Thermoplastics. PLASTICS that may be heated beyond their melting point and cooled again any number of times without any appreciable change in their properties (see ACRYLIC RESIN, CELLULOSE ACETATE, NYLON, POLYETHYLENE, POLYMETHYL METHACRYLATE, POLYSTYRENE, POLYTETRAFLUOROETHYLENE, and POLYVINYL CHLORIDE).

Thermosetting Plastics. PLASTICS in which a chemical reaction occurs when they are moulded under heat and pressure (see ALKYD RESIN, EPOXY RESIN, PHENOLIC RESIN, POLYESTER RESIN, SILICONES and UREA RESIN).

Thermostat. An apparatus for automatically controlling temperature, either at a predetermined value or within specified limits, by adjusting the flow of heat into or out of the enclosure, body, or fluid, whose temperature it is controlling. A thermostat

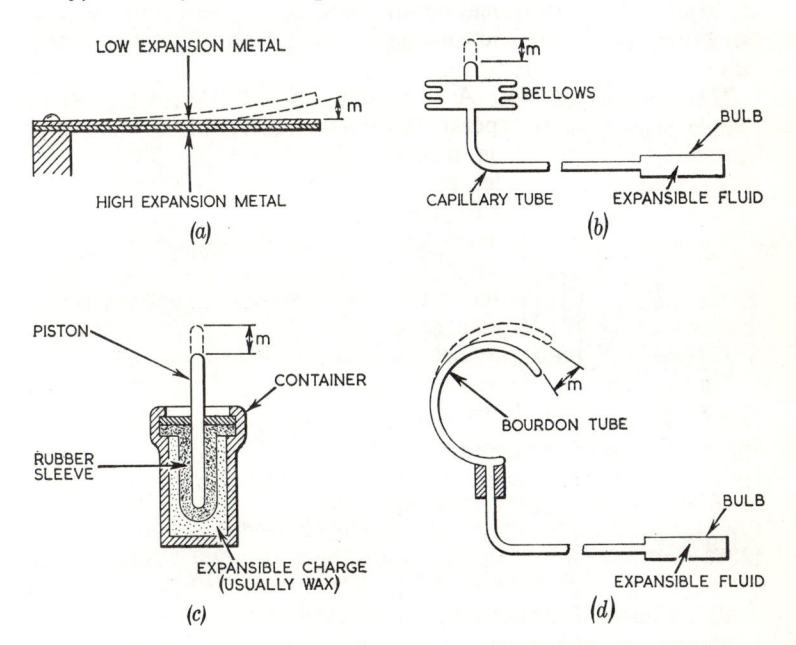

FIG. T-3. TEMPERATURE DETECTING ELEMENTS.

(*a*) Cantilever bimetal elements. (*b*) Bellows volumetric expansion element with remote bulb. (*c*) Piston volumetric expansion element. (*d*) Bourdon tube volumetric expansion element with remote bulb. m is the movement generated by temperature change.

Thévinin's Theorem

comprises basically (*a*) a temperature-sensitive element generating a mechanical force or electrical signal, and (*b*) a regulator which is either an electric switch or a fluid valve, or some comparable device (see Fig. T–3). The regulator directly or indirectly adjusts the flow of the heating or cooling medium in accordance with the response it receives from the detecting element.

Thévenin's Theorem. See HELMHOLTZ-THÉVENIN THEOREM.

Thomson Effect. A thermoelectric effect. If different parts of the same metallic conductor are maintained at different temperatures, a temperature-gradient will exist and a heat flow will take place in the metal. If now an electric current is passed through the conductor, the temperature distribution will be disturbed. The accompanying evolution or absorption of heat (other than the Joulian heat) comprises the Thomson effect. It is reversible, changing sign with the direction of the current, and it is proportional to the product of current and temperature-gradient. It is also known as *Kelvin Effect*.

Three-Ammeter Method. A method of measuring small single-phase powers. In Fig. T–4 the load has a phase-angle ϕ and

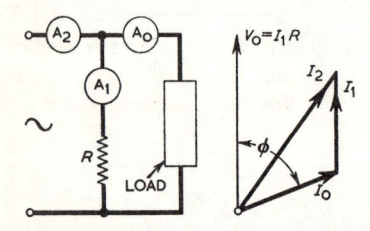

FIG. T–4. THREE-AMMETER METHOD OF MEASURING SMALL SINGLE-PHASE POWERS.

takes I_0 amperes at V_0 volts. By using three instruments, and ignoring any of their impedance effects, the load power P_0 and power factor $\cos \phi$ can be obtained. From the complexor diagram,

$$I_2{}^2 = I_0{}^2 + I_1{}^2 + 2I_0(V_0/R) \cos \phi$$

Hence

$$P_0 = V_0 I_0 \cos \phi = \tfrac{1}{2}(I_2{}^2 - I_0{}^2 - I_1{}^2)R$$

and the power factor is

$$\cos \phi = \tfrac{1}{2}(I_1{}^2 - I_0{}^2 - I_1{}^2)I_0 I_1$$

See also THREE-VOLTMETER METHOD.

Three-Phase Four-Wire System. A system of distribution employing four conductors, three of which are connected to the three-phase supply and the fourth having its termination on a neutral point in the source of supply.

Three-Phase System. A POLYPHASE SYSTEM in which the number of phases M is 3, and the phase displacement is $2\pi/3$. It is the system of distribution, transmission and generation in almost universal use for electrical energy supply in bulk, the phase voltages being as far as possible balanced.

Three-Voltmeter Method. A method of measuring small single-phase powers, similar to the THREE-AMMETER METHOD (q.v.). With the same symbols as in the latter case (Fig. T–5),

$$V_2^2 = V_0^2 + V_1^2 + 2V_0 I_0 R \cos \phi,$$

giving

$$P_0 = V_0 I_0 \cos \phi = \tfrac{1}{2}(V_2^2 - V_0^2 - V_1^2)/R$$

and

$$\cos \phi = \tfrac{1}{2}(V_2^2 - V_0^2 - V_1^2)/V_0 V_1$$

The three-meter methods have the disadvantages that, by requiring the difference between squared values, the errors in the instruments are magnified.

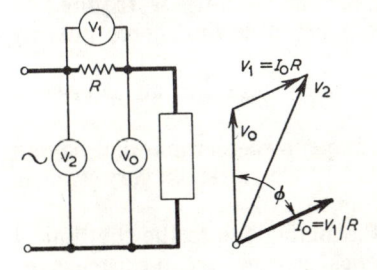

FIG T–5. THREE-VOLTMETER METHOD OF MEASURING SMALL SINGLE-PHASE POWERS.

Three-Wire System. A distribution system, for direct current or single-phase alternating current, employing two conductors and a neutral wire. The supply is taken between one conductor and the neutral, the latter being usually earthed and carrying only the difference current.

Thrustor. An electrohydraulic mechanism that converts electrical energy into a straight-line, constant thrust. A centrifugal pump at the bottom of an oil-filled tank forces a piston upwards with a constant force. It is a control device, and may be used on a mechanical brake.

Thury System. A system of high-voltage d.c. transmission. A number of series-wound generators are connected in series to supply a number of loads also in series, the loads usually being

Thyratron

motors driving auxiliary generators in substations. The voltage of the main generators is varied to maintain the current constant.

Thyratron. A gas-filled triode or tetrode. It is now being replaced for many applications by the THYRISTOR (q.v.).

Thyristor. A controlled rectifier capable of blocking current in the forward direction until a firing pulse at the gate switches the device on. Once conduction has begun, the gate has no further control over the current, and the device can be switched off only by reversing the polarity of the applied voltage or by reducing the forward current below the "holding" current (by increasing the resistance in the external circuit). The thyristor is particularly suited to controlling the exchange of power between a source and load. When operating from an a.c. source, the thyristor can be switched on at any time during the positive half cycle of the supply voltage (by timing the firing pulse). It will remain conducting until the end of the cycle when it will switch off by itself as the polarity reverses (*natural commutation*). The amount of power fed to the load can thus be controlled by adjustment of the firing pulse. When the power source is d.c., further circuitry is required to switch the device off (*forced commutation*) as the source polarity is always positive.

Typical circuits using thyristors are those for an INVERTER, CONVERTER, or RECTIFIER.

Ticonal. Trade name for a permanent-magnet material having the composition Al 8, Ni 14, Co 24, Cu 3. It is very similar to, and prepared in the same way as, ALCOMAX.

Tidal Power. Power resources available due to the rhythmical movements of the seas. A few practical schemes are in existence for converting this power into electrical energy, and it has been estimated that about 30 TWh/year might be developed from all suitable sites.

Tie Line. (1) An interconnecting line in a distribution system.

(2) A link between two or more private automatic telephone exchanges.

Time Constant. The time in which a function, such as the charge on a capacitor or current in an inductor, falls to $1/e$ of its initial value ($1/e \simeq 0 \cdot 368$). If C is the capacitance, L the inductance, and R the total series resistance, the time constants of capacitive and inductive circuits are RC and L/R respectively.

Time-Control Acceleration, Starting. A method of automatically controlling the acceleration of a resistance-started electric motor during the starting period. Use is made of one or more accelerating relays of the time-delay type for determining the period that elapses between the closing of successive accelerating contactors. The acceleration period of the motor is independent of the load conditions.

Time/Current-Control Starting. A method of automatically controlling the starting of a resistance-started motor. It is similar to TIME-CONTROL STARTING above but incorporates a timing relay designed to be influenced by the magnitude of the motor current during acceleration in such a manner that the time delay increases with motor load. Thus, a motor lightly loaded would have the accelerating contactors closed more rapidly than if it were heavily loaded.

Time of Recovery. In an automatically regulated circuit, the time required after a specified disturbance for the voltage or current to recover to 95% of its original value.

Tip Speed Ratio. An important design figure for wind-power generating plant. It is the circumferential speed of the rotor blade tips divided by the wind speed, and usually lies between 5 and 10 for such plant.

Tirrell Regulator. An AUTOMATIC VOLTAGE REGULATOR of the vibrating-contact type. A regulating field rheostat is rapidly switched in and out of circuit to correct deviations from the nominal value of voltage.

Tong Tester. A portable test equipment for measuring current. It consists of a secondary winding wound upon one leg of a rectangular frame core and usually connected to an integral ammeter. One corner of the core consists of a butt joint, and one side adjacent to the butt joint is hinged so that the core can be opened and closed around a cable or busbar carrying the current to be measured.

Toroidal Winding. Alternative name for RING WINDING (q.v.).

Torque-Amplifier Selsyn. A transmitter SELSYN having a special arrangement of windings, involving both a commutator and sliprings, whereby the torque reaction at the transmitter is largely neutralized. The unit combines the functions of a torque amplifier and transmitter Selsyn, requiring much less manual effort.

Torque Angle. Alternative name for LOAD ANGLE (q.v.).

Torque Motor. A motor that makes a part of a revolution when current is applied, to exert a torque against some controlling force such as a spring. It is used for the operation of brakes, rheostats, etc.

Total Break Time

Total Break Time. Of a circuit-breaker, the time elapsing between the application of tripping power and the extinction of the arc caused by the opening of the contacts.

Totally Enclosed Machine. Machine in which there is no access whatever to the inside of the casing, so that cooling must be effected by conduction to and subsequent dissipation from the *external* surface.

Tough-Rubber Sheath. Protective covering applied to an insulated cable, consisting of a rubber mixed with hardening substances and suitably vulcanized to make it waterproof and corrosion and abrasion resistant.

Tower. See LATTICE TOWER.

Townsend Discharge. Suggested mechanism by which a spark occurs between electrodes in a gas. It depends on the production of an electron at a point near the cathode, to precipitate an electron avalanche. As this develops across the gap, secondary sources of ionization appear, and the current between the electrodes develops to the point of breakdown.

Tracking. The formation, along the surface of an insulating material, of carbon conducting paths caused by surface stresses. Fibrous materials such as paper, cotton or asbestos are generally very susceptible to tracking, but their strength is greatly influenced by the impregnant or bonding medium. Melamine impregnants generally give the best resistance to track formation.

Traction Battery. Battery for use in the power unit of a battery-driven vehicle.

Traction Motor. A motor designed for use in electric traction. A SERIES-CHARACTERISTIC MOTOR is inherently ideal for this work.

Train Line. A cable extending the length of a railway coach and terminating in sockets at each end. Couplers between each coach enable electrical continuity to be maintained through the length of the train.

Transducer. A device which, under the influence of a change in energy level of one form or in one system, produces a corresponding change in energy level of another form or in another system. The ideal correspondence between the input and output is usually considered to be an output proportional to the input, or to its derivative or integral. In such a case the device is a *linear* transducer. In a *reversible* transducer the action is reversible, i.e. the correspondence may be in either direction. See Fig. T-6.

Transformation Ratio

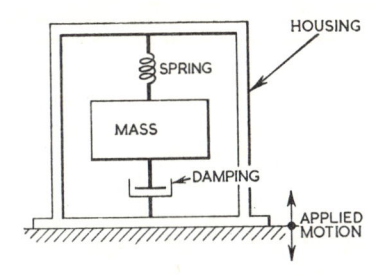

FIG. T–6. A SEISMIC TRANSDUCER.

Displacement of the mass may be made proportional to the applied displacement, velocity or acceleration. This displacement may be measured by capacitance or resistance strain gauges, or by inductive, piezoelectric or e.m.f. generation effects.

Transductor. The basic amplifying element of a MAGNETIC AMPLIFIER consisting of iron-cored coils. It is also known as a *saturable reactor*.

Transfer Admittance. The ratio of the current flowing in any part of a network to the electromotive force applied elsewhere in the network that produced that current.

Transfer Function. A type of SYSTEM FUNCTION. It relates the response at a given pair of terminals to a source or stimulus at another pair of terminals. Its usual form is $K.G(p)$, where K is a multiplier or "steady-state gain", and $G(p)$ is a function of time in terms of $p = d/dt$ and the parameters of the network.

Transfer Impedance. The ratio of the voltage applied across two given terminals in a network to the current flowing at some other point in the network.

Transfer Instrument. An instrument so constructed that it can be calibrated on direct current and used to indicate a.c. quantities without sensible error. A example is an electrodynamic instrument, in which the working principle is inherently independent of the kind of voltage or current concerned. Alternatively, a transfer instrument may embody a thermal method of comparison, using a potentiometer with a thermocouple to compare the heats developed by alternating and direct current sources.

Transfer Switch. Switch for transferring load current from one tap-selector switch on a transformer to another, so that each is not carrying a current when it operates.

Transfer Tripping. See INTERTRIPPING.

Transformation Ratio. (1) Of a power transformer, the ratio of the voltage between the terminals of the higher-voltage winding to that between the terminals of the lower-voltage winding at no load.

Transformer

(2) Of an instrument transformer, the ratio of the measured quantity (current or voltage) in the primary winding to the current or voltage in the secondary winding.

See also TURNS RATIO.

Transformer. A static electromagnetic device consisting fundamentally of two electric windings or circuits (Fig. T-7) magnetically interlinked by an iron core, whereby an alternating

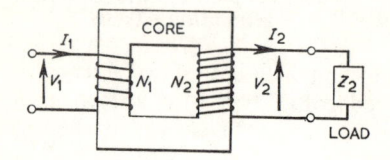

FIG. T-7. ELEMENTARY TRANSFORMER.

$$\frac{V_1}{V_2} = \frac{N_1}{N_2} = \frac{I_2}{I_1}$$

electromotive force applied to one of the windings produces, by electromagnetic induction, a corresponding e.m.f. in the other winding. The winding to which the e.m.f. is applied is known as the *primary* winding, the other being the *secondary* winding. Transformers are employed for converting energy received at one voltage to energy transmitted at a different voltage; for separating electrical circuits while permitting power flow between them; and for matching impedances in communication circuits so that maximum power is obtained from a source (see also INSTRUMENT TRANSFORMER, CURRENT TRANSFORMER, VOLTAGE TRANSFORMER).

Transformer E.M.F. An electromotive force induced by transformer action in a coil or other conductor that is linked with an alternating flux.

Transformer Hum. See NOISE.

Transformer Kiosk. A weatherproof enclosure for a power transformer and its associated switchgear, with no additional space.

Transformer Oil. See INSULATING OIL.

Transformer Ratio. See TRANSFORMATION RATIO.

Transformer Substation. A substation installed at a point on a network where a change of voltage has to be carried out. It is therefore to be found on sites adjacent to generation stations where power is received at generation voltage and is stepped up to the grid voltage of, usually, 132 or 275 kV. These stations are also important switching points where feeders from a number of directions may be marshalled through circuit-breakers on to the common busbars. At other points on the grid network, transformer substations are needed where supplies from the grid are delivered to the Area Boards responsible for distribution to the public. Here reduction in voltage takes place to the subtransmission voltage required by the Board concerned, usually 33 kV. On the subtransmission network there may sometimes be a further intermediate stage of voltage reduction before reaching the many local distribution centres where the final reduction to consumer voltage takes place.

Transformer Tank. A container in which a transformer is immersed in a cooling and insulating medium. It may carry fins to provide a heat-dissipating surface for a transformer with natural oil cooling.

Transformeter. Combination of an integrating meter and associated current transformers in a single unit. It is available for a single-phase or polyphase circuit.

Transient. The variation of a system quantity (such as current, flux, velocity) in a physical system following a disturbance. The variation persists, normally, until the system has settled down to a new energy state. If only one energy-storage component is concerned, a *single-energy transient* results; two such components produce a *double-energy transient*, and so on.

Transient Reactance. Reactance of the armature winding of a synchronous machine, caused by the leakage flux.

Transient Stability. STABILITY following sudden disturbances, such as sudden load increases on a power system, or switching operations or faults.

Transistor. The solid-state counterpart of the thermionic valve. Electrons or holes flow from the emitter electrode through the solid semiconductor crystal to the collector. Basically, the transistor can do much the same work as the thermionic valve, and in many cases it has distinct advantages. For instance, it has no filament and so needs no warm-up time; it is very efficient, and robust; and it can be much smaller. Low-power units are about one-twentieth of the volume of the modern miniature valve.

Transistor Tetrode. TRANSISTOR in which arrangement is made to apply a transverse voltage across the base layer, so that the emitter current is restricted to a smaller area, decreasing collector capacitance and increasing maximum frequency.

Transmission Dynamometer. A device for measuring the mechanical output of a prime mover or electric motor transmitted to another equipment. Simple mechanical transmission dynamometers comprise an epicyclic gear or a belt in the transmitted drive.

Transmission Line

An electrical dynamometer comprises a d.c. generator that converts the output of the machine under investigation to electrical power, this being then dissipated in resistors, or delivered to a suitable power network. The field system of the generator is mounted on trunnions, and the torque is determined from the force needed to keep it stationary.

Transmission Line. A conductor for the transmission of electrical energy from a generating station or substation to other stations or substations.

Transport Number. In electrolysis, the fraction of the total current carried by the migrating anions or by the cations.

Transposition. Variation of the relative position of parallel conductors to reduce or eliminate unwanted inductive effects. In the case of *power lines*, the harm is principally due to the resulting assymmetry in line voltage drops by which the terminal voltage may be affected, possibly leading to a unbalance. To avoid this, the three (or six) conductors of an overhead circuit are transposed in the course of the line run so that each phase conductor becomes in turn the upper conductor. For overhead *telephone lines*, if the relative positions of the pairs remains unchanged there will be interference and cross-talk. Transposing is simply carried out by taking each two pairs and rotating them cyclically so that the phase of the induced voltages is changed by 180° at each stage.

Parasitic eddy currents are developed in the slot conductors of large a.c. machines, and a large solid conductor must be subdivided into strips sufficiently shallow to limit the eddy-current loss. However, the residual loss of a given strip depends upon its position in

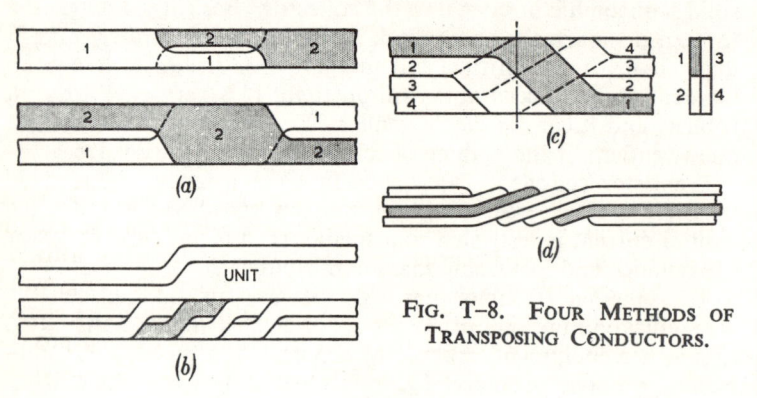

FIG. T–8. FOUR METHODS OF TRANSPOSING CONDUCTORS.

Trembling Bell

the slot, the loss being greatest in the strip nearest the slot-opening. To maintain the advantage gained, either the subdivision must be maintained throughout the winding, or the conductors themselves must be twisted within the slot so that each strip occupies all possible positions. Four methods of transposition are shown in Fig. T-8.

Travelling Cable. On an electric lift car, a flexible cable providing electrical connection between the car and a fixed point.

Travelling Wave. A SURGE propagated along a conductor.

Travelling-Wave Accelerator. A LINEAR ACCELERATOR for the production of high-energy electrons.

Trembling Bell. Bell with a mechanism giving a trembling action. It is the most widely used type for alarm and general

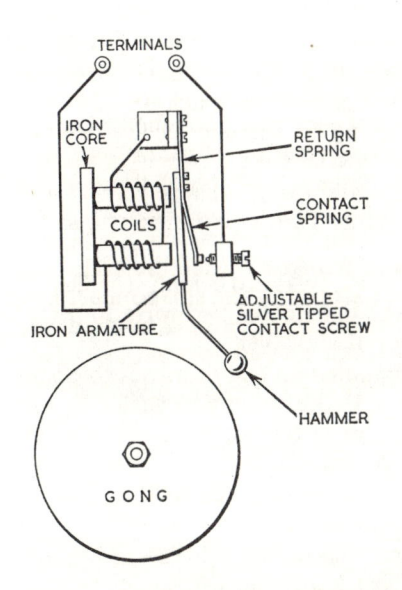

FIG. T-9. PRINCIPLE OF OPERATION OF A SIMPLE TREMBLING BELL.

purposes; its principle of operation is illustrated in Fig. T-9. When the external circuit is closed, current passes through the exciting coils of the electromagnet; the armature is therefore attracted towards the pole pieces and the hammer strikes the gong, which action breaks the circuit at the interrupter contacts. The electromagnet flux having ceased, the armature returns to its *off* position by the action of the supporting spring, thereby remaking the circuit

Triboelectricity

at the contacts. The same cycle is continually repeated until the external circuit is finally opened at the pushbutton or switch.

Triboelectricity. Electricity produced by friction. Substances may be arranged in a *triboelectric series* so that any substance will acquire a positive charge if rubbed with another substance below it in the series, or a negative charge if rubbed with a substance above it.

Trickle Charge. A steady, low-current charge designed to maintain a battery in a fully charged condition.

Trifurcating Box. A closed box in which the cores of a three-core cable can be connected to external conductors.

Triode. A three-electrode electronic tube consisting of a cathode, an anode and a control grid, or a similar semiconductor device. A gas-filled triode is a THYRATRON.

Triple-*n* Harmonic. A HARMONIC of order that is a multiple of 3, usually an odd multiple (e.g. 3rd, 9th, 15th . . .). Such harmonics are of importance in three-phase systems and machines.

Tripping Fuse. A fuse containing an auxiliary device that will trip other apparatus, such as a switch.

Trolley. Fitting for an electrically driven vehicle so that contact can be provided with an overhead wire by means of a *trolley wire*.

Trolley Frog. A device used in overhead supply for electric traction. It is incorporated at the junction of two wires to permit the passage of a current-collector along either wire.

Tropical Switch. A switch mounted on feet or bosses to ensure an air space between its base and the mounting surface. This safeguards the switch against the effects of excessively damp climates. It is also known as a *feet switch*.

Troughing. Preformed channel designed to accommodate cables and protect them from mechanical damage.

Truck-Type Switchgear. Switchgear in which the circuit-breaker is mounted on a wheeled truck together with ancillary items such as current and voltage transformers, relay instruments and control switches. The truck rolls on the floor and is guided into a housing containing busbars and circuit connections which are selected by means of plugs.

Trunk Feeder. A FEEDER connecting two sources of electrical energy, or two networks. It is also known as a *trunk main* or *interconnecting feeder*.

Turbulator

Trunking. Cable casing, constructed of metal, wood or insulating material, and usually of rectangular cross-section. One side is normally removable or hinged for the whole of its length to allow cables to be inserted.

Tube-Sinking. Method of applying an aluminium sheath to a cable. It consists essentially of pulling a cable core into a long aluminium tube, and then pulling both through a sinking die so that the sheath is brought close to the core. The cold-working makes the metal harder and less ductile. Tubes sufficient to cover a 250 m length of cable can be produced.

Tufnol. Trade name for a type of SYNTHETIC-RESIN-BONDED PAPER (q.v.).

Tumbler Switch. A small single-pole switch with a quick-break action operated by a lever handle pivoted near the face of the switch. It is commonly used in low-power circuits such as those controlling individual lamps.

Tuned Circuit. Network comprising an inductive and a capacitive element, one or both of which can be varied to alter the RESONANT FREQUENCY.

Tungsten. A metal with a very high melting point, used widely as a lamp filament. It is also employed in some thermionic valves and for sparking points and contacts. *Tungsten steels* were used for permanent magnets.

Tungsten Arc. (1) An arc between tungsten electrodes, the radiation being principally from the incandescent electrodes.

(2) An arc through tungsten vapour, having a characteristic emission.

Turbine. See STEAM TURBINE, WATER TURBINE.

Turbo-alternator. An alternator designed for high-speed operation, coupled to steam or gas turbines. It has few poles and forms the most important class of machine for power generation in the world today. A typical air-cooled turbo-alternator is shown in Fig. T–10.

Turbovisory Equipment. Steam-turbine supervisory equipment. Modern turbines require special instrumentation, particularly for starting. This is arranged to indicate and/or record the following parameters: (*a*) shaft eccentricity; (*b*) differential expansion between rotating and stationary parts; (*c*) overall expansion of the turbine; (*d*) vibration of bearing pedestals.

Turbulator. Trade name for a side-vented EXPLOSION POT having a number of magnetic inserts embedded in the pile of

Turnbuckle

insulating plates adjacent to, and surrounding, the vents. As the arc is drawn out, it comes under the influence of the field set up by these inserts and is moved towards and seals the throats of the vents, allowing pressure to build up inside the chamber. At cur-

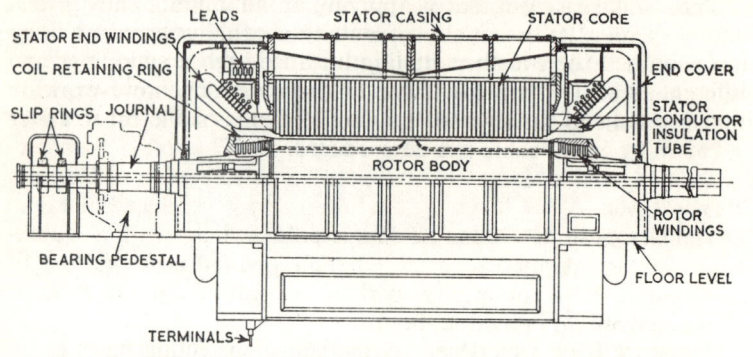

FIG. T–10. TYPICAL AIR-COOLED TURBO-ALTERNATOR.
(*G.E.C. Ltd.*)

rent zero the arc dies down and the gases are ejected through the vents by the pressure inside the pot. This process is repeated until the turbulent flow of gases during current zero produces sufficient cooling and de-ionization of the arc path to prevent restriking taking place.

Turnbuckle. Fitting for adjusting the tension of a wire. It has a bolt at each end, one having a right-hand and the other a left-hand thread.

Turns Ratio. The ratio of the number of turns in the phase winding of a transformer that is associated with the higher voltage to the corresponding number of turns in the lower-voltage winding.

Two-Fluid Cell. A voltaic cell in which the anode and cathode are immersed in different electrolytes.

Two-In-Hand Winding. A method of winding a lap-connected armature in which two circuits are wound simultaneously (see also LAP WINDING).

Two-Part Tariff. A TARIFF that involves a fixed standing charge plus a charge which is proportional to the number of units of energy supplied.

Two-Rate Tariff. A TARIFF in which different prices are charged for energy supplied at various times of the day.

Two-Reaction Theory. An analysis of the behaviour of a rotating electrical machine by considering the magnetomotive forces acting along the direct axis and the quadrature axis. The former is the direction in which the field winding magnetizes, the latter is electrically in space quadrature to it. With the m.m.f. of any given winding resolved into axis components produced by corresponding axis windings, there is no mutual inductance between windings on different axes. The flux due to windings on one axis, however, will induce e.m.f.s of rotation in moving elements of the other axis.

Two-Way Switch. Alternative name for a LANDING SWITCH (q.v.).

Type Test. A test taken on the first machine produced of each type or design to determine its characteristics.

U

Ultrasonic Flaw Detection. A method of detecting flaws in a specimen, based upon impulse reflection. A crystal probe is placed upon the specimen, usually with a coupling medium such as oil between. Ultrasonic impulses enter the specimen from the point contact in a closely knit beam and are reflected back by flaws such as cavities, cracks or inclusions. These are reproduced on a cathode-ray screen and compared with the wave reflected from the back of the specimen.

Ultrasonic Machining. Methods of removing metal, employing SPARK MACHINING or IMPACT GRINDING. Ultrasonic drilling has been applied to the production of dies in tungsten carbide and hardened steel; the dicing of germanium for transistor manufacture; the cutting of sapphires and agates for instrument bearings; engraving of glass and ceramics; and the threading of hard and brittle materials.

Ultrasonics. The study and application of acoustic waves of frequencies above the range of human hearing. It is also known as *supersonics*.

Ultra-violet Radiation. Electromagnetic radiation of frequencies higher than those of visible light, and wavelengths of 3,900–50 Å. Ultra-violet waves affect photographic plates, can produce fluorescence and are able to ionize gases. They may also be employed in ELECTROTHERAPY.

Unbalance Factor. In a three-phase system, the ratio of the negative phase-sequence current to the positive. It may also be

Undercompensation

applied to the ratio of the appropriate voltages, but the term UNSYMMETRY FACTOR is more usual in the latter case.

Undercompensation. The result of a COMPENSATING WINDING having less effect than the armature reaction it opposes. The output voltage will tend to decrease as the load current increases.

Underground Distribution. System of distribution in general practice in urban and suburban areas where overhead systems are likely to be dangerous or unsightly or to cause too much obstruction. Both high-voltage and low-voltage systems may be laid underground, utilizing insulated cable of suitable type and laid with necessary safeguards.

Undervoltage Release. Device that opens a switch when the supply to a circuit falls below a given voltage. A trip-coil is incorporated for this purpose in such equipment as hand-operated motor starters so that the motor does not re-start when an interrupted supply is restored. It is also known as a *low-volt release*.

Unearthed System. A system with no conductive connection to earth. It has the advantage that should an earth fault occur on any line conductor there is little interference to the system operation on load. But should a second earth fault develop on another line there will be effectively a short-circuit which will require some means of protection to cut off the machine, by fuses, circuit-breakers, etc.

When a communication circuit such as a telephone line is unearthed (i.e. does not rely on an earth connection for signal currents) there is less chance of interference or cross-talk between adjacent circuits, which are otherwise liable to interference through their common earth impedance.

Unexcited Synchronous Motor. A single- or three-phase SYNCHRONOUS MOTOR having no form of d.c. excitation.

Unijunction Transistor. A TRANSISTOR using only one *p–n* junction on a bar with ohmic contacts at the two ends. See *Glossary of Semiconductor Terms* in Appendix.

Unipolar Field Effect Transistor. TRANSISTOR in which the flow of electrons along a bar is controlled by *p–n* junctions on the sides of the bar.

Unipolar Machine. See HOMOPOLAR MACHINE.

Unit. See C.G.S. SYSTEM, M.K.S. SYSTEM, SYSTÈME INTERNATIONAL, and under the names of individual units.

Unit Capacitor. An INDUSTRIAL CAPACITOR in which the impregnated-paper dielectric and electrodes are encased, usually in rectangular containers, so that the elements substantially fill the whole of the space in the main container, providing good terminal contact between the elements and the container to aid in dissipating the heat from the dielectric. Connection between the capacitor terminals and the electrode foils of the elements is made by soldering connections either directly to the projecting edges of wide electrode foils, or to metal tabs inserted in the element winding. Unit capacitor ratings are lower than those of TANK CAPACITORS.

Unit Function. A function of time, such that its value is zero for all values of t up to $t = 0$ and is unity for all t from $t = 0$ to $t = \infty$. Its principal application in transient analysis is as a switching function. Thus, a lightning surge may be idealized into unit-function shape; and closing the switch separating a d.c. source from a circuit is equivalent analytically to the application of a unit-function voltage to the circuit.

Unit Magnetic Pole. A magnetic pole such that, when it is one centimetre from a similar pole in a vacuum, each exerts a force of one dyne on the other.

Unit Testing. A form of direct testing applied to a unit, or group of units, of a multi-unit circuit-breaker. By this means, it is possible to assess the behaviour of the complete circuit-breaker up to its rated making and braking capacity from tests made on one or more of its units.

Universal Bridge. A commonly used laboratory bridge (usually self-contained) employing three built-in arms, viz. two resistors P and Q, and a parallel capacitor-resistor combination (Fig. U–1).

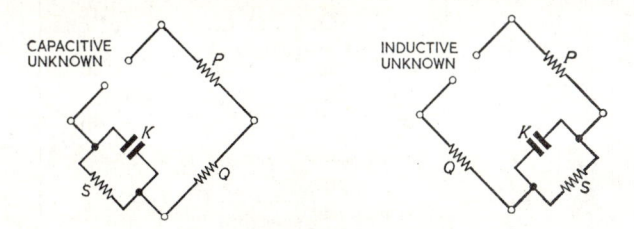

FIG. U–1. A UNIVERSAL BRIDGE.

It can be switched to a real-ratio form for measuring capacitance, and to a real-product form for the measurement of inductive unknowns (see A.C. BRIDGE), in each case giving also an indication of power factor from the balance relation of the capacitance and

Universal Motor

resistance arms. P and Q are adjustable so that the range can be extended up and down. Taking the balance conditions respectively for (i) a parallel CR and (ii) a series LR unknown, the balance conditions are:

(i) $$\frac{R}{1+j\omega CR} = \frac{P}{Q} \cdot \frac{S}{1+j\omega KS}$$

Whence
$$R = S(P/Q) \quad \text{and} \quad C = K(Q/P)$$

(ii) $$R + j\omega L = PS[(1/Q) + j\omega K]$$

Whence
$$R = S(P/Q) \quad \text{and} \quad L = K(PS)$$

Both balances are independent of frequency.

Universal Motor. A commutator machine designed to work on d.c. and single-phase a.c. supplies without alteration. It normally has a circuit identical to that of a d.c. series motor, and its construction is the same except that it is necessary to laminate the field structure as well as the armature core.

Universal Test Set. A TEST SET for use particularly for radio and other low-current circuits. It measures direct and alternating quantities and resistance. Such instruments use a rectifier for alternating current, with provision for excluding the rectifier when direct current is measured.

Unsymmetry Factor. In a three-phase system, the ratio of the negative phase-sequence voltage to the positive. See also UNBALANCE FACTOR.

Upset Butt Welding. A variety of BUTT WELDING in which pressure is first applied to the joint and then current passed until welding temperature is reached and the weld is "upset". It is also known as *slow butt welding*.

Urea Resin. Moulding (thermosetting) material, produced by the controlled reaction of urea and formaldehyde. Its characteristics are similar to those of PHENOLIC RESIN, though in general the moulding temperature is rather lower and the rate of cure faster. Urea resins produce mouldings possessing hard, durable surfaces with excellent resistance to abrasion, and they have good electrical properties with excellent resistance to tracking. This anti-tracking characteristic plus the ability of urea mouldings to be produced in an unlimited range of colours, including white and pastel shades, make them popular with some manufacturers for the production of plugs, sockets, ceiling roses, housings, switchplates, etc.

V

Vacuum. Space from which gas has been removed. A perfect vacuum, in which the gas pressure is zero, is unattainable, but a DIFFUSION PUMP is capable of producing down to 10^{-7} mmHg. Vacua of the order of 10^{-6} mmHg are employed in high-vacuum valves and other electron tubes. Vacuum pressures are commonly measured in conventional millimetres of mercury or in torrs. 1 mmHg = 1·0000001 torr. 1 torr = 1/760 atm. (1 mmHg = 133·32 Nm^{-2}.)

Vacuum Impregnation. The filling of voids in a dielectric material by reduction of pressure, to increase breakdown strength and permittivity. Mineral oil, paraffin wax, petroleum jelly, etc. may be used to fill voids. A typical vacuum impregnation process would involve 50 hours at a pressure corresponding to 0·5 mmHg. This treatment, by removing most of the air, would result in a capacitor having an acceptable breakdown voltage and other properties suitable for d.c. operation. If, however, the capacitor is intended for a.c. operation, then a still lower pressure might have to be applied during the vacuum impregnation, to prevent ionization effects which would result in local overheating and a consequent reduction in useful life.

Valve. A term first applied to a device which allowed a current to pass through it in one direction only; later extended to cover developments of the original valve whereby it was enabled to amplify signal voltages and to generate electrical oscillations. The envelope of an electron valve may be either evacuated or gas-filled. High-vacuum valves have thermionic cathodes; gas-filled valves may have thermionic or cold cathodes. Valves may be classified by the number of electrodes they contain as *diode* (2), *triode* (3), *tetrode* (4), *pentode* (5), *hexode* (6), *heptode* (7), *octode* (8), *nonode* (9), etc.

Valve Rectifier. See THERMIONIC RECTIFIER.

Valve Voltmeter. A voltage-measuring device in which the alternating voltage to be measured is rectified by a thermionic valve and the resultant direct voltage measured by a conventional d.c. instrument. Its input impedance can be made very high so that the instrument imposes no appreciable load on the source to be measured, and it can be used to measure alternating voltages of any frequency with equal accuracy.

Van de Graaff Generator. An electrostatic generator capable of providing a direct potential of up to about 8 MV of either

Var

polarity which can be used to accelerate electrons or ions down an evacuated accelerator tube. The mechanical work expanded in turning an endless belt against electrostatic force is converted into electric-field energy (Fig. V–1). Charge is sprayed on to the belt from "spray-set" needle-points at about 50 kV. The charge is

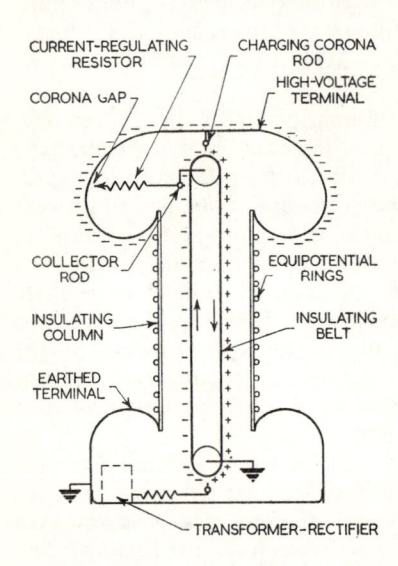

FIG. V–1. PRINCIPLE OF A VAN DE GRAAFF GENERATOR.

Electrostatic charge is carried on the endless belt of insulating material to accumulate on an insulated sphere.

carried upwards to the interior of the h.v. electrode, a metal sphere to which it is transferred by means of a second spray-set. Operation is inside a tank filled with pressurized insulating gas, e.g. nitrogen-freon mixture at 1,700 kN/m^2. See PARTICLE ACCELERATOR.

Var. Abbreviation for REACTIVE VOLT–AMPERES (q.v.).

Variable-Block Tariff. A form of BLOCK-RATE TARIFF in which the quantities of units forming each block are variable in accordance with factors such as maximum demand or size of premises.

Variable-Speed Motor. A motor, the speed of which can be varied while remaining independent of the load.

Variable-Voltage Control. Control scheme that provides a variable-voltage d.c. supply for the armature of a shunt motor (see GENERATOR-FIELD CONTROL, WARD LEONARD CONTROL).

Variac. A trade name applied to an AUTO-TRANSFORMER which is wound with enamelled wire on an iron core, usually toroidal in shape. A sliding contact moves on a bared track over the turns so that the voltage changes in steps corresponding to the volt-drop of a single turn. A double-wound arrangement with a bare secondary winding can also be employed. Losses are small, but the sliding contact (usually a carbon brush) limits the current collected to about 20 A. Output voltages above the supply voltage can be obtained if the supply voltage is fed into a fraction of the winding as shown in Fig. V–2.

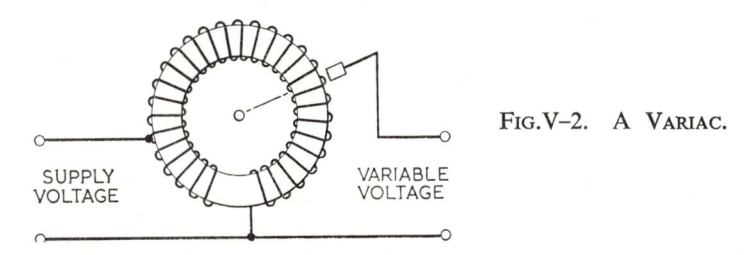

FIG. V–2. A VARIAC.

Varley's Loop Test. A cable fault-localization test. Measurements of resistance are made with a resistance bridge, first with the fault forming one junction of the bridge, and second with the cable conductor resistance measured directly.

Varmeter. Instrument for the measurement of the reactive component of the volt–amperes in a circuit, and calibrated in REACTIVE VOLT–AMPERES.

Vector. A quantity having both magnitude and direction. It may be represented by a straight line of length equivalent to the magnitude, and drawn in a direction corresponding to that of the quantity. A sine-wave quantity may be represented by a vector rotating about a fixed point. Vectors may be added or subtracted as if they represented coplanar forces.

Vector Field. See FIELD.

Vectormeter. An instrument having two pointers. One moves horizontally in proportion to power load, the other vertically in dependence on the wattless load. The intersection of the two pointers indicates the load on a machine.

Ventilation. Of an electrical machine, the provision of access for cooling air while preventing the entry of dirt and moisture. Types of ventilating enclosure are specified in B.S. 2613. See OPEN MACHINE, SCREEN-PROTECTED MACHINE, PIPE-VENTILATED MACHINE, TOTALLY ENCLOSED MACHINE.

Vertical Plugging

Vertical Plugging. Method of switchgear busbar selection in which the circuit-breaker is mounted on a carriage arranged to move both horizontally and vertically. The sockets for duplicate busbars and circuit connections are arranged in a horizontal plane. The required busbar is selected by moving the carriage horizontally and then plugging the circuit-breaker vertically upwards into the appropriate sockets.

Vibrating-Reed Electrometer. A vibrator unit in which a small capacitor has its capacitance modulated by the steady vibration of one of its electrodes. If the capacitor is charged from some source, an alternating component of the charge produced by the variation can be readily amplified for final conversion to direct current for indication.

Vibration Damper. Device attached to overhead-line conductor to prevent excessive vibration being caused by the wind, etc.

Vibration Galvanometer. An instrument resembling a direct-current-operated galvanometer, but producing an alternating torque. The reflected light from a small mirror on the moving part traces out a band of light, which reduces to a stationary spot when the current is reduced to zero (as in the balance condition of an a.c. bridge network). The inertia of the movement restricts the width of the light band, and thus the sensitivity, unless mechanical resonance is used to increase the response. In practice the frequency of mechanical resonance is made equal to that of the alternating current to be indicated by adjusting the tension of the coil supports. The vibration galvanometer is a useful bridge detector for industrial frequencies, and low audio frequencies up to about 250 Hz.

Vibrator. A device for generating a mechanical vibration, in particular by electrical means. Fig. V-3 shows the essential features of an electromagnetic vibrator. When the coil, suspended in the annular gap of the permanent magnet, is supplied with alternating current of a suitable frequency, the force developed in it is transferred to a shaft that provides linear vibrational movement.

Vibrator Power Pack. A device primarily intended to convert direct current at low voltage into direct current at high voltage without rotating plant. The basic feature is a vibrating reed with appropriate contacts.

Villari Effect. Magnetostrictive effect appertaining to changes in the magnetization of a specimen when it is mechanically stressed longitudinally or transversely.

Voltage Divider

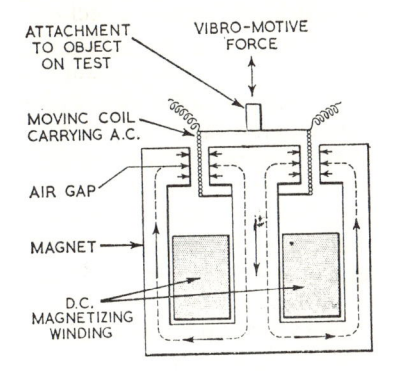

FIG. V-3. AN ELECTRO-MAGNETIC VIBRATOR.

Virtual Value. See EFFECTIVE VALUE.

Volt. The practical unit of electric potential, electromotive force, and potential difference; also the absolute unit of these quantities in the M.K.S. system (abbreviation: V). The volt is that potential difference which, associated with a current of one ampere, transfers energy at the rate of one watt. More precisely, a charge of one coulomb moved between points having a potential difference of one volt acquires or loses a potential energy of one joule.

Volt–Ampere. The apparent power VI given by the product of voltage V and current I in r.m.s. values (abbreviation: VA). It is normal to rate a.c. electrical apparatus in VA (or kVA or MVA) either as a matter of convenience (as with switchgear or rectifiers), or because such a quantity better describes the service value of the device.

Volta Effect. The phenomenon occurring when two dissimilar metals are placed in mutual contact in air, one acquiring a positive potential with respect to the other.

Voltage. The value of a potential difference or electromotive force expressed in volts. The term is also used more loosely for the difference of potential.

Voltage Coefficient. Of a d.c. generator, the ratio of the e.m.f. to the product of armature speed, number of conductors, and flux.

Voltage Divider. A device to permit a fixed or variable fraction of a given supply voltage to be obtained; often, although ambiguously, called a POTENTIOMETER. Two forms are generally available, the most common being a resistor, across the ends of which the supply or reference voltage is applied. By two tapped

Voltage Doubler

connections, or by one fixed connection and a slider (or similar) mechanism, a fraction of the reference voltage can be picked off. For alternating voltages a capacitor chain can be employed, with the units chosen so that the appropriate fraction of the voltage is available across one or more adjacent units of the chain.

Voltage Doubler. An arrangement of rectifiers that rectifies separately each half cycle of an applied alternating voltage, and adds the two resulting rectified voltages to produce a direct voltage that has an amplitude approximately double that of the peak amplitude of the applied voltage (see LATOUR DOUBLER, SCHENKEL DOUBLER; also COCKROFT–WALTON MULTIPLIER).

Voltage Gradient. A term used interchangeably with *electric field intensity*, and measured in volts per unit length within an electric field region, such as an insulator subjected to voltage stress. Strictly, the voltage gradient is a vector with direction opposite to that of the electric field.

Voltage-Regulating Relay. Relay used for automatic control of ON-LOAD TAP CHANGING. When the voltage varies the relay initiates a tapping change.

Voltage Regulation. (1) The variation of the terminal voltage of a generator or the secondary circuit of a transformer (see VOLTAGE REGULATOR).

(2) The maintenance of the voltage of an m.v. or l.v. distribution system within $\pm 6\%$ of the declared voltage.

Voltage Regulator. A device that, supplied at a given input voltage, will provide an adjustable output voltage. Common types of regulator are the resistance divider (Fig. V–4), AUTO-TRANSFOR-

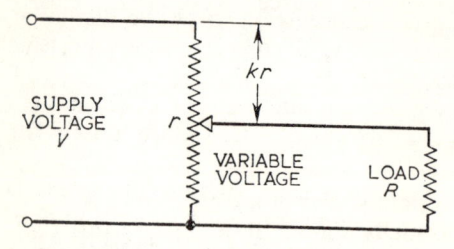

FIG. V–4. A RESISTANCE-DIVIDER VOLTAGE REGULATOR.

For smooth voltage variation, r should never exceed $3R$.

MER and MOVING-COIL REGULATOR. A transformer with tap-changing may be used for the same purpose, but then the output voltage is varied in steps rather broader than those obtainable from other voltage-regulating devices. See also INDUCTION

Voltmeter

Regulator, Variac, Booster Transformer; also Automatic Voltage Regulator.

Voltage Stabilizer. Device incorporated in a circuit to maintain a constant output voltage from a poorly regulated power supply and/or a varying output load. The simplest form of voltage stabilizer makes use of the substantially constant voltage characteristic of a cold-cathode glow-discharge tube when connected as shown in Fig. V–5, provided the operating limits of the tube are not exceeded.

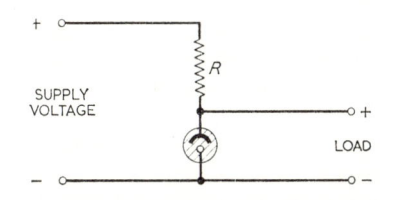

Fig. V–5. Voltage Stabilizer Employing a Cold-Cathode Glow-Discharge Tube.

Voltage Transformer. An Instrument Transformer for the transformation of voltage. Its operation and design are similar to those of an ordinary power transformer. It may be single-phase or three-phase. If single-phase, a two-limb core, with one or both limbs wound, or a shell-type core is used. If three-phase, either a three-limb or a five-limb core can be used. The primary winding, which is the winding connected into the circuit that is to be measured or controlled, is continuously excited at more or less constant voltage. It may be known alternatively as a *potential transformer*.

Voltaic Cell. A self-contained source of electrical energy consisting of two electrodes immersed in an electrolyte (see Primary Cell).

Voltaic Current. Electric current resulting from chemical action.

Voltameter. Instrument for measuring quantity of electricity; also known as a Coulometer (q.v.).

Voltmeter. An instrument for measuring the voltage or potential difference between two points, one of which may be at earth potential. The value is normally shown on a scale. Most voltmeters actually measure a current passing through a fixed resistance, the instrument then comprising a milliammeter or microammeter with a high multiplier resistance in series, the combination being connected across the points to be measured. An *electrostatic voltmeter* measures the voltage directly. The repulsion between

Volume Resistivity

similarly charged electrodes, or the attraction between oppositely charged electrodes, provides the small working force or torque. The electrostatic voltmeter will read direct and r.m.s. alternating voltage (see also ELECTROMETER). In the simple triode VALVE VOLTMETER, the low voltage to be measured is applied to the grid of the valve; a milliammeter then measures the resulting change in anode current.

Voltmeters for d.c. use include permanent-magnet types such as the *moving-coil*, and the now little-used *polarized moving-iron* type; being polarized they only read unidirectional voltages and will show polarity. Voltmeters that will read on both d.c. and a.c. circuits are the *moving-iron* and *electrodynamic*, the latter being generally similar to the moving-coil type except that the permanent magnet is replaced by one or two fixed coils, with or without a ferromagnetic core. A voltmeter for a.c. use only is the *induction* instrument in which a magnetic field rotates or pulsates in a fixed iron system, causing induced currents in a light aluminium disc or cylinder in the air-gap. The resulting current reacts with the line-current magnetization to provide a force or torque.

Volume Resistivity. The resistance offered to a constant and uniformly distributed current by a conductor of unit length and unit cross-sectional area. Dimensionally, resistivity has the units Ω-m, Ω-cm, Ω-in., etc., depending on the unit length chosen to express the dimensions of the cubic-shaped conductor.

Vulcanized India Rubber. RUBBER treated with sulphur or sulphur compounds to modify its physical properties. The product, which is used as a low-voltage-cable insulant, is largely unaffected by temperature changes and is elastic.

W

Wagner Earth. A device to eliminate false balance conditions in a.c. bridges caused by the effects of capacitance at higher audio frequencies between the bridge arms and earth. If, as in Fig. W–1, the bridge supply is not earthed but the bridge is so arranged that points B and D are at earth potential, then the detector will also be at earth potential and no stray capacitance currents will flow in it. Further, all branch capacitances to earth will be stabilized. The Wagner earth method secures this condition by use of two additional impedances Z_5 and Z_6, the junction E of which is solidly earthed. Z_5 and Z_6 must be of the same type as either Z_3 and Z_4,

Ward Leonard Control

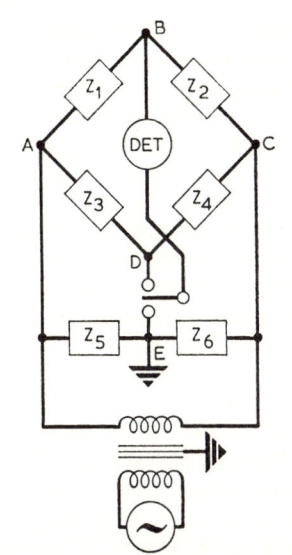

FIG. W-1. APPLICATION OF A WAGNER EARTH TO AN A.C. BRIDGE.

The ratio-arms Z_5 and Z_6, connected to earth, enable perfect balance to be obtained.

or Z_1 and Z_2. If balance is achieved for both positions of the switch, then points B and D must have the same potential to earth as E, which is zero.

Walker Advancer. A PHASE ADVANCER that resembles the LEBLANC ADVANCER with the addition of a stator winding in series with the armature winding. The electromotive force generated can be adjusted as regards its phase angle by altering the brush position relative to the stator winding.

Ward Leonard Control. A form of motor control. The armature of the motor is supplied from its own generator, usually driven by an a.c. motor, as shown in Fig. W-2. The field of the motor is

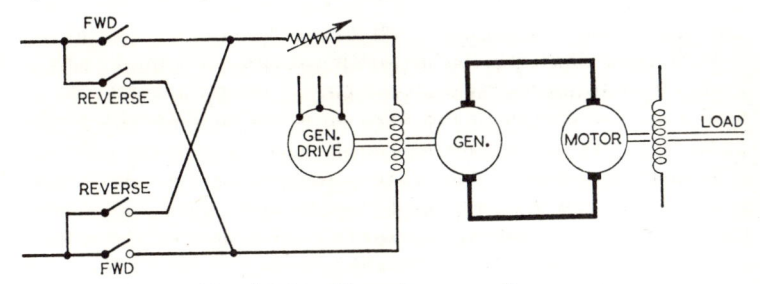

FIG. W-2. WARD LEONARD DRIVE.

energized from another, and sometimes constant, source. To vary the motor armature voltage, the voltage of the generator is varied by controlling its field current. This can be done directly by a rheostat in the generator field circuit or by other more refined methods. To reverse the armature voltage it is only necessary to reverse the generator field. The generator rotation is not reversed.

The Ward Leonard system gives a smooth control of armature voltage, and as the armature loop-circuit resistance remains constant, the slope of the speed/torque characteristic is substantially the same at all speeds. The speed is also less sensitive to torque variations, the natural speed drop usually being less than 10% at 100% full-load torque.

Ward Leonard/Ilgner Control. A form of WARD LEONARD CONTROL incorporating a load-equalizing device consisting of a flywheel added to the main motor-generator set. This flywheel is arranged to give up some of its stored kinetic energy by slowing the set down on the occurrence of high peak loads, the set running up to speed again when the load is removed.

Water Turbine. A prime-mover utilizing hydraulic power, e.g. for driving electric generators. The design is affected by the head of water available. An *impulse* turbine is the PELTON WHEEL; a *reaction* type is the FRANCIS TURBINE. A development of the reaction turbine is the PROPELLER TURBINE or KAPLAN TURBINE.

Watt. The practical and M.K.S. unit of POWER (abbreviation: W). It is the rate of work, or energy transfer, at one joule per second. The JOULE has a mechanical definition, so that the watt is primarily also a mechanical unit. It is electrically defined as the rate of transfer or conversion of energy when a current of one ampere flows between points having a difference of potential of one volt.

Watt-Hour. A unit of energy, being the work done by 1 watt acting for 1 hour. It equals 3,600 joules.

Watt-Hour Efficiency. The ratio between the amount of output energy in watt-hours drawn from a battery during a test discharge and the input energy in the same units required to charge the battery.

Watt-Hour Meter. An INTEGRATING METER for measuring energy. This is usually expressed in kilowatt-hours. It is alternatively known as an *integrating wattmeter* or a *recording wattmeter*.

Wattless Component. Alternative name for REACTIVE COMPONENT (q.v.).

Wattmeter

Wattless Current. Alternative name for the REACTIVE COMPONENT of an alternating current.

Wattmeter. An instrument for the measurement of the power in an electrical circuit. Two common forms are the INDUCTION INSTRUMENT and the ELECTRODYNAMIC INSTRUMENT. A wattmeter normally contains a current coil, carrying either the whole or a definite fraction of the load current, and a voltage coil carrying a current proportional of the voltage. In measuring the power in a single-phase or a d.c. circuit, the simplest form of connection is that given in Fig. W–3(a). The current coil is connected in one line and the voltage coil is connected between the lines. Where the

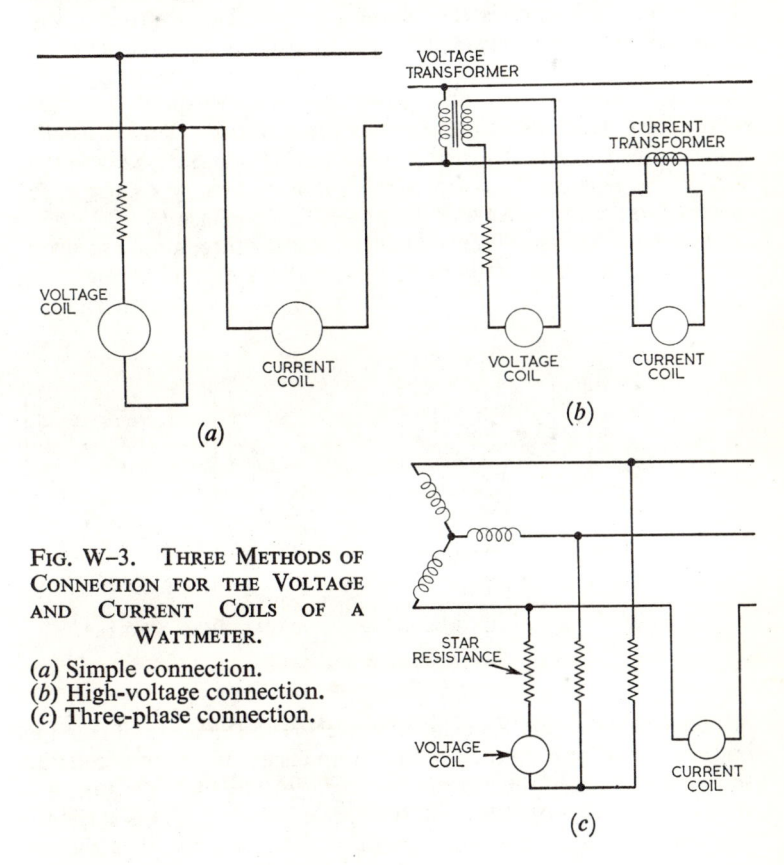

FIG. W–3. THREE METHODS OF CONNECTION FOR THE VOLTAGE AND CURRENT COILS OF A WATTMETER.

(a) Simple connection.
(b) High-voltage connection.
(c) Three-phase connection.

355

Wauchope Starter

current is too high to pass directly through the coils of the instrument, and in the case of high-voltage circuits, current and voltage transformers are used as shown in (*b*).

In a balanced three-phase circuit the total power is three times the power in one phase, and it is only necessary to measure the power in one of the phases and to multiply the result by three. One way of doing this is shown in Fig. W–3(*c*). Three equal resistances are connected in the form of a star, the voltage coil being included in one of them. The current coil is connected in the same line as that to which the voltage coil is connected. The voltage coil thus carries a current proportional to the phase voltage, and the current coil carries the phase current. The wattmeter thus measures the phase power, and multiplication by three gives the total power. If the load is unbalanced it can be shown that, with a three-wire circuit, the total power can be measured with two wattmeters, the individual readings being added. This is usually done automatically by using a DOUBLE-ELEMENT WATTMETER. For a three-phase four-wire circuit it is necessary to use three wattmeters, each measuring the power in the phase.

Wauchope Starter. A star/delta starter for three-phase squirrel-cage induction motors which achieves a smooth acceleration approaching that of a slip-ring motor. Fig. W–4 shows the changes

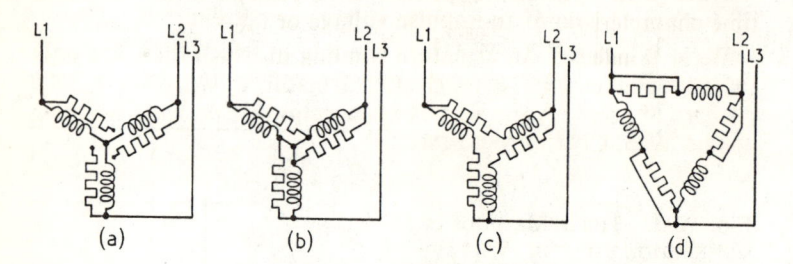

FIG. W–4. SWITCHING SEQUENCE OF A WAUCHOPE STARTER. (*Allen West & Co. Ltd*).

of connections in the Wauchope starter. The sequence of starting is: (*a*) The motor is connected in star and is permitted to accelerate to a stable speed in the normal manner. (*b*) The three resistors are then connected in parallel with the motor windings. This is simply a preparatory step and is in operation for a fraction of a second

Wave Winding

only. (c) The star point of the windings is now opened. It will be seen that the resistors have been so connected that the motor windings are in delta, with a resistor in series with each winding. At this stage the voltage across the motor windings in increased, giving a corresponding increase in torque, and the motor accelerates to a higher steady speed. (d) The resistors are then short-circuited and the motor windings connected in delta across full line voltage.

Wave Filter. A QUADRIPOLE with a characteristic that enables it to discriminate between signals of various frequencies. It is a recurrent network (or artificial line) designed to transmit signals in a desired band (or bands) of frequency and attenuate those of other frequencies.

Wave-Front. The rising portion of a voltage (or current) time characteristic of an impulse voltage or current.

Wave Impedance. The CHARACTERISTIC IMPEDANCE of a transmission line, or a measure of the impedance of a waveguide.

Wave-Shape. The voltage or current/time characteristic of a surge. It may be defined by the DURATION of the wave-front (T_1) and the time to half-value of the wavetail (T_2), and hence is known as a T_1/T_2 wave.

Wave-Tail. The falling portion of a voltage/time or current/time characteristic of an impulse voltage or current.

Wave Winding. An armature winding in which there are only two parallel circuits through the armature, regardless of the number of poles. A simple wave connection in five slots is shown in Fig. W–5. Cf. LAP WINDING.

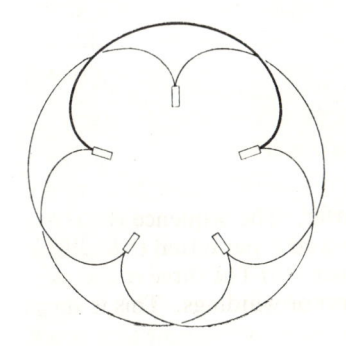

FIG. W–5. SIMPLE WAVE CONNECTION.

The diagram shows a wave-wound armature with five slots and five bars.

Waveform

Waveform. The shape of the curve produced when the magnitude of a quantity that varies (usually in a periodic manner) with time is plotted on a time base.

Waveform Analysis. The reduction of a complex alternating or other waveform, normally repetitive, to its component fundamental and HARMONIC sine waves.

Waveguide. A tube of conducting material, usually copper, used as an alternative to a coaxial feeder for conveying electromagnetic radiation. The commonest shape is rectangular.

Way. A space in a multiple duct unit to contain a cable.

Wayleave. Permission to carry transmission lines or telephone wires over private land or buildings; also the rent or charge for such permission.

Weber. The M.K.S. unit of magnetic flux (abbreviation: Wb). From the theory of electromagnetic induction, the flux must be measurable in terms of volt-seconds, and 1 Wb = 1 Vs. A *weber per square metre* is the unit of magnetic induction. $1 \text{ Wb/m}^2 = 10^4$ gauss; $1 \text{ Wb} = 10^8$ maxwells.

Weber Dynamometer. Early type of dynamometer, in which a small moving coil is hung from a bifilar suspension inside a fixed coil.

Weld. A union between metal surfaces that have been heated to a plastic or liquid condition. It may involve the addition of more metal, and the application of pressure. See ARC WELDING, RESISTANCE WELDING.

Wenner Configuration. An arrangement of earth electrodes consisting of four electrodes in a straight line at equal intervals.

Werren Overlap Test. A modification of the MURRAY LOOP TEST for locating faults in cables, designed for use when the insulation of the return conductor has a low value, such as may often be found in a multicore cable where all cores are affected by the fault.

Weston Cell. A primary STANDARD CELL, most commonly used. It has electrodes of mercury and cadmium amalgam in an electrolyte of cadmium sulphate with mercurous sulphate as depolarizer (Fig. W–6). Its e.m.f. is taken as 1·108 V at 20°C, and the rate of change with temperature is closely known.

Wet Cell. An electrolytic cell, in which the electrolyte is a liquid. Wet primary cells are the DANIELL CELL and LECLANCHÉ CELL.

Wetherill Separator. A MAGNETIC SEPARATOR for removing a number of materials from a mixture. The material to be treated is passed on a conveyor belt between the poles of a magnet. The success of the machine is dependent on the shape of the magnet poles: the lower magnet face is flat, and the upper pole is arranged

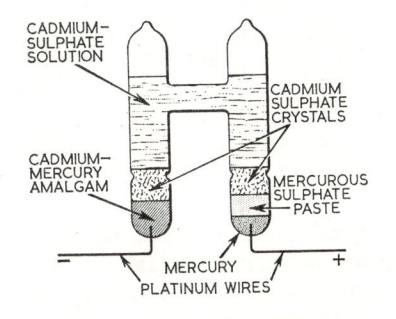

FIG. W-6. WESTON STANDARD CELL.

with a ridge to provide a concentration of the field. When the material passes under the magnets, any particular particle which is affected jumps towards the upper pole and is intercepted by a second, take-off, belt. This belt carries it off to the side and discharges it into a chute. In practice a number of magnets are employed; the process is continuous.

Wheatstone Bridge. A BRIDGE circuit consisting of four resistance arms, a galvanometer and a battery or other d.c. source. If three of the resistances are known, the fourth may be determined. In Fig. W-7, no current will flow in the galvanometer branch G if

$$Pi_1 = Si_2 \quad \text{and} \quad Qi_1 = Ri_2$$

whence $PR = QS$ is the *balance condition*.

White Radiation. Radiation consisting of a continuous range of frequencies. It is also known as *heterogeneous* or *continuous* radiation.

Wiedemann Effect. Magnetostrictive effect in which a specimen twists when magnetized longitudinally and circumferentially at the same time.

Wilson Cloud Chamber. Apparatus for viewing and recording nuclear events as indicated by the tracks of particles. It consists of an observation chamber in which moist air can be made to expand

Wimshurst Machine

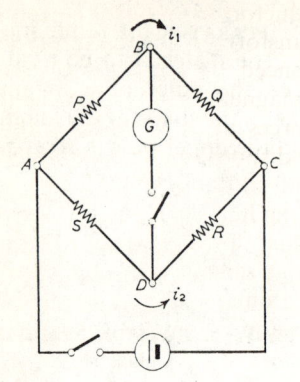

FIG. W-7. WHEATSTONE BRIDGE CIRCUIT.

The galvanometer is undeflected if the ratio of P to Q equals that of S to R.

rapidly. If ions are present they form nuclei on which the vapour molecules condense. Since ions are produced along the tracks of fast-moving charged particles, it follows that such tracks are rendered visible by the vapour condensation from the supersaturated mixture.

Wimshurst Machine. An electrostatic generator, in which two coaxial discs carrying metal strips revolve in opposite directions. Charges induced on the strips are collected by brushes.

Wind-Power Generator. Generator that employs the wind as a source of power. Many types of windmill have been used for various purposes, but only two have been seriously considered for power production, the propeller type and the ANDREAU GENERATOR. For a *propeller* unit, the power in the wind is $P = 0.0000053\ AV^3$ kilowatts, where A is the area in square feet swept by the rotor, and V is the wind speed in miles per hour. Theoretically, it is possible to extract a maximum of 59.3% of the power in the wind but, in practice, this extraction is not likely to be much greater than 40% and may be as low as 10%. Full power output is produced at a *rated wind speed* which is usually about 30 mile/h for a very windy site and 25 or 20 mile/h for less windy places. For wind speeds above the rated value the power output is controlled to the full design capacity.

Windage Loss. The power absorbed in a machine through incidental disturbance of the atmosphere by moving parts.

Winder. An electrically driven engine for hoisting a cage up the shaft of a mine or similar vertical shaft.

Winding. A system of insulated conductors forming the current-carrying element of a machine or transformer and designed to produce a magnetic field or be influenced by one. An electric machine operates as a result of the magnetic flux set up in its magnetic circuit by magnetomotive forces arising from currents flowing in groups of windings suitably disposed on the stator and rotor. The flux usually sets up an e.m.f. in the winding owing to the conductors of the winding cutting the flux, or the turns of the winding being linked with a varying flux. The interaction of the m.m.f.s of the stator and rotor windings sets up a torque. Both e.m.f.s and torques are usually produced simultaneously, but in the generator it is the e.m.f. and in the motor the torque that is the more obvious.

Winding Ends. The leads brought out from the two ends of a phase winding.

Wiping Gland. A watertight gland that is achieved by the use of a *wiped joint*, or joint around which molten solder is wiped by hand with a cloth pad.

Wire Gauge. A means of specifying the size of circular-section wire or the thickness of sheet. Several sets of gauges are in common use, but the practice of specifying diameters in millimetres is growing, particularly in Europe. In Britain the legal standard is the Standard Wire Gauge (S.W.G.), but there is nevertheless a Birmingham Wire Gauge (B.W.G.) commonly used for steel. There is no legal standard in the U.S.A., the custom with electrical conductors being to employ the American Wire Gauge (A.W.G.), alternatively called the Brown and Sharp gauge (B. & S.). With the S.W.G. the even-numbered gauges are the primary preferred sizes, the odd being secondary sizes. Half sizes are available but are not recommended.

Withstand Test. A test applied to equipment designed to operate at high voltages. A specified alternating voltage is applied to the test object for a specified time. This voltage must be withstood without flashover or puncture and may be performed with the test object dry or subjected to artificial rain (*wet test*). It may be followed by a FLASHOVER TEST.

Work Function. Potential difference traversed by an electron in performing sufficient work to permit it to leave the surface of a conductor. The work function varies with the material.

Wound-Rotor Motor. Alternative term for a SLIP-RING MOTOR (q.v.).

X

X-Amplifier. Thermionic amplifier for amplifying the voltage producing the horizontal deflection of the beam in a CATHODE-RAY TUBE.

X-Plates. Pair of flat parallel electrodes mounted vertically, side by side in a cathode-ray tube. A difference of potential applied between the two plates produces horizontal deflection of the beam.

X-Ray. Electromagnetic radiation forming part of the electromagnetic spectrum and occupying a band beginning at the short wavelength end of the ultra-violet spectrum. The rays are generated in an X-RAY TUBE as the result of the impact of a stream of electrons of high energy on a suitable metal target.

X-Ray Crystallography. The study of the sub-microscopic structure of materials by the production of interference patterns resulting from the diffraction of X-rays by atomic planes (*X-ray diffraction*).

X-Ray Spectrometer. Instrument for determining the wavelengths of X-rays and the relative intensities of various wavelengths in an X-ray spectrum.

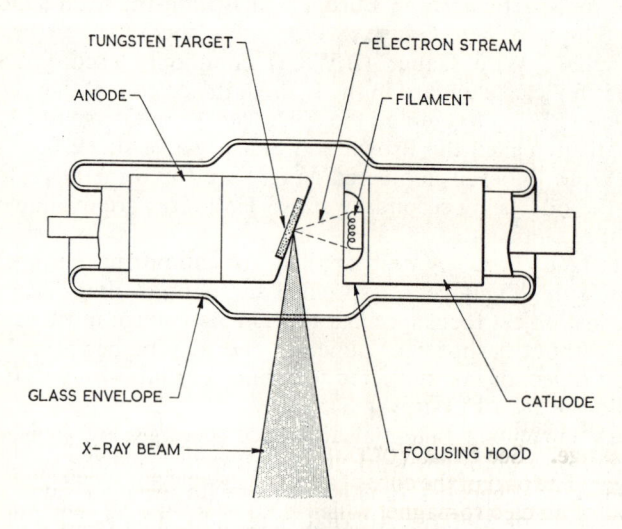

FIG. X–1. X-RAY TUBE.

X-Ray Tube. A two-electrode vacuous device (internal pressure less than 10^{-6} mmHg). It is fitted with a pure tungsten filament, from which electrons are thermionically emitted, as the main component of the cathode, and an anode of suitable design to serve as a target for the stream of electrons passing from cathode to anode when a high voltage of appropriate polarity is applied between the two electrodes (Fig. X–1). The high-velocity electrons impinging on the metal target cause X-rays to be emitted.

X-Unit. A unit for wavelengths of electromagnetic waves. It is approximately 10^{-3} Å.

Xenon. A rare-gas element, atomic weight 131·3. It is present in the atmosphere in one part in 170 million, by volume. It is employed in some types of gas-filled electron tubes or discharge lamps.

Xerography. A dry photographic process based on ELECTROSTATICS and TRIBOELECTRICITY, in direct contrast to other methods that require wet chemical processing for image development and fixation.

Xeroradiography. The production of an X-ray image by XEROGRAPHY.

Y

Y-Amplifier. Thermionic amplifier for amplifying the voltage producing vertical deflection of the beam in a CATHODE-RAY TUBE.

Y-Class Insulation. One of seven classes of insulating materials for electrical machinery and apparatus defined in B.S. 2757 on the basis of thermal stability in service. Class Y insulation is assigned a temperature of 90°C. It consists of materials such as cotton, silk or paper without impregnation. It was formerly known as Class O.

Y-Connection. An American term meaning a three-phase STAR CONNECTION (q.v.).

Y-Plates. Pair of flat, parallel electrodes mounted horizontally, one above the other in a cathode-ray tube. A difference of potential applied between the two plates produces vertical deflection of the beam.

Y-Voltage. See STAR VOLTAGE.

Yoke. The part of the core of a transformer or the ferromagnetic material of an electromagnet which is not surrounded by the windings. It serves to join the magnetic poles or to complete the magnetic circuit.

Yoke Permeameter

Yoke Permeameter. Instrument for measurement of the magnetic characteristics of a ferromagnetic material. A bar of the material is placed in a made-up magnetizing coil, and the magnetic circuit is completed by either one or two yokes of laminated, high-permeability material of large cross-section. B and H are measured ballistically by means of search-coils close to the specimen. The compensated instrument shown in Fig. Y–1 contains, in addition

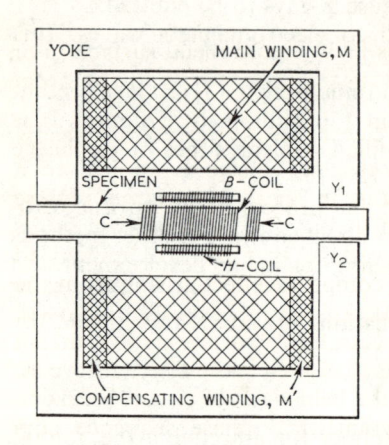

FIG. Y–1. A TYPICAL COMPENSATED YOKE PERMEAMETER.

to the main magnetizing winding M, a compensating winding M′ in two halves, one on each side of the winding M. For each reading the compensating current in M′ is adjusted to give uniform magnetization along the length covered by the search-coils. An uncompensated permeameter is the FAHY SIMPLEX PERMEAMETER.

Yoke Suspension. A method of mounting a traction motor on an electric truck, more commonly known as BAR SUSPENSION (q.v.).

Youngstown Switch. See PALMER LIMIT SWITCH.

Z

Zener Breakdown. A field-emission effect occuring when a high electric-field intensity exists across the depletion layer of a SEMICONDUCTOR. This field effectively reduces the forbidden energy gap, so that electrons can pass more easily from the valence band to the conduction band (see Fig. Z–1). It is, in fact, an example of the quantum-mechanical *tunnel effect* and is analogous

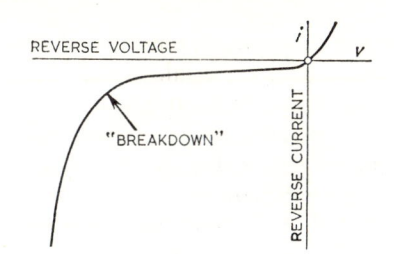

FIG. Z-1. ZENER BREAKDOWN.

to the field emission of electrons from a cold metal surface. For germanium and silicon the critical field intensity is of the order of 0·2 MV/cm.

Zener Diode. See *Glossary of Semiconductor Terms* in Appendix.

Zener Effect. When a high reverse voltage is applied to a *p–n* junction in an SEMICONDUCTOR there is a large reverse "breakdown" current, as shown in Fig. Z-1. Breakdown limits the magnitude of the reverse voltage that can be applied to rectifiers and transistors. It is a composite effect arising from the phenomena of AVALANCHE BREAKDOWN and ZENER BREAKDOWN.

Zero-Pause. The momentary cessation of an alternating current when passing through a zero value between successive half-cycles. The zero-pause is significant in the behaviour of circuit-breakers.

Zero Phase Sequence. A symmetrical phase sequence corresponding to three equal co-phasal currents. Any unbalanced system can be resolved into three symmetrical systems, of POSITIVE PHASE SEQUENCE, NEGATIVE PHASE SEQUENCE and zero phase sequence, which are mutually independent.

Zero-Type Dynamometer. Dynamometer in which the electrical forces are balanced by mechanical forces to bring the indicating pointer to zero.

Zeta. Zero-energy thermonuclear assembly; a device for the study of PLASMAS. It is toroidal and contains deuterium at low pressure. A heavy discharge current can cause a narrow thread of plasma to form by a pinch effect. This plasma is unstable and lasts for only a few thousandths of a second. It was hoped that a stable, dense plasma would be formed that would be hot enough to enable atomic *fusion* to take place. This would be an efficient source of electrical energy.

Zigzag Connection. A method of STAR CONNECTION in which each branch of the star contains contributions from two different phases.

Zigzag Leakage

Zigzag Leakage. The leakage flux produced in and near the gap between stator and rotor of an electrical machine. It results from the "solenoidal" combination of the stator and rotor ampere-conductors, which are normally in approximate phase opposition and so magnetize circumferentially along the air-gap. The leakage flux tends to cross and re-cross the gap between the stator and rotor tooth-tips if the gap-length is relatively short: hence the term zigzag.

Zinc–Air Battery. A dry primary battery using zinc as the anode material and oxygen as the other electrochemical reactant at the cathode. The cathode shown in Fig. Z–2 consists of several layers held in an external plastics frame; the battery is supplied in a sealed package as it starts to deteriorate when exposed to the air.

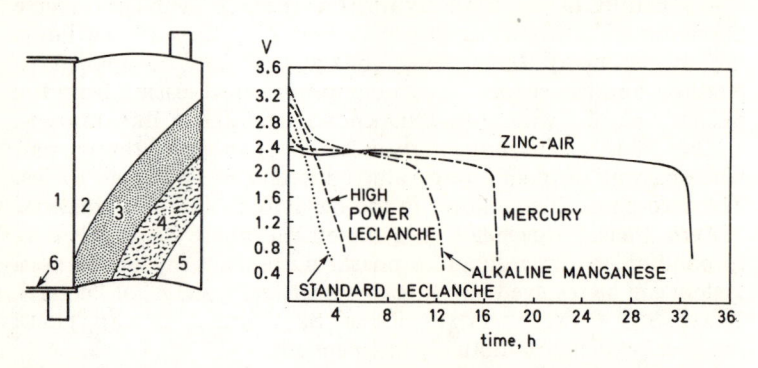

FIG. Z–2. THE CATHODE STRUCTURE OF A ZINC–AIR CELL, AND TYPICAL COMPARATIVE DISCHARGE CURVES AT 100mA.
1. Highly porous zinc.
2. Permeable separator.
3. Metal mesh current collector.
4. Catalyst.
5. Microporous p.t.f.e. film.
6. Plastic outer case.

(*Crompton Parkinson Ltd.*)

The system combines high energy density with high current output. In some designs, the active life can be extended by replacement of the zinc anode when this is exhausted. Research is still progressing on the development of an economic fully rechargeable secondary cell.

Zirconium. Metallic element of atomic weight 91·22, melting point about 1,700°C. Its high absorption rate for oxygen and nitrogen makes it of use in the manufacture of electronic valves.

APPENDIX

S.I. Primary and Derived Units
Decimal Prefixes
General Abbreviations
Letter Symbols
Semiconductor Terms

Appendix

UNITS, ABBREVIATIONS AND SYMBOLS

Basic S.I. Units

The following are the basic units in the system:

Physical quantity	Unit	Symbol
length	metre	m
mass	kilogram(me)	kg
time	second	s
electric current	ampere	A
temperature	kelvin	K
luminous intensity	candela	cd
amount of substance	mole	mol

Supplementary Dimensionless Units

plane angle	radian	rad
solid angle	steradian	sr

Derived S.I. Units in terms of the Basic Units

area	square metre	m^2
volume	cubic metre	m^3
speed, velocity	metre per second	m/s
acceleration	metre per second squared	m/s^2
density, mass density	kilogramme per cubic metre	kg/m^3
concentration (of amount of substance)	mole per cubic metre	mol/m^3
activity (radioactive)	1 per second	1/s
specific volume	cubic metre per kilogramme	m^3/kg
luminance	candela per square metre	cd/m^2

Appendix

Derived S.I. Units with Special Names

Physical quantity	Unit	Symbol	Definition of unit
energy	joule	J	$kg\,m^2\,s^{-2}$
force	newton	N	$kg\,m\,s^{-2} = J\,m^{-1}$
power	watt	W	$kg\,m^2\,s^{-3} = J\,s^{-1}$
electric charge	coulomb	C	$A\,s$
electric p.d.	volt	V	$kg\,m^2\,s^{-3}\,A^{-1} = WA^{-1}$
electric resistance	ohm	Ω	$kg\,m^2\,s^{-3}\,A^{-2} = VA^{-1}$
electric capacitance	farad	F	$A^2\,s^4\,kg^{-1}\,m^{-2} = CV^{-1}$
magnetic flux	weber	Wb	$kg\,m^2\,s^{-2}\,A^{-1} = V\,s$
magnetic flux density	tesla	T	$kg\,s^{-2}\,A^{-1} = Wb\,m^{-2}$
inductance	henry	H	$kg\,m^2\,s^{-2}\,A^{-2} = VsA^{-1}$
luminous flux	lumen	lm	$cd\,sr$
illumination	lux	lx	$cd\,sr\,m^{-2}$
frequency	hertz	Hz	s^{-1}
pressure	pascal	Pa	$kg\,m^{-1}\,s^{-2} = Nm^{-2}$
conductance	siemens	S	$A^2\,s^3\,kg^{-1}\,m^{-2} = \Omega^{-1}$
viscosity, dynamic	poiseuille	Pl	$kg\,m^{-1}\,s^{-1} = N\,s\,m^{-2}$

Other Derived S.I. Units

electric field strength	volt per metre	V/m	$m\,kg\,s^{-3}\,A^{-1}$
electric charge density	coulomb per cubic metre	C/m³	$s\,Am^{-3}$
electric flux density	coulomb per square metre	C/m²	$S\,A\,m^{-2}$
permittivity	farad per metre	F/m	$s^4\,A^2\,m^{-3}\,kg^{-1}$
current density	ampere per square metre	A/m²	$A\,m^{-2}$
magnetic field strength	ampere per metre	A/m	$A\,m^{-1}$
permeability	henry per metre	H/m	$m\,kg\,s^{-2}\,A^{-2}$

Appendix

Decimal Prefixes

The recommended prefixes to be used should be those indicating multiples and sub-multiples that differ from a unit in steps of 10^3, i.e. prefixes representing 10 raised to a power which is a multiple of ± 3.

Prefix	Symbol	Factor	
tera	T	10^{12}	$= 1\ 000\ 000\ 000\ 000$
giga	G	10^9	$= 1\ 000\ 000\ 000$
mega	M	10^6	$= 1\ 000\ 000$
kilo	k	10^3	$= 1\ 000$
hecto*	h	10^2	$= 100$
deca*	da	10^1	$= 10$
deci*	d	10^{-1}	$= 0.1$
centi*	c	10^{-2}	$= 0.01$
milli	m	10^{-3}	$= 0.001$
micro	µ	10^{-6}	$= 0.000\ 001$
nano	n	10^{-9}	$= 0.000\ 000\ 001$
pico	p	10^{-12}	$= 0.000\ 000\ 000\ 001$
femto	f	10^{-15}	$= 0.000\ 000\ 000\ 000\ 001$
atto	a	10^{-18}	$= 0.000\ 000\ 000\ 000\ 000\ 001$

*These prefixes should be strictly limited in use to occasions where the recommended prefixes are inconvenient.

Appendix

Abbreviations

absolute	abs.	medium voltage	m.v.
after-diversity demand	a.d.d.	melting point	m.p.
air circuit-breaker	a.c.b.	metre-kilogramme-	
alternating-current	a.c.	second	M.K.S.
atmospheric	atm.	miles per hour	mile/h or m.p.h.
availability factor	a.f.	mineral-insulated copper-	
boiling point	b.p.	covered (cable)	m.i.c.c.
centimetre-gramme-		minimum	min.
second	C.G.S.	oil circuit-breaker	o.c.b.
characteristic impedance		paper braided jute	
ratio	c.i.r.	(insulated cable)	p.b.j.
current transformer	c.t.	paper insulated (power	
degree	deg.	cable)	p.i.
direct-current	d.c.	peak-to-peak	p-p
diversity factor	d.f.	potential difference	p.d.
electromagnetic unit	e.m.u.	power factor	p.f.
electromotive force	e.m.f.	rate of rise of restriking	
electrostatic unit	e.s.u.	voltage	r.r.r.v.
equation	eqn.	revolutions per	
freezing point	f.p.	minute	rev/min or r.p.m.
high-breaking (rupturing)-		root-mean-square	r.m.s.
capacity	h.b.c.	short-circuit ratio	s.c.r.
high pressure	h.p.	silicon controlled rectifier	s.c.r.
high voltage	h.v.	standard temperature	
horsepower	hp	and pressure	s.t.p.
brake horsepower	bhp	synthetic-resin-bonded	
effective horsepower	ehp	paper (insulation)	s.r.b.p.
indicated horsepower	ihp	Systèms International d'Unités	S.I.
nominal horsepower	nhp	temperature	temp.
shaft horsepower	shp	transformer	xfmr.
infra-red	i.r.	ultra-violet	u.v.
load factor	ld.f.	vacuum	vac.
low pressure	l.p.	voltage transformer	v.t.
low voltage	l.v.	volume	vol.
magnetomotive force	m.m.f.	weight	wt.
maximum	max.		
maximum continuous		which see	q.v.
rating	m.c.r.	compare	cf.

Appendix

Letter Symbols

- A magnetic vector potential
- a 120° operator
- B magnetic flux density
    susceptance
- C capacitance
- c speed of light
- D electric flux density
- E electromotive force
    electric force
- e charge of electron
- F magnetomotive force
    Faraday's constant
- f frequency
- G conductance
- H magnetizing force
- h 120° operator
    Planck's constant
- I current
    intensity of magnetization
- i instantaneous value of current
- J current density
- j 90° (complex) operator
- L inductance
    self-inductance
- L_{mn} mutual-inductance
- M mutual-inductance
    intensity of magnetization
    magnetomotive force
- m mass of electron
- N number of turns
- P power
    electric polarization
- Q electric charge
    reactive power
    Q-factor
- R resistance
    reluctance
- r resistivity
- S apparent power
    reluctance
    Poynting vector
- V electromotive force
    potential
    difference of potential
- v instantaneous value of potential
    phase velocity of electromagnetic waves
- W electrical energy
- X reactance
- Y admittance
- Z impedance
- α attenuation coefficient
- β phase-change coefficient
    Bohr magneton
- γ propagation coefficient
    conductivity
- δ dielectric loss angle
- ϵ, ε permittivity, dielectric constant
- κ magnetic susceptibility (volume)
- Λ permanence
- λ wavelength
    load angle, torque angle
- μ electric or magnetic dipole moment
    permeability
- ν reluctivity
- ρ resistivity
    charge density (volume)
- σ conductivity
    charge density (surface)
- Φ magnetic flux
- ϕ electronic exit work-function
    phase difference
- $\cos \phi$ power factor
- χ magnetic susceptibility (mass)
- Ψ electric flux
- ω angular frequency

Appendix

GLOSSARY OF SEMICONDUCTOR TERMS

Diac. A five-layer two-terminal gateless device as shown in the figure. When the applied voltage is high enough, the device will begin to conduct as a normal untriggered thyristor (through layers $p_2 n_2 p_1 n_1$). See THYRISTOR in main dictionary. When the polarity of the supply reverses, a similar situation occurs, but this time conduction is via layers $p_1 n_2 p_2 n_3$. The device is available with a range of breakover voltages, but for voltages above about 30V the device is generally known as a *sidac*.

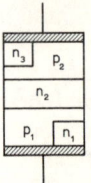

F.E.T. (Field-Effect Transistor). A three-terminal device in which current carriers (holes or electrons) flow from source to drain along a p- or n-type channel. The rate of flow of carriers is controlled by an electric field applied across the channel, the field being proportional to the gate (input) voltage. The device is therefore voltage controlled, and its high input impedance and low noise figure in the audio range make it useful as a first stage in audio 'hi-fi' amplifiers.

Integrated Circuit. A single crystal of semiconductor in which is fabricated a complete electronic circuit, which may consist of several transistors, diodes and resistors. Typical dimensions of such a 'chip' are $1\frac{1}{4}$ mm square by $\frac{1}{4}$ mm thick. This is enclosed in a case, about $18 \times 6 \times 3$ mm, with, say, 14 pins, so that it can conveniently be connected into electronic equipment.

Light-Emitting Diode. A pn junction which emits light at a fixed frequency when a sufficiently high current is flowing across the junction. Silicon and germanium do not have suitable characteristics to produce visible light by this means, the most common material in use at present being gallium arsenide. These devices have longer lives, greater efficiency, and are more reliable than tungsten filament lamps.

M.O.S.T. (Metal-Oxide-Semiconductor Transistor). A type of f.e.t. (insulated-gate f.e.t.) having a layer of oxide between the metallic gate and the channel of semiconducting material. It has a higher input impedance and greater bandwidth than a comparable junction-gate f.e.t.

S.C.S. (Silicon Controlled Switch). A four-layer device similar to a THYRISTOR but with both gate leads brought out. It can therefore be switched on by a positive pulse on the cathode gate (as in a normal thyristor), or by a negative pulse on the anode gate.

Triac. A five-layer device similar to a DIAC (see above) but with a gate electrode, producing a bidirectional controllable device.

Tunnel Diode (Esaki Diode). A heavily doped semiconductor diode that exhibits negative resistance over part of its I/V characteristic. It is used in oscillators, and low noise amplifiers.

Unijunction Transistor. A two-layer device consisting of a p-type rod placed asymmetrically in an n-type base. There are two contacts to the n region, base 1 being nearer to the emitter than base 2. The device will only conduct at a certain value of emitter-base 1 voltage (the value of which can be varied by the biasing arrangements). It is therefore a triggerable device which is used in many types of oscillator circuit.

Varactor Diode. A pn junction device in which the depletion zone capacitance is nonlinearly related to the applied voltage. The nonlinearity is achieved by uneven impurity doping of the semiconductor materials. These diodes are used as tuning diodes in oscillator circuits, and in frequency multiplier systems.

Zener Diode. A diode designed to be operated under reverse breakdown conditions when the voltage across the device is practically constant, the current being limited by external circuit resistance to prevent it exceeding the allowable dissipation. The devices, which are available with breakdown voltages up to several hundred volts, are used for voltage stabilisation.